Utilization and Management of Bioresources

Sadhan Kumar Ghosh
Editor

Utilization and Management of Bioresources

Proceedings of 6th IconSWM 2016

Springer

Editor
Sadhan Kumar Ghosh
Department of Mechanical Engineering
Jadavpur University
Kolkata, West Bengal, India

ISBN 978-981-13-5373-4 ISBN 978-981-10-5349-8 (eBook)
DOI 10.1007/978-981-10-5349-8

Printed on acid-free paper

This Springer imprint is published by Springer Nature
The registered company is Springer Nature Singapore Pte Ltd.
The registered company address is: 152 Beach Road, #21-01/04 Gateway East, Singapore 189721, Singapore

Preface

Healthy environment and sustainable living conditions are the basic goal for mankind. To achieve such goals, we need renewable energy resources, sustainable technology and excellent waste management leading towards a circular economy. It is also our duty to ensure that we pose least harm to the environment by preventing soil, air and water pollution. Since the inception, the International Society of Waste Management, Air and Water (ISWMAW) has been working towards its commitment to sustainable waste management and environmental protection.

The IconSWM movement under the aegis of the International Society of Waste Management, Air and Water (ISWMAW) was initiated for better waste management and environmental protection in the year 2009 through generating awareness and bringing all the stakeholders together from all over the world in a bracket for discussion and formulating actions. It establishes research projects across the country in collaboration with several institutes worldwide and the Consortium of Researchers in International Collaboration (CRIC). IconSWM has become significantly one of the biggest platforms in India for knowledge sharing, awareness generation and encouraging the Urban Local Bodies (ULB), government departments, researchers, industries, NGOs, communities and other stakeholders in the area of waste management and resource conservation. The conference attracted huge interest from academics, practitioners and policymakers around the world because of its important theme areas and the conference's timeliness in addressing the need of resource utilization.

Biomass and bioenergy production is now imperative towards a sustainable future. To meet the high global energy crisis, bioenergy is destined to play a crucial role. While the energy demand is a concern, climate change is another issue that can also be tackled with the help of bioenergy. Biohydrogen and biomethane generation is of current interest among the researchers. Other than the energy sector, biotechnologists and biochemical engineers have develop different routes to produce chemicals from different biomasses. Worldwide, many biorefineries are operating in full capacity. Technologies have been developed for valorization of waste streams from these biorefineries. Despite the renewable nature, bioenergy production is a very fast-growing area, and still it occupies only 5 % of the total energy market share.

Nonetheless, some of the initiatives of commercial bioenergy production include biodiesel prepared from agricultural crop oils and waste oils, bioethanol extraction from sugarcane and cornstarch, biogas production from organic fraction of waste materials, etc.

However, the economy of the biorefineries is highly stochastic and affected by the feedstock price fluctuations and supply chain sustainability. Development of new or modified technologies is a current requirement. Some of the developing areas include conversion of lignocellulosic materials into biofuels and other chemicals, biohydrogen production, etc. Nanotechnology is considered to be an important area of research that could be a way forward in producing bioenergy in more effective means through process intensification. IconSWM aimed to achieve the realization of these ideas, initiatives and processes through stakeholder's participation, interaction and networking. In the 6th IconSWM, nearly 20 % of the abstracts received from 30 different countries were based on biomass, biofuels and bioenergy. We have segregated the biomass- and bioenergy-related research articles and covered them in several tracks including biomass and biowaste utilization, biogas, biofuels and bioenergy, applied biotechnology and bioenergy systems and applied nanotechnology. The conference offered both the academics and practitioners the opportunity to share knowledge and experience relevant to the utilization and management of bioresources. The overarching question was how we collaborate to facilitate further development in these emerging areas. This book unveils the selected papers from the conference.

The 6th IconSWM received more than 320 abstracts and 280 full papers from 40 different countries. Each full paper was put to the review process and was reviewed by at least two experts. This book includes only 30 accepted full papers based on biomass and bioenergy and is organized in 5 (five) parts as follows:

Part I: Biomass and Biowaste Utilization
Part II: Biogas
Part III: Biofuels and Bioenergy
Part IV: Applied Biotechnology and Bioenergy Systems
Part V: Applied Nanotechnology

Kolkata, India
3 September 2017

Sadhan Kumar Ghosh

6th IconSWM 2016 Technical Committee

Prof. Sadhan Kumar Ghosh
Editor

Members

Prof. Agamuthu Pariatamby
University of
Malaya, Malaysia

Mr. C. R. C. Mohanty
Advisor, UNCRD,
Japan

Prof. Allen H. Hu
National Taipei
University of Technology

Prof. Deben Baruah
Tezpur University,
India

Mr. K. D. Bhardwaj
National Productivity
Council, India

Prof. JinhuiLi
Tsinghua University,
R. P. China

Dr. Rodrigo Lozano
University of Gävle,
Sweden

Dr. Marisa Soares Borges
Federal University
of Parana, Brazil

Prof. Prasanta K. Dey
Aston University, UK

Dr. Nguyen Trung Thang
ISPONRE, Vietnam

Prof. Mont Michael Nelles
University of Rostock,
Germany

Dr. Ronald L. Mersky
Widener University,
USA

Dr. Suneel Pandey
TERI, India

Prof. S. T. ElSheltawi
University of Cairo,
Egypt

Dr. H. N. Chanakya
Indian Institute of Science,
India

Prof. Seo Yong Chil
Yonsei University;
Rep. of Korea

Prof. Ramkrishna Sen
IIT Kharagpur, India

Dr. Sunil Herat
Griffith University,
Australia

Dr. Tomas Ramos
Universidade Nova de
Lisboa, Portugal

Prof. J. W. C Wong
Hong Kong Baptist
University

Acknowledgement

ISWMAW and IconSWM express gratitude to Jadavpur University, Kolkata, and the Centre for Quality Management System. ISWMAW and IconSWM are indebted to the United Nations Centre for Regional Development (UNCRD) for providing support.

ISWMAW and IconSWM are thankful to the co-organizers, namely, the Centre for Sustainable Technologies at the Indian Institute of Science Bangalore, Indian Institute of Technology Kharagpur, The Energy Research Institute (TERI) and Consortium of Researchers in International Collaboration (CRIC). The gratitude is extended to the supporting organizations, namely, the International Partnership for Expanding Waste Management Services of Local Authorities (IPLA), Japan; the Institute of Strategy and Policy on Natural Resources and Environment (ISPONRE), Vietnam; Deutsches Biomasse forschungs zentrum gemeinnützige GmbH (DBFZ), Germany; Aston University, UK; the University of Rostock, Germany; and the UK and India Education Research Initiative (UKIERI). IconSWM Committee Members gratefully acknowledges the help provided by the sponsoring organizations and exhibitors, namely, the Ministry of Urban Development, Government of India; Department of Environment, Government of West Bengal; BBMP, DMA, Karnataka; Hiland Group, Kolkata; Oil India Ltd. and IOCL (R&D); Geocycle Ltd.; SUEZ, UK; and CMAK, Bangalore, and the active participation of the authors and attending delegates.

The committee expresses gratitude to Mr. Praveen Prakash, IAS, JS (W), UDD, Government of India; Mr. Biswajit Roy, director (BD and HR) of Oil India Ltd.; Mr. Arnab Roy, IAS, principal secretary, Environment, Government of West Bengal; Dr. N. Visal, IAS, DMA, Government of Karnataka; Mr. M. P. Singh, GM R&D, IOCL; Ms. Chikako Takase and Mr. CRC Mohanty, UNCRD, Japan; Mrs. Dipsikha Deka, Oil India Ltd.; Mr. N. K. Belani, president, CREDAI; Mr. Manjunath Prasad, IAS, commissioner, BBMP; Mr. G. Kumar Nayak, IAS, MD, Karnataka Power Corp. Ltd.; Mr. Khalil Ahmed, IAS, commissioner, KMC; Prof. Seo Yong Chil, Yonsei University; Prof. S. T. El Sheltawi, Cairo University; Mr. C. P Marak, chairman, MSPCB; and Mr. B. K. Barua, ASPCB, for their support.

ISWMAW and IconSWM are thankful to all the members in the core group, the International Scientific Committee, the Organising Committee and the Secretariat, the Technical Committee members and many of the authors included in this book who served as the referees. Thanks go to all who provided constructive and comprehensive reviews. The active participation of all the team members in IconSWM across the country is acknowledged with gratitude with special mention of Dr. Suneel Pandey and his team in TERI; Ms. Sheetal Singh and her team in CMAK; Prof. Deben Barua in Tezpur University, Assam, and his team; Dr. V. Kirubakaran; Mr. Gautam Ghosh and his team; Mr. Tapas Ghatak; Dr. H. N. Chanakya; Prof. Amit Ghosh; Ms. Sampriti Kakati; Mrs. Pranati Ghosh; Mr. Biswajit Debnath; Mr. Rahul Baidya; Mr. Sannidhya Kumar Ghosh; Mr. Suresh Mondal; Mr. Bisweswar Ghosh; Mr. Gobinda Debnath; and the research team members in the Mechanical Engineering Department and ISWMAW, Kolkata, HQ.

Special thanks go to the team in Springer (India) Private Limited, in particular Mr. Aninda Bose and Ms. Kamiya Khatter, whose contribution throughout the process of this publication has been invaluable. In closing, the insights and excellent contributions of all the authors to this book are gratefully acknowledged.

Contents

Contributors

A. Bhattacharya School of Bioscience and Engineering, Jadavpur University, Kolkata, India

A. Brosowski Deutsches Biomasseforschungszentrum GmbH (German Biomass Research Centre), Leipzig, Germany

A. Ghosh Bose Institute, Kolkata, West Bengal, India

A. Halder Glass Division, CSIR-Central Glass and Ceramic Research Institute, Kolkata, India

A. Mallick Heritage Institute of Technology, Kolkata, India

Aniruddha Mukhopadhyay University of Calcutta, Kolkata, India

A. Schüch Professorship Waste Management and Material Flow, University of Rostock, Rostock, Germany

A.K. Mandal Glass Division, CSIR-Central Glass and Ceramic Research Institute, Kolkata, West Bengal, India

A.O. Coker Department of Civil Engineering, Faculty of Technology, University of Ibadan, Ibadan, Nigeria

Adeshola Oluremi Openibo Department of Vehicle Inspection Service, Lagos State Ministry of Transportation, Lagos, Nigeria

Amit Ganguly CSIR Central Mechanical Engineering Research Institute, Durgapur, India

Amrita Phukon Assam Agricultural University, Jorhat, India

Anita Rajor School of Energy and Environment, Thapar University, Patiala, Punjab, India

Annapurna Jetty CSIR-Indian Institute of Chemical Technology, Hyderabad, Telangana, India

Anubrati Chatterji Department of Food and Nutrition, Barrackpore Rastraguru Surendranath College, Barrackpore, Kolkata, West Bengal, India

Anup D. Chahande Department of Biotechnology, PIET, Nagpur, India

Avijit Bhowal Department of Chemical Engineering, Jadavpur University, Kolkata, India

Barkhang Brahma Department of Energy, Tezpur University, Napaam, Assam, India

Bulbul Gupta Biotechnology Branch, University Institute of Engineering and Technology, Panjab University, Chandigarh, India

C. Thamaraiselvi Department of Biotechnology, Mother Teresa Women's University, Kodaikanal, Tamil Nadu, India

C.G. Achi Department of Civil Engineering, Faculty of Technology, University of Ibadan, Ibadan, Nigeria

Dipa Ray University of Limerick, Limerick, Ireland

D.C. Baruah Department of Energy, Tezpur University, Napaam, Assam, India

D.S.N. Goutham CSIR-Indian Institute of Chemical Technology, Hyderabad, Telangana, India

Dipam Patowary Department of Energy, Tezpur University, Napaam, Assam, India

E. Faedsura School of Biochemical Engineering and Technology, Sirindhorn International Institute of Technology, Thammasat University, Bangkok, Thailand

E.A. Abdel Kader Chemical Engineering & Pilot Plant Department, National Research Center, Cairo, Egypt

Esha Bala Maulana Azad National Institute of Technology, Bhopal, India

G. El Diwani Chemical Engineering and Pilot Plant Department, National Research Center, Cairo, Egypt

G. Morscheck Professorship Waste Management and Material Flow, University of Rostock, Rostock, Germany

Gunajit Dev Sarma Department of Energy, Tezpur University, Napaam, Assam, India

H. Sudrajat School of Biochemical Engineering and Technology, Sirindhorn International Institute of Technology, Thammasat University, Bangkok, Thailand

H.S. Hussein Chemical Engineering and Pilot Plant Department, National Research Center, Cairo, Egypt

Helen West School of Biosciences, University of Nottingham, Sutton Bonington Campus, Leicestershire, UK

Hemantajeet Medhi Tezpur University, Napaam, Assam, India

Hoang Le Phuong Thai Nguyen University of Technology, Thai Nguyen Province, Vietnam

I. Mallick Bose Institute, Kolkata, West Bengal, India

J. Senophiyah Mary Department of Environmental Engineering, Government College of Technology, Coimbatore, India

Jaspreet Kaur Biotechnology Branch, University Institute of Engineering and Technology, Panjab University, Chandigarh, India

K.G. Ahila Department of Biotechnology, Mother Teresa Women's University, Kodaikanal, Tamil Nadu, India

Kailash Hazarika Department of Nematology, Assam Agricultural University, Jorhat, Assam, India

Kangkana Saikia Tezpur University, Napaam, Assam, India

Kashyap Porag Bezbaruah Department of Soil Science, Assam Agricultural University, Jorhat, Assam, India

Kunal School of Energy and Environment, Thapar University, Patiala, Punjab, India

L. Ramrakhiani Ceramic Membrane Division, CSIR-Central Glass and Ceramic Research Institute, Kolkata, West Bengal, India

L. Weerasundara Faculty of Applied Sciences, University of Sri Jayewardenepura, Nugegoda, Sri Lanka

L.M. Pandey Center for the Environment, Indian Institute of Technology, Guwahati, India

The Department of Biosciences & Bioengineering, Indian Institute of Technology, Guwahati, India

M. Basu Post Graduate Department of Microbiology, Barrackpore Rastraguru Surendranath College, Barrackpore, Kolkata, West Bengal, India

M. Mukhopadhyay University of Calcutta, Kolkata, India

M. Nelles Professorship Waste Management and Material Flow, University of Rostock, Rostock, Germany

M. Vasanthy Department of Environmental Biotechnology, Bharathidasan University, Trichy, Tamil Nadu, India

M. Vithanage Faculty of Applied Sciences, University of Sri Jayewardenepura, Nugegoda, Sri Lanka

M.K.C. Sridhar Department of Environmental Health Sciences, Faculty of Public Health, University of Ibadan, Ibadan, Nigeria

M.M. Panda Utkal University, Bhubaneswar, Odisha, India

Michèle Clarke School of Geography, University of Nottingham, University Park, Nottingham, UK

Modhuleena Mandal Department of Food Technology & Biochemical Engineering, Jadavpur University, Kolkata, West Bengal, India

N.K. Attia Chemical Engineering and Pilot Plant Department, National Research Center, Cairo, Egypt

N.N. Deshavath Indian Institute of Technology Guwahati, Guwahati, Assam, India

Nguyen Thi Kim Thai National University of Civil Engineering, Hanoi City, Vietnam

Nilay Borah Department of Soil Science, Assam Agricultural University, Jorhat, India

Nurudeen Adekunle Raji Department of Mechanical Engineering, Faculty of Engineering, Lagos State University, Lagos, Nigeria

P. Basak School of Bioscience and Engineering, Jadavpur University, Kolkata, India

P. Datta Center for the Environment, Indian Institute of Technology, Guwahati, India

P. Dutta Post Graduate Department of Microbiology, Barrackpore Rastraguru Surendranath College, Barrackpore, Kolkata, West Bengal, India

P. Ghosh Department of Food Technology & Biochemical Engineering, Jadavpur University, Kolkata, West Bengal, India

P. Kumarathilaka Faculty of Applied Sciences, University of Sri Jayewardenepura, Nugegoda, Sri Lanka

P.K. Amritha National Institute of Technology, Calicut, India

P.P. Anilkumar National Institute of Technology, Calicut, India

Pallavi Roy Tezpur University, Napaam, Assam, India

Papita Das Department of Chemical Engineering, Jadavpur University, Kolkata, India

Piyush Nanda Department of Biotechnology, Indian Institute of Technology, Kharagpur, India

Pradip Kumar Chatterjee CSIR Central Mechanical Engineering, Research Institute, Durgapur, India

Pranay A. Raut Department of Biotechnology, PIET, Nagpur, India

Pranjal Das Tezpur University, Napaam, Assam, India

Prassana Kumar Pathak Department of Soil Science, Assam Agricultural University, Jorhat, Assam, India

Priyabrata Banerjee CSIR- Central Mechanical Engineering, Research Institute, Durgapur, India

R. Chakraborty Jadavpur University, Kolkata, West Bengal, India

R. El-Araby Chemical Engineering and Pilot Plant Department, National Research Center, Cairo, Egypt

Rajen Barua Department of Soil Science, Assam Agricultural University, Jorhat, Assam, India

Ramalingam Dineshkumar Department of Biotechnology, Indian Institute of Technology, Kharagpur, India

Ramkrishna Sen Department of Biotechnology, Indian Institute of Technology, Kharagpur, India

Richard Blanchard Loughborough University, Leicestershire, UK

Rupam Patowary Department of Energy, Tezpur University, Napaam, Assam, India

S. Arayawate School of Biochemical Engineering and Technology, Sirindhorn International Institute of Tech, Thammasat University, Bangkok, Thailand

S. Babel School of Biochemical Engineering and Technology, Sirindhorn International Institute of Technology, Thammasat University, Bangkok, Thailand

S. Biswas School of Bioscience and Engineering, Jadavpur University, Kolkata, India

S. Chatterjee Jadavpur University, Kolkata, West Bengal, India

S. Ghosh Ceramic Membrane Division, CSIR-Central Glass and Ceramic Research Institute, Kolkata, West Bengal, India

S. Mahanta Ramjas College, University of Delhi, New Delhi, India

S. Majumdar Ceramic Membrane Division, CSIR-Central Glass and Ceramic Research Institute, Kolkata, West Bengal, India

S. Mayakaduwa Faculty of Applied Sciences, University of Sri Jayewardenepura, Nugegoda, Sri Lanka

Sunanda Sain University of Calcutta, Kolkata, India

S. Tiwari The Department of Biosciences & Bioengineering, Indian Institute of Technology, Guwahati, India

S.K. Sahoo Utkal University, Bhubaneswar, Odisha, India

Sachankar Buragohain Department of Energy, Tezpur University, Napaam, Assam, India

Sampriti Kataki Department of Energy, Tezpur University, Napaam, Assam, India

Siddhartha Datta Department of Chemical Engineering, Jadavpur University, Kolkata, India

Shubhalakshmi Sengupta Department of Chemical Engineering, Jadavpur University, Kolkata, India

Sutripta Sarkar Department of Food and Nutrition, Barrackpore Rastraguru Surendranath College, Barrackpore, Kolkata, West Bengal, India

T. Bandara Faculty of Applied Sciences, University of Sri Jayewardenepura, Nugegoda, Sri Lanka

T. Meenambal Department of Environmental Engineering, Government College of Technology, Coimbatore, India

Uma Ghosh Department of Food Technology & Biochemical Engineering, Jadavpur University, Kolkata, West Bengal, India

Uttariya Roy Department of Chemical Engineering, Jadavpur University, Kolkata, India

V.V. Dasu Indian Institute of Technology Guwahati, Guwahati, Assam, India

V.V. Goud Indian Institute of Technology Guwahati, Guwahati, Assam, India

Vidyadhar V. Gedam WTM Division, CSIR-NEERI, Nagpur, India

Y. Jayawardhana Faculty of Applied Sciences, University of Sri Jayewardenepura, Nugegoda, Sri Lanka

Yogesh P. Moharkar Department of Biotechnology, PIET, Nagpur, India

About the Editor

Dr. Sadhan Kumar Ghosh Professor and former head of the Mechanical Engineering Department and founder coordinator of the Centre for QMS at Jadavpur University, Kolkata-700032, India, is a renowned personality in the field of waste management, green manufacturing and TQM. He served as the director of CBWE, Ministry of Labour and Employment, Government of India, and at his initial stage of career in the L&T Ltd. and Gramophone Company of India Ltd. Professor Ghosh is the founder and chairman of the IconSWM and the president of the International Society of Waste Management, Air and Water and the chairman of the "Indian Congress on Quality, Environment, Energy and Safety Management Systems (ICQESMS)". He has been awarded the Distinguished Visiting Fellowship 2012 by the Royal Academy of Engineering, UK, for work on "Energy Recovery from Municipal Solid Waste". He received the Boston Pledge and NABC 2006 award for the most eco-friendly innovation "Conversion of plastics and jute waste to wealth" in the ESP/50 K Business Plan Competition at Houston, Texas, USA. He received Indian patent on "eco-friendly plastics recycling machine and the process thereof" (Patent no.202532 dated 02/03/2007) and "Automatic High Speed Jute Ribboning Machine" in Bangladesh (Patent no. 1005146 dated 17/02/2014) innovated by him, patent in India is awaited. His research interests include circular economy, 3R, waste management (MSW, e-waste, C&D Waste, biomass, biowaste, etc.), sustainable development, supply chain management, QEES Management System standards and SMEs performance.

Professor Ghosh is the international expert of the (a) United Nations Centre for Regional Development under the UN in 3R Forum (Adelaide, 2016) and (b) Asian Productivity Organization (APO), Japan and China Productivity Council (CPC) on Green Manufacturing and Green Factory (Taiwan 2014, 2015, 2016). His papers on waste management, namely, on biowaste, e-wastes and MSW received Excellence Awards in many countries. His very recent international research projects as India Leader are (a) Erasmus+ with Aston University, UK, and ATI and UTCC, Thailand, 2015–2016; (b) Research on GP for the Base of the Pyramid for Sustainable Development in APO Member Countries, sponsored by APO, Japan, and CPC, Taiwan, 2016–2017; (c) UKIERI 2011–2013 Research Project on "Climate Change

Issues and Environmental Performance of Small and Medium-Sized Enterprises (SMEs) in India and the UK"; (d) UKIERI 2012–2014 Research Project on "Waste to Energy from Municipal Waste – design of DSS for planning and implementation"; (e) UKIERI funding on "Developing Low Carbon SMEs through Lean and Green Manufacturing" for 2015–2016; and (f) Royal Society – DST fund for "UK-India Seminar of Bio Energy" held in February 2015 with UK partners, and he acted as PIr of more than 20 research projects.

He has been serving as the editor, associate editor, editorial board member and member of international steering committees and served as the keynote/plenary speaker in many international conferences in Australia, China, Germany, Hong Kong, Italy, Laos, Malaysia, the USA, the UK and India. He was the organizing chair of the 4th 3RINC, by Materials Cycles Society, Japan, and the chair elect of the 7th IconSWM 2017. He has initiated the movement for worldwide sustainable waste management with partners in many countries through the "Task Force for International Cooperation on Waste Utilization" (TF4ICWU) and Consortium of Researchers in International Collaboration (CRIC). He was the convener of ISO/TC 61 WG24, was a member in the Indian Mirror Committee of ISO/TC 207 and chaired the ISO/TC 61 WG2 meeting in several countries. He wrote 9 books, more than 24 edited volumes, more than 150 national and international publications and book chapters. He was the State Level Advisory Committee Member of Plastics Waste (Management and Handling) Rules, 2011. He is available at sadhankghosh9@gmail.com.

Part I
Biomass and Biowaste Utilization

Biogenic Waste and Residues in Germany: Amount, Current Utilization and Perspectives

M. Nelles, A. Brosowski, G. Morscheck, and A. Schüch

Abstract The utilization of biogenetic waste and residues draws more and more attention worldwide, driven by the fight against climate change and the efforts to save greenhouse gas emissions. Biodegradable waste and residuals already contribute to energy supply and are used in new biorefinery concepts. The potential of remaining available wastes and residues is compared to the small energy crop potential but important for climate protection goals. In total 98.4 million tons dry mass represent the technical potential that means it is available for utilization in Germany. 30.9 million tons of this technical potential is currently not in use. The biogas sector in Germany is highly developed but has to be further developed as well as the other bioenergy sectors to fulfil the future tasks. The German energy transition has yet not reached its ambitious goals; the process stagnated in certain areas. Biomass and waste biomass must contribute to development a bioeconomy by the combined material and energy utilization in biorefineries and contribute to the energy transition in combination with the other renewable energies. Examples for enhanced energetically utilization of biowaste as well as biorefinery concepts are described.

Keywords Biodegradable wastes and residues • Recycling • Biorefinery • Bioeconomy • Bioenergy

M. Nelles (✉) • A. Schüch
Professorship Waste Management and Material Flow, University of Rostock, Rostock, Germany

Deutsches Biomasseforschungszentrum GmbH (German Biomass Research Centre), Leipzig, Germany
e-mail: michael.nelles@uni-rostock.de

A. Brosowski
Deutsches Biomasseforschungszentrum GmbH (German Biomass Research Centre), Leipzig, Germany

G. Morscheck
Professorship Waste Management and Material Flow, University of Rostock, Rostock, Germany

S.K. Ghosh (ed.), *Utilization and Management of Bioresources*,
DOI 10.1007/978-981-10-5349-8_1

1 Introduction

The utilization of biogenetic waste and residues draws more and more attention worldwide, driven by the fight against climate change and the efforts to save greenhouse gas emissions. The federal government wants to significantly reduce the emissions of greenhouse gases in Germany. This objective only can be achieved by a sustainable energy economy – saving energy, renewable energy generation and efficient use of energy.

Bioenergy, inclusive the biogenetic waste and residues, contributes to this goal driven by the renewable energy law (EEG). The climate protection goal of Germany for the years 2008–2012 was to decrease the greenhouse gas emissions up to 21% compared to 1990. This goal was achieved by the saving of 130 million tons carbon dioxide equivalents (CO_{2eq}) in 2011. More than the half of this was reached by bioenergy (61 million tons CO_{2eq}). In 2015 bioenergy saved nearly 66 million tons CO_{2eq}, 39% of total contribution of the renewable energies (UBA 2016; BMWi 2016).

To meet the climate protection goals to save 40% in 2020, respectively, 80–95% in 2050 (compared with 1990), further developments are necessary; bioenergy is indispensable for that.

Today in Germany the waste management inclusive of the energetic use and recycling of organic waste saves annually about 56 million tons of carbon dioxide equivalent compared to 1990 and contributes significantly to the achievement of the climate protection goals. That was achieved by more than two decades in Germany since the establishment of waste separation in the households. The separate collection of biowaste and green waste takes a leading position in comparison across Europe. In response to the EU's waste framework directive, the Waste Management Act of 2012 (KrWG) in § 11 paragraph 1 obligates waste producers and mandated waste management authorities to collect biowaste separately at the latest as of January 1st 2015. The term "biowaste" in § 3 KrWG comprises yard, park and landscape management waste as well as food and kitchen waste. This will result in an increased available biowaste flow, to be used sustainably.

Also the combined energetic and material utilization of agricultural waste and residues as farmyard manure contributes to the saving of greenhouse gases. The use of loggings and other woody biomass residues in biomass heat and power plants is established in Germany and produces a considerable amount of renewable energy as mentioned before.

The switch from a substitution system towards more market-oriented options for renewable energies, together with increasing prices and limited availability of energy crops, faces the bioenergy sector with economic challenges. New solutions are needed. The greenhouse gas saving by bioenergy is considerable but only one advantage of biomass. The sector has to be transferred into a bioeconomy to combine the energetic and material use of biomass. The aim is to create added value and to make the biomass utilization more sustainable inclusive the economic performance. Focus of politics and science is on the biogenetic waste and residues.

2 Potential of Biogenic Waste and Residues

To estimate the unused technical biomass potential, the investigation of the theoretical, the technical and the already used technical potential is necessary. For Germany this was done for biomass waste and residues in 2015 by the German Biomass Research Centre (DBFZ). In this study the following biomass categories are investigated: woody and forestry by-products (8 biomasses, e.g. bark, sawdust, etc.); agricultural residuals and by-products (18 biomasses, e.g. intercrops, straw, dung, etc.); municipal waste (17 biomasses, e.g. biowaste collected by bin, organic share of the residual waste, food waste, sewage sludge, etc.); industrial residuals (29 biomasses, e.g. slaughterhouse waste, food processing waste, etc.); and residuals from other areas (21 biomasses, e.g. landscape cultivation or protection material, communal green waste, etc.). For 77 of the 93 biomass kinds could collect consistent data. Caused by deficient data sources, the following biomasses were not considered: winter and summer intercrops; horticulture residuals; beet leaves; straw of rape, grain maize, sunflower and grain legumes; liquid poultry manure; dung of horses, sheep and goats; riverside green; and water plants and nutshells (Brosowski et al. 2015).

As main result of the mentioned study, a theoretical biogenic waste and residues potential of about 151.1 million tons dry matter (t dm) in Germany was estimated (Fig. 1). Due to restrictions 43.1 million t dm is not utilizable. For another 9.7 million t dm the database is unclear. In total 98.4 million t dm represents the technical potential that means it is available for utilization (Fig. 1). Two thirds of this potential is currently used in various material flows. One third is unused. Around 95% of the unused technical potential is logging residues, farmyard manure and cereal straw (Brosowski et al. 2015).

The municipal organic waste represents with 8–12% and the industrial organic wastes with 9% a small share of the theoretical biomass residues potential (Brosowski et al. 2015). Only 35% (green waste and biowaste, 3.4 and 3.9 million tons) of the theoretical biowaste potential (21.1 million tons) has been separately collected by the public waste management authorities in bins, bags or containers in 2010, while roughly 23% (4.8 million tons) was collected with the residual waste. Another significant amount was disposed of in private backyards composting and in non-recorded private business biowaste treatment facilities (Morscheck et al. 2015).

In 2050 the yearly German primary energy demand of 6891 PJ could be covered to 26% by local biomass. In the bioenergy scenario 2050 the Agency for Renewable Resources (Fachagentur Nachwachsende Rohstoffe FNR) calculated together with experts in this field a total potential of 1819 PJ/a. Energy plants are considered with 740 PJ (from 4 million ha; in 2015, 247 PJ from 2.2 million ha), biogenetic waste and residues with 134 PJ (FNR 2016).

The use of available biogenetic waste and residues is important for climate protection goals and regional added value and development. Besides the contribution to energy supply, it is used more and more in new biorefinery concepts. An example for a biorefinery concept is the use of straw and other organic residues from the

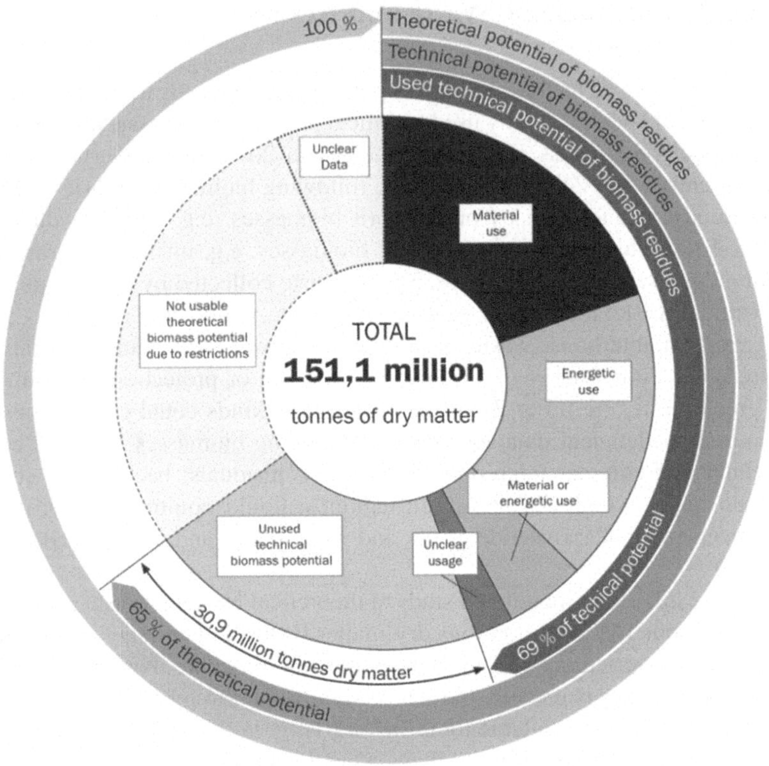

Fig. 1 Theoretical and technical biomass potentials from waste and residues and utilization – status quo in Germany (based on Brosowski et al. 2015, FNR-research project FKZ 22020114)

ethanol production for biogas, which decrease the total footprint of the ethanol production.

3 Utilization

Currently 12.5% (2015) of the primary energy consumption in Germany comes from renewable energy sources (BMWi 2016); about 79% of this is provided by biomass inclusive from wastes and residues dominated by solid biomass for heating purposes (Bloche-Daub et al. 2015). Globally renewable energy contributes with 10% (≈53 EJ) to the total primary energy consumption (≈550 EJ). Biomass is with 62% (33 EJ), the most used renewable source (Bloche-Daub et al. 2015). Following the current pathways and aimed future waste, biomass utilization options in Germany are described.

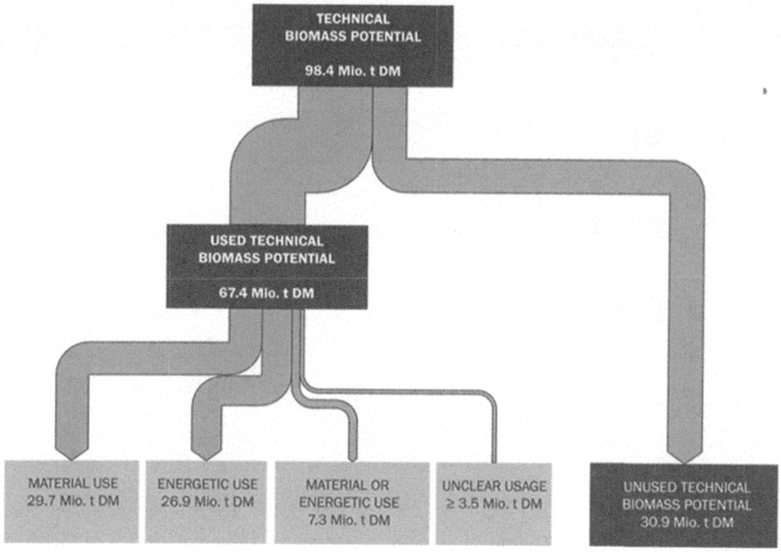

Fig. 2 Current utilization of biogenetic waste and residues (Brosowski et al. 2015, figure adapted)

3.1 Current Pathways

Figure 2 shows the used and unused share of the technical biomass potential of biogenetic waste and residuals in Germany. For 3.5 million tons dry matter, the utilization is unclear. Figure 3 gives a more detailed overview about different biomass categories.

Agricultural residues and by-products have with 17.6 million tons per year the highest unused technical potential in Germany. Currently 15% of the technical potential of this biomass stream is recycled mostly as organic fertilizer; 16% is energetically used in biogas or incineration plants (Fig. 3). A high percentage, 68%, is already unused despite the fact that technical solutions are available. Reasons for that are high requirements regarding operating permission, emission limitations and partly economic reasons.

While the material use of woody and forestry by-products outweighs in the past, currently the use for energetic purposes amounts 66%. The technical potential is exhausted up to 71%. Only parts of the forest residues (needles, small-diameter wood, etc.), with an amount of 11.9 million tons, are available for further purposes. Industrial residues are completely used for energetic or material utilization pathways, whereby the material use outweighs (Fig. 3). Municipal organic waste and residues are mainly recycled by composting, while the industrial biowaste and residuals are often treated in anaerobic digestion plants (AD). Since 2012 more than eight existing composting plants are equipped with an additional AD stage. New AD plants with post-composting are established also to build capacities for the enhanced

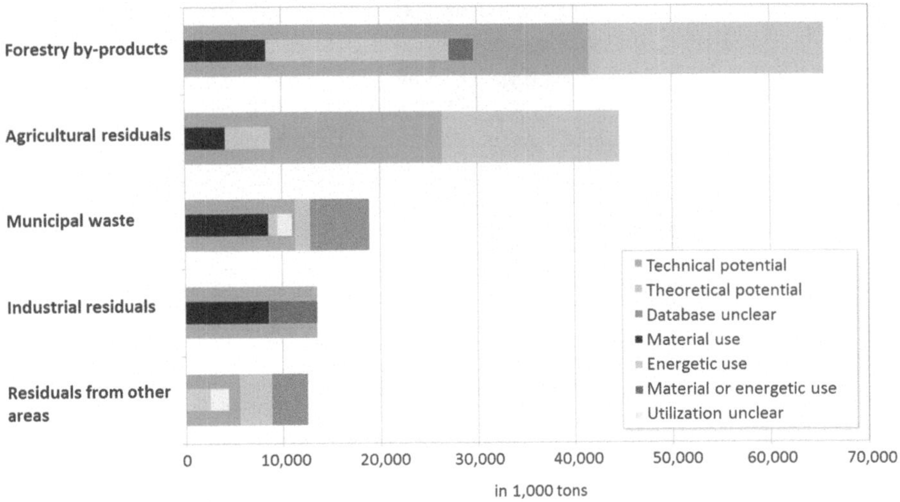

Fig. 3 Current utilization of different biogenetic waste and residue categories in Germany, dry mass per year (Brosowski et al. 2015, figure adapted)

separate collection of municipal biowaste in the next years. For example, since 2012 13 of the 17 newly built waste treatment (AD) plants use mainly municipal biowaste (Schüch et al. 2014; Nelles et al. 2015).

For the smallest considered biomass category "biogenetic residuals from other areas," the database regarding the potentials as the current utilization is unclear (Fig. 3).

3.2 *Future Utilization Options*

A sustainable management of biogenic material flow combined material and energy recovery paths (nutrient and carbon recycling, energy supply, reducing carbon dioxide emissions by replacing fossil fuels, reducing the peat demand and lower treatment costs with extended regional added value). For a future utilization of the residual and waste biomass potential, different options are promising.

The optimization of already established material flows, e.g. the mentioned combination of anaerobic digestion (AD) and composting, is one option. Compost and digestate are used as organic fertilizer or replace peat in potting soil and plant substrates. But also the energetic part of this pathway has to be developed. Caused by efficiency and economic reasons, the utilization of the waste heat draws more attention. Promising small technical solutions for solid biomass are backup stoves with integrated water pockets and micro/mini-combined heat and power installations (Lenz 2016).

Bioenergy, inclusive this from biogenic waste and residues, has to provide positive and negative electrical capacity to avoid the retaining at fossil energy sources in a more and more renewable energy system (Holzhammer et al. 2015; Mauky et al. 2014). The increasing share of fluctuating renewable energies as wind and solar power needs to be balanced in the future in a way that substitutes the most relevant fossil fuels (Holzhammer et al. 2013). This could be implemented in a so-called smart bioenergy concept (Thrän et al. 2016). Biogas plants are today able to meet this flexibility demand. Pilot project showed that a demand-oriented energy production is also possible for solid biomass conversion plants (Hoffstede 2013).

A sustainable cascading can mitigate greenhouse gas emissions. The "biorefinery" concept addresses economic biomass supply chains to make unused biomass potentially accessible. New technologies with low emissions and economic performance and new products are developed (BMBF 2014). Biogenic residues and waste biomass have to contribute to a bioeconomy development by the combined material and energy utilization in biorefineries and contribute to the energy transition in combination with other renewable energies. Generation of new products from biogenetic waste streams is topic in research and innovation to make these conversions more attractive and economically feasible. These developments will help to explore the available biomass potential. Second-generation biorefinery concepts are focused on residues as straw, grass, waste wood or other biogenetic wastes.

Examples in research and praxis show that these future utilization options of biogenetic waste and residues are ongoing. The integration of anaerobic and aerobic treatment of organic waste is often considered in revised waste management concepts of the waste management authorities. For plants with a capacity of more than 15,000 tons per year and waste treatment prices from at least 100 €/ton, this option seems to be economically feasible. So in the last decade, a number of compost plants are equipped with a fermentation stage or are planned.

The combination of anaerobic digestion and thermal biomass utilization brings additional positive synergy effects as shown in the following example of the AVA Augsburg: The biogas plant with plug flow technology was started in December 2013 with an approved yearly throughput of 55,000 tons biowaste. Currently the plant produces about 35 million kWh per year and provides the heat demand of 3,000 and the electricity demand of 4000 households, respectively, the fuel demand (CNG) of 3500 private cars (with each 15,000 km mileage). Furthermore yearly 12,000 tons compost and 14,000 tons liquid fertilizer is produced. Synergy effects, regarding emissions and saved investments (for biofilters, washers), are reached by the use of exhaust gas streams (up to 30,000 m^3/h) of the biowaste treatment plant for the incineration of the neighbouring already existing plant. The investment cost amounts 17.4 million euros (Bauer et al. 2014; AVA 2016).

A smaller plant concept was implemented in Schwerin by the SAS GmbH, owned by the city of Schwerin and the Remondis GmbH & Co. KG: here the biogas plant with an approved yearly throughput of 18,000 tons biowaste was built at a new industrial area of Schwerin. In 2015, 15,943 tons biowaste is treated, 6000 tons (without Christmas trees and green waste) of this was collected at the City Schwerin, and the other mass was provided by the Remondis Group from other regions. A plug flow reactor (constructed by the company Eisenmann) and connected CHP provide more than 650 households (four persons per household) with in total up to 6.2 million kWh electricity. The plant operates since the end of 2014 and produces 7000 tons compost and 9000 tons liquid fertilizer. For this plant, 6.9 million euros is invested. The starting phase was attended by intensive public relations activities and the change to larger collection bins with ident system, to enhance the biowaste quality and to increase the collected amount (Lange 2016). With a flexible operation of biomass power plants, the electricity generation can be adjusted to the electricity demand and the fluctuating wind and solar energy production. In addition, the yields of biomass/biogas power plants can be increased by the exploitation of price fluctuations.

The flexible electricity production was implemented in a demonstration project (FlexHKW) at the biomass heat and power plant "Bioenergie Wächtersbach." This, with woodchips-fed plant, has a capacity of 1.2 $MW_{el.}$, respectively, 4.9 MW_{th} and is equipped with an ORC and an 6 MW_{th} oil-fired peak load boiler. The plant provides a 19 km district heat network with private customers, municipal buildings and industrial consumers. In the project a bypass around the turbine was used and some modifications in the system control made. Thus, it is now possible in times with lower power demand to bypass a portion of the steam around the turbine and to use for heating the network. The costs amounted below 10,000 euros. These investments have been paid back after 1 year by the provision of minutes reserve and (negative) secondary standard service for electricity (Krengel 2015; Hoffstede 2013).

The Fraunhofer Center for Chemical-Biotechnological Processes (CBP) is the heart of the cluster of excellence: "Bioeconomy." Sixty partners from the region pool hear their competences to build an optimum utilization chain around the local beech wood – through a targeted integration of chemical, paper, pulp, automotive, construction and textile industries. Other partners in the cluster "Bioeconomy" are Linde, Total and Vattenfall as well as many medium-sized companies as, for example, Homatherm. Also the German Biomass Research Centre (DBFZ) is integrated as a research institution in the network (BMBF 2014).

The awarded concept of VERBIO biorefinery is based on system with closed material loops and the use of the whole plant (low-quality grain and straw) for the production of biofuels. The coupling of a biomethane, bioethanol and organic fertilizer production plant is worldwide the first biorefinery realized at large scale. By the double utilization of the raw materials a particularly efficient and CO_2-saving process was achieved. The raw materials used are converted to 90% into the energy sources bioethanol and biomethane (verbiogas). The residues are going back to the crop area as fertilizer and contribute thereby to a sustainable agriculture (Verbio 2016).

4 Conclusion

Despite the fact that Germany has a leading position in separate collection and treatment of organic waste and biogas sector, there is room and need for further developments. Bioenergy has to play its role to transfer the energy system away from fossil towards renewable energy sources. The German energy transition has yet not reached its ambitious goals; the process stagnated in certain areas (Thrän et al. 2016). The position of the bioenergy has to be changed from a quantitative contribution to a higher qualitative contribution. Positive examples and results of research and developments give positive impulses for this development. However the missing political endorsement create insecurity to the bioenergy branch and obstructs the implementation activities.

References

AVA (2016) Abfallverwertung Augsburg, Energie, Biogas, http://www.ava-augsburg.de/energie/biogas/. Last access 20 June 2016

Bauer WP, Kroner T, Meier G (2014) Kombination von Vergärung mit Abfallverbrennung – energetische und wirtschaftliche Synergien. In: Strategie Planung Umweltrecht 8 (2014), TK Verlag – Fachverlag für Kreislaufwirtschaft (January 2014), pp 101–109

Bloche-Daub K, Witt J, Lenz V, Nelles M (2015) Märkte und Trends von regenerativen Energien weltweit, in der EU und in Deutschland. In: Herbes C, Friege C (Hrsg) Marketing Erneuerbarer Energien – Grundlagen, Geschäftsmodelle, Fallbeispiele, S 29–59, ISBN 978-3-658-04967-6. doi:10.1007/978-3-658-04968-3_2

BMBF (2014) Bioökonomie in Deutschland Chancen für eine biobasierte und nachhaltige Zukunft, Hrsg.: Bundesministerium für Bildung und Forschung (BMBF) and Bundesministerium für Ernährung und Landwirtschaft (BMEL), Berlin 2014

BMWi (2016) Zeitreihen zur Entwicklung der erneuerbaren Energien in Deutschland, Bundesministerium für Wirtschaft und Energie (BMWi, Hrsg.) unter Verwendung von Daten der Arbeitsgruppe Erneuerbare Energien-Statistik (AGEE-Stat), Stand: Februar 2016

Brosowski A, Mantau U, Mahro B, Thrän D, Adler P, Erdmann G, Stinner W, Blanke C, Hering T, Reinholdt G (2015) Biomassepotenziale von Rest- und Abfallstoffen – Status quo in Deutschland, Schriftenreihe Nachwachsende Rohstoffe, Bd. 36, Fachagentur Nachwachsende Rohstoffe e.V., 2015, ISBN 978-3-942147-29-3

FNR (2016) Einheimische Bioenergie kann 2050 bis zu 26 Prozent unseres Primärenergiebedarfs decken, press release 27.04.16. http://www.fnr.de. Last access at 17 June 2016

Hoffstede U (2013) FlexHKW – Flexibilisierung des Betriebes von Heizkraftwerken. In: Konferenzreader 5. Statuskonferenz "Energetische Biomassenutzung", 14./15. November 2013 in Leipzig, DBFZ, Leipzig, pp 496–499

Holzhammer U, Nelles M, Scholwin F (2013) Auswirkungen der flexiblen Stromproduktion aus Biogas auf den konventionellen Kraftwerkspark und dessen CO_2-Emissionen, In: Kern M, Raussen T (Hrsg) Optimierte Erfassung und Verwertung von Bioabfall, Reihe Witzenhausen-Institut, Neues aus Forschung und Praxis, S. 145–167, Dezember 2013, ISBN 3-928673-65-3

Holzhammer U, Stelzer M, Nelles M, Scholwin F (2015) Die neue Flexibilität der Stromwirtschaft und die zukünftige Rolle der Abfallwirtschaft, In: Erich-Schmidt-Verlag (Hrsg) Müll und Abfall, 02.15, S. 79–89. ISBN 978-3-503-12493-0

Krengel U (2015) Flexible Stromproduktion mit Heizkraftwerken möglich, IWES press Release 29. Oktober 2015, pp 1–2

Lange A (2016) Quo vadis BAVA Schwerinum? Bioabfallverwertungsanlage Schwerin – erst Betriebserfahrungen. In: Conference proceedings of the 17. DIALOG Abfallwirtschaft MV 15. June 2016, ISBN 978-3-86009-432-7, pp 113–124

Lenz V (2016) SmartBiomassHeat – Wärme aus Biomasse im Verbund der erneuerbaren Energien. In: Conference proceedings of the 10. Rostocker Bioenergieforum, 16.-17.06.2016 in Rostock, ISBN 978-3-86009-433-4, pp 83–96

Mauky E, Jacobi HF, Liebetrau J, Nelles M (2014) Flexible biogas production for demand-driven energy supply – feeding strategies and types of substrates. Bioresour Technol 178:262–269. doi:10.1016/j.biortech.2014.08.123

Morscheck G, Schüch A, Lemke A, Nelles M (2015) Biowaste recycling in Germany – further challenges. In: Conference proceedings of the 5th international conference on Solid Waste Management (5 IconSWM2015) from 25–27 November 2015, at the Indian Institute of Science, Bangalore, India

Nelles M, Grünes J, Rensberg N, Schüch A (2015) Utilization of organic waste and residues in Germany. Appl Mech Mater 768:714–721. doi:10.4028/www.scientific.net/AMM.768.714. ISSN 1662-7482

Schüch A, Daniel-Gromke J, Liebetrau J, Nelles M (2014) Stand und Perspektiven der Abfall- und Reststoffvergärung in Deutschland. In: Fachverband Biogas (Hrsg) BIOGASJournal 2/2014, S. 34–38, ISSN 1619-8913

Thrän D, Seitz ST, Wirkner R, Nelles M (2016) Die Rolle der Bioenergie in der Energiewende – das „Smart Bioenergy"-Konzept. In: Conference proceedings of the 10. Rostocker Bioenergieforum, 16.-17.06.2016 in Rostock, ISBN 978-3-86009-433-4, pp 15–27

UBA (2016) Netto-Bilanz der vermiedenen Treibhausgas-Emissionen durch die Nutzung erneuerbarer Energien (2015), Umweltbundesamt (UBA), Emissionsbilanz erneuerbarer Energieträger, unter Verwendung von Daten der AGEE-Stat, Stand: März 2016

Verbio (2016) Bioraffinerie. http://www.verbio.de/nachhaltigkeit/bioraffinerie/. Last access at 17 June 2016

Recent Trends in Catalytic Hydrolysis of Waste Lignocellulosic Biomass for Production of Fermentable Sugars

S. Chatterjee and R. Chakraborty

Abstract This article presents a comprehensive comparative assessment of the reaction conditions employed in the heterogeneous and homogeneous catalytic hydrolysis of waste lignocellulosic biomass (WLB) for the production of fermentable sugar (FS) for its subsequent conversion to renewable bioethanol. The effects of catalyst type and reaction conditions on the selectivity of FS in catalytic hydrolysis of low-cost WLB have been meticulously assessed. Moreover, representative radar plots demonstrating FS (substrate for bioethanol) yield in both homogeneous and heterogeneous catalytic protocols have been elucidated. An intensive global attention has recently been paid for the improvement of catalytic technologies pertaining to efficient pretreatment and hydrolysis for conversion of WLB to FS. Cellulose [$(C_6H_{10}O_5)_n$], the foremost component in WLB materials, is a biodegradable polymer of simple carbohydrates, consisting of β (1, 4)-linkage of D-glucose units, which can be depolymerized to FS for the subsequent sustainable synthesis of renewable biofuels. In this article, a critical assessment of the production of FS through catalytic pretreatment and subsequent hydrolysis of WLB resources has been elucidated. The abundant presence of low-cost WLB and their potential application for synthesis of FS (D-glucose) and other derivatives (xylose) for subsequent bioethanol, biobutanol, bio-H_2 production can provide an economically sustainable and environmentally benign avenue to mitigate energy crisis and global climate change.

The present study reveals the effects of important process parameters, viz. hydrolysis time, catalyst concentration, temperature and water to WLB ratio on the selectivity of D-glucose in both homogeneous and heterogeneous catalytic hydrolysis of WLB along with various advanced pretreatment intensification protocols. In order to improve the existing drawbacks, recent efforts have been made to develop advanced methods through utilization of ionic liquid, microwave, and infrared irradiation as well as ultrasonication to make the overall process more efficient and environmentally benign.

Keywords Lignocellulosic biomass • Hydrolysis • Fermentable sugar • Heterogeneous catalysts • Homogeneous catalysts • Process intensification

---

S. Chatterjee • R. Chakraborty (✉)
Jadavpur University, Kolkata, West Bengal, India
e-mail: ch.rajat@gmail.com

S.K. Ghosh (ed.), *Utilization and Management of Bioresources*,
DOI 10.1007/978-981-10-5349-8_2

1 Introduction

The major constituents of natural waste lignocellulosic biomass (WLB) are lignin, cellulose and hemicelluloses (Somerville et al. 2004). The percentage of the constituents may vary according to the nature of the WLB (Carrier et al. 2011). Owing to the widespread networks of inter- and intramolecular hydrogen bonding, cellulose is insoluble in water. It is, therefore, problematic to process and hydrolyze in solution (Murakami et al. 2007). Thus, it is necessary to carry out economic and effective pretreatment of WLB to convert it into accessible cellulose and hemicelluloses for subsequent catalytic hydrolysis leading to the formation of fermentable sugar (FS) for subsequent bioethanol production (Kumar et al. 2009; Mosier et al. 2005).

The main objectives of the pretreatment process are (Somerville et al. 2004) to disrupt the crystalline structure of the cellulose present in WLB, (Carrier et al. 2011) to break down the complex polymer chains of lignin as well as hemicelluloses and (Murakami et al. 2007) to increase the porosity and the surface area of cellulose for better accessibility and high productivity of desired product for subsequent hydrolysis step (Lynd et al. 1999; Li et al. 2010). In this article, both homogeneous and heterogeneous catalytic pretreatment processes have been discussed. Moreover, recent pretreatment process intensification protocols such as the application of microwave radiation, ionic liquids (ILs) and ultrasound wave have been meticulously reviewed.

Furthermore, the catalytic conversion of WLB to valuable chemicals faces additional difficulties owing to the inert chemical structure and complex molecular distribution of carbon, oxygen and hydrogen present in the WLB. In this context, it is necessary to develop new highly activated, recyclable low-cost eco-friendly heterogeneous solid catalysts for the yield of desired product FS (D-glucose) (Abbadi et al. 1998; Dhepe et al. 2005; Stocker 2008). To date, most of the research works related to kinetics of heterogeneous catalyzed conversion of WLB are based on the lignocellulosic model compounds such as synthetic crystalline cellulose, commercial cellobiose and xylan (Bootsma and Shanks 2007; Lai et al. 2011; Rick et al. 2012).

The present review mainly concerns about the production of FS (D-glucose) through catalytic pretreatment and subsequent hydrolysis of naturally available WLB resources, incorporating the significant effects of corresponding process parameters. The abundant presence of low-cost WLB and their potential application for synthesis of FS can provide sustainable environmentally benign avenue overcoming energy crisis and mitigating global environmental problems. Additionally, growing endeavours are being made to integrate the multistep batch pretreatment and hydrolysis processes into a continuous conversion of WLB employing well-designed multifunctional low-cost environmentally friendly heterogeneous solid catalysts to achieve the desired sustainable yields of valuable products (Yan et al. 2006).

2 Pretreatment of WLB

In last few years, many procedures have been employed for cellulose pretreatment prior to hydrolysis (Hamelinck et al. 2005; Sousa et al. 2009); selected works have been discussed in the following sections. In Fig. 1, a schematic for pretreatment and subsequent processes for FS synthesis has been depicted. Moreover, various new advanced pretreatment processes and their different technical aspects have been highlighted in comparison with the existing processes in the following sections.

2.1 Homogeneous Acid-Catalyzed Pretreatment

In recent years, in addition to physical pretreatments, e.g. grinding and milling, many researchers applied physicochemical homogeneous acid (HCl, H_3PO_4)-catalyzed pretreatment on various types of WLB (corn stover, *Achyranthes aspera*, *Sida acuta*) to produce FS (86.2 wt. %, 85.4 wt. %) which typically involves moderate reaction time of 40 min to 1 h at a moderately high temperature range of 60–120 °C (Siripong et al. 2016; Zu et al. 2014). Nonetheless, owing to requirements of elevated temperature (Kim et al. 2014; Kundu and Lee 2015) and lengthy time, the process involved high energy consumption in addition to equipment corrosion and generation of the massive acidic waste stream and difficulties in product separation. Thus, to augment the efficiency of the pretreatment process, heterogeneous catalyst(s) must be applied to overcome the difficulties pertaining to homogeneous acid-catalyzed pretreatment processes.

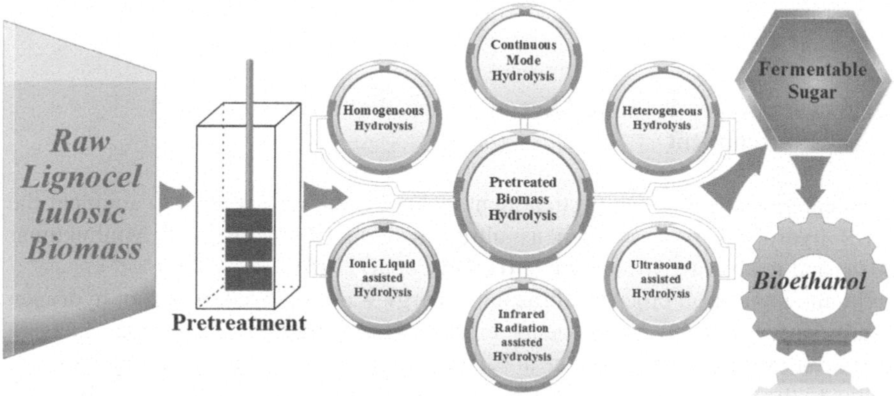

Fig. 1 Schematic for the hydrolysis of WLB to synthesize FS and subsequent bioethanol production

2.2 Heterogeneous Acid-Catalyzed Pretreatment

Recently, Tan and Lee (2015) reported maximum 99.8 wt. % FS (D-glucose) from macroalgae cellulosic residue using Dowex™ Dr-G8 as a heterogeneous solid catalyst at 120 °C for 30 min. Although solid-catalyzed pretreatment demonstrated remarkable performance, however, all these catalysts were synthesized using reagent-grade chemicals rendering expensive catalyst, which, in turn, made the process cost-intensive.

2.3 Intensification of Pretreatment Using Recent Protocols

Recently, several researchers have reported improved new processes such as ultrasonication and applications of the microwave to intensify the pretreatment of the raw WLB; these are precisely discussed in the following sections.

2.3.1 Pretreatment Using Ionic Liquid (IL)

Pretreatment employing ILs has gained much interest in the field of WLB hydrolysis (Perez-Pimienta et al. 2016). Recently, Farahania et al. (2016) reported the pretreatment of poplar biomass with IL (1-ethyl-3-methyl-imidazolium acetate) at a lower temperature of 50 °C for 24 h rendering significant (80 wt. %) FS (glucose) yield. However, the application of ILs in WLB hydrolysis is not economically attractive due to its high cost, handling difficulties and problems in product separation.

2.3.2 Ultrasonication- and Microwave Irradiation (MI)-Assisted Pretreatment

In recent years, new green-pretreatment technologies were applied to augment the WLB hydrolysis process such as ultrasonication (Ramadoss and Muthukumar 2014) and microwave irradiation (MI) (Diaz et al. 2015).

In ultrasonication process, ultrasound energy was employed to disrupt hydrogen bonds between WLB components enabling mass transport and, thus, causing improved WLB digestion (Li et al. 2015). In spite of being a green technology, owing to the higher power consumptions (400–600 W) and relatively poor FS (glucose) yields (Chen et al. 2011) made the overall process economically less attractive.

On the other hand, another new eco-friendly pretreatment technique, viz. MI, has been applied over the last few years to intensify and enhance FS and subsequent bioethanol production (Ninomiya et al. 2014). Recently, Gabhane et al. (2014) used banana agricultural waste for obtaining maximum FS yield around 47.33 wt. % at 50 °C in 15 min. However, the high power requirement of MI system made the overall process economically unattractive for scale-up purposes.

3 Hydrolysis of WLB

The next step after pretreatment in the synthesis of FS is hydrolysis of oligosaccharides, cellulose and hemicelluloses components. Various industrially important chemicals like FS, 5-HMF, furfural, etc. are synthesized in this step. To intensify the hydrolysis of WLB and to augment the desired product selectivity, various advanced cutting-edge technologies like ultrasonication, microwave radiation and different types of green solid heterogeneous solid catalysts (synthesized or commercially available) have been employed in the hydrolysis. The characteristics of the emerging hydrolysis processes have been enunciated in following sections.

3.1 Homogeneous Acid-Catalyzed Hydrolysis of WLB

Many industrially important fermentative products such as bioethanol, levulinic acid, furfural, etc. can be obtained from waste WLB hydrolysate (Jönsson et al. 2013). The most well-established, extensively used procedure, e.g. acid hydrolysis (Gütsch et al. 2012), has been found effective for WLB hydrolysis. Nonetheless, homogeneous acid hydrolysis has several limitations such as elevated reaction temperature (170–240 °C), equipment corrosion, difficulty in product separation and high reaction time (Taherzadeh and Karimi 2007) making the overall process problematic.

3.2 Heterogeneous Solid Acid-Catalyzed Hydrolysis of WLB

Over recent past, the applications of green solid acid heterogeneous catalysts such as supported metals, acid resins, H-form zeolites, carbonaceous acids, functionalized silica, metal oxides, etc. had gained widespread interest in the WLB hydrolysis process. For WLB hydrolysis, Amberlyst-15, a solid acid catalyst (Meena et al. 2015), was found very much effective (Onda et al. 2008; Pang et al. 2010), and owing to the presence of the SO_3H group, the catalyst could selectively allow penetration of hydrogen ions of reactants during the hydrolysis reaction.

Notably, most of the heterogeneous solid acid catalysts applied for WLB hydrolysis were developed from cost-intensive reagent-grade chemicals besides requiring prolonged hydrolysis time and relatively elevated temperature in comparison with homogeneous WLB hydrolysis. Notably, the economic sustainability of hydrolysis process is greatly dependent on the efficacy, reusability and cost-effectiveness of the heterogeneous catalyst. In this context, low-cost catalysts which have been preliminary derived from waste biomass resources like carbon-based solid acid catalysts (CBSAC) are noteworthy (Suganuma et al. 2008). This catalyst rendered hydrolysis of untreated cellulose yielding 4 wt. % glucose at 100 °C for 3 h.

However, most of the previously mentioned research works have been performed on lignocellulosic model compounds, i.e. commercially available microcrystalline cellulose, synthetic cellobiose and xylan (Lai et al. 2011; Rick et al. 2012). Recently, Nata et al. (2015) achieved approximately 19.91 mg/mL of FS from hydrolysis of corn starch at elevated temperature (150 °C) over prolonged (6 h) reaction time using C_4-SO_3H as the solid acid catalyst. Very recently, Hu et al. (2016) have prepared a magnetic carbonaceous solid acid containing chlorine (−Cl) groups as cellulose-binding sites and sulfonic (−SO_3H) groups as cellulose-hydrolyzing sites; the catalyst was used for hydrolysis of rice straw in presence of ionic liquid 1-butyl-3-methylimidazolium chloride ([BMIM]Cl) at high reaction temperature of 130 °C for 4 h resulting in maximum 78.5 % of FS. Additionally, Zhong et al. (2015) and Sakdaronnarong et al. (2016) also investigated the efficacy of various carbon-based catalysts on hydrolysis of wheat straw and sugarcane bagasse, respectively.

The effects of reaction conditions on the selectivity of FS are presented in Table 1 for heterogeneous solid acid-catalyzed systems. The yield of FS varies significantly from 9 wt. % to 50 wt. % through application of different types of heterogeneous catalysts. A poor selectivity of FS was observed using Nafion/silica catalyst, whereas using sulfonated silica/carbon nanocomposites, considerable improvement in sugar selectivity could be achieved.

On the other hand, the usage of AC-SO_3H, as well as $HNbMoO_6$, exhibited a moderate selectivity to FS. However, in general, high temperature and long reaction time remain as the major drawbacks to solid-catalyzed hydrolysis. Figure 2 demonstrates the deviation in FS selectivity in accordance with select references.

Table 1 Overview of reaction conditions and selectivity of FS using heterogeneous catalysts

Catalyst	Reaction condition Temperature (K)	Time (h)	FS selectivity (%)	References	Remarks
AC-SO_3H	423	24	40.5	Onda et al. (2008)	Elevated reaction temperature; long reaction time; poor FS selectivity
Sulfonated silica/carbon nanocomposites	423	24	50.0	Vyver et al. (2010)	
$HNbMoO_6$	403	12	21.0	Takagaki et al. (2008)	
Nafion/silica	463	24	9.0	Hegner et al. (2010)	

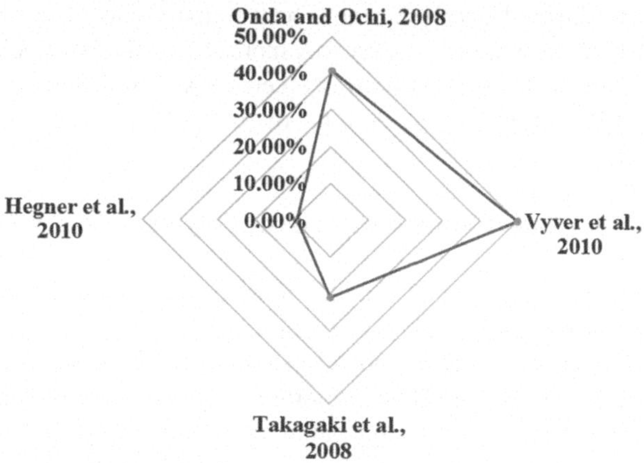

Fig. 2 Radar plot depicting selectivity of fermentable sugar through the heterogeneous catalytic protocol in concurrence with select references

3.3 Intensification of Hydrolysis Using Recent Protocols

In recent years, several research works have been reported involving applications of the microwave, ultrasound and ILs to intensify the hydrolysis of the WLB for possible enhancement of FS yield. These have been briefly discussed in the following sections.

3.3.1 Ultrasound-Assisted (US) Hydrolysis of WLB

US hydrolysis of WLB has gained remarkable attention in the recent past. Werle et al. (2013) reported acid hydrolysis of waste palm leaves (*Roystonea oleracea*) by ultrasound obtaining maximum 74 % yield of FS at 65 °C for 300 min using phosphoric acid as catalyst. Silva et al. (2015) reported the application of ultrasound in enzymatic hydrolysis of sugarcane bagasse attaining maximum FS of 217 g kg^{-1} for 4 h at 47 °C. On the other hand, Borah et al. (2016) reported ultrasound-assisted enzymatic hydrolysis of various types of WLB such as *Eichhornia crassipes* and *Saccharum spontaneum*, to achieve maximum 40.02 % of FS. Nonetheless, the overall process was economically unattractive due to the high equipment and energy cost for ultrasonication.

3.3.2 Microwave Irradiation (MI)-Assisted Hydrolysis of WLB

In recent past, MI was applied to WLB hydrolysis to enhance the hydrolysis rate and to augment the yield of FS and consequent bioethanol production (Xue et al. 2011). Recently, Villière et al. (2013) reported maximum 46 wt. % FS yield from wet potato sludge (industrial waste) in 2 h at 60 °C using combined MI and ultrasonication and sulphuric acid as catalyst. Moreover, the research group also stated that, during hydrolysis, rapid heat transfer was observed due to the application of MI, whereas mass transfer was increased at the interfacial boundary layers of solid-liquid hydrolysis system. Nevertheless, high microwave energy requirements (1 kW) might be a concern for the overall process economy.

3.3.3 Ionic Liquid (ILs)-Assisted Hydrolysis of WLB

Recently, Wang et al. (2014) reported hydrolysis of bamboo biomass using six different types of metal ions including Mg^{2+}, Ca^{2+}, Na^{+}, K^{+}, Cu^{2+} and Fe^{3+} obtaining 67.1 wt. % of FS at 100 °C in 4 h with $CuCl_2$ as co-catalyst. On the other hand, Ramli and Amin (2014) reported usage of 1-butyl-3-methylimidazolium bromide (BMIMBr) as ILs for oil palm frond and empty fruit bunch direct hydrolysis that resulted in maximum 27.4 wt. % and 24.8 wt. % FS yield (120 °C,4 h) using solid Fe/HY catalyst. Nonetheless, the process has limitations in terms of use of high-cost ILs and difficult product separation with lower FS yield.

3.3.4 Hydrolysis WLB Carried Out in Continuous Mode Reactors

Most of WLB hydrolyses were carried out in batch reactors. Some pioneer research works have been conducted in continuous-type reactors (CTR) resulting augmented yield of FS at mild reaction conditions using shorter residence time (Kumakura and Kaetsu 1978; Church and Wooldrldge 1981). Kim et al. (2005) reported hydrolysis of glucans from pretreated corn fibre obtaining around 90 wt. % of glucose at 160 °C in 3.5 min residence time in a packed-bed reactor using a cation exchanger. Importantly, they observed that the cost of catalyst had a significant influence on the overall production cost of FS and subsequent bioethanol production. To date, very scanty reports are available for conversion of WLB into FS in CTR (Kim et al. 2005) using heterogeneous catalysts. Future works should focus on enhancing the capacity of the FS production using continuous catalytic reactor.

3.3.5 Fast Pretreatment-Hydrolysis WLB Using Energy-Efficient Infrared Radiation

More recently, our research group (Chatterjee et al. 2016) has successfully achieved a promising high 89.87 mol% glucose (FS) yield from waste watermelon (*Citrullus lanatus*) peel at 60 °C applying energy-efficient far-infrared radiation in presence of heterogeneous Amberlyst-15 catalyst in one-pot pretreatment-hydrolysis system in much shorter time.

4 Prospects and Conclusion

Waste lignocellulosic biomass is the most abundant renewable bioresource on earth. It has been considered as one of the most significant natural resources for the production of industrially valuable products such as glucose, fructose, 5-HMF as well as bioethanol and biobutanol. In the past years, several pretreatments and subsequent hydrolysis processes have been developed. Generally, the heterogeneous pretreatment and hydrolysis processes are preferable over homogenous processes mainly due to the case of product separation and reduced corrosion, leaching and handling problems. However, the invention of the heterogeneous solid catalysts with high catalytic activity for both pretreatment and subsequent hydrolysis processes still faces a challenging issue mainly owing to the solid-solid contact between the insoluble waste lignocellulosic biomass and solid catalyst.

However, the application of ionic liquids as a solvent or as a catalyst can be considered as an alternative solution; nonetheless, high cost and difficulties associated with separation and recovery have made such process economically unattractive. On the other hand, other eco-friendly intensification protocols such as the application of microwave and ultrasound on pretreatment and consequent hydrolysis steps rendered superior yields of desired products at comparatively milder operating conditions in comparison with conventional heated reactors. Nevertheless, microwave and ultrasound processes usually require high energy inputs; the consequent increment in production cost makes the overall process economically unpleasant. Furthermore, the most economically sustainable avenue in terms of feedstock should be augmented utilization of waste lignocellulosic biomass rather than pure cellulose.

The present study reveals that the usage of both homogeneous and heterogeneous catalysts in conventional pretreatment and hydrolysis steps of waste lignocellulosic biomass involve high reaction time, temperature and significantly low selectivity towards desired product. In order to improve the existing protocols, effort must be made to develop newer advanced technology involving process intensification protocol such as energy-efficient infrared radiation. Besides, more research needs to be conducted on continuous mode of catalytic pretreatment and subsequent hydrolysis for enhanced throughput of fermentable sugar in such a way that the overall process remains environmentally benign as well as energy-efficient and economically viable to render a sustainable protocol.

References

Abbadi A, Gotlieb KF, Bekkum H (1998) Study on solid acid catalyzed hydrolysis of maltose and related polysaccharides. Starch/Staerke 50(1):23–28

Bootsma JA, Shanks BH (2007) Cellobiose hydrolysis using organic–inorganic hybrid mesoporous silica catalysts. Appl Catal A 327:44–51

Borah AJ, Agarwal M, Poudyal M, Goyal A, Moholkar VS (2016) Mechanistic investigation in ultrasound induced enhancement of enzymatic hydrolysis of invasive biomass species. Bioresour Technol 213:342–349

Carrier M, Loppinet-Serani A, Denux D, Lasnier JM, Ham-Pichavant F, Cansell F, Aymonier C (2011) Thermogravimetric analysis as a new method to determine the lignocellulosic composition of biomass. Biomass Bioenergy 35(1):298–307

Chatterjee S, Barman S, Chakraborty R (2016) Far infrared radiated energy-proficient rapid one-pot green hydrolysis of waste watermelon peel: optimization and heterogeneous kinetics of glucose synthesis. RSC Adv 6:74278–74287

Chen WH, Tu YJ, Sheen HK (2011) Disruption of sugarcane bagasse lignocellulosic structure by means of dilute sulfuric acid pretreatment with microwave assisted heating. Appl Energy 88:2726–2734

Church JA, Wooldrldge D (1981) Continuous high-solids acid hydrolysis of biomass in a 11/2-in. Plug flow reactor. Ind Eng Chem Prod Res Dev 20:378–382

Dhepe PL, Ohashi M, Inagaki S, Ichikawa M, Fukuoka A (2005) Hydrolysis of sugars catalyzed by water-tolerant sulfonated mesoporous silicas. Catal Lett 102(3–4):163–169

Diaz AB, Moretti MMS, Bussoli CB, Nunes CCC, Blandino A, Silva R, Gomes E (2015) Evaluation of microwave-assisted pretreatment of lignocellulosic biomass immersed in alkaline glycerol for fermentable sugars production. Bioresour Technol 185:316–323

Farahania SV, Kimb Y, Schalla C (2016) A coupled low temperature oxidative and ionic liquid pretreatment of lignocellulosic biomass. Catal Today 269:2–8

Gabhane J, William SPMP, Gadhe A, Rath R, Vaidya AN, Wate S (2014) Pretreatment of banana agricultural waste for bio-ethanol production: individual and interactive effects of acid and alkali pretreatments with autoclaving, microwave heating and ultrasonication. Waste Manag 34:498–503

Gütsch JS, Nousiainen T, Sixta H (2012) Comparative evaluation of autohydrolysis and acid-catalyzed hydrolysis of Eucalyptus globulus wood. Bioresour Technol 109:77–85

Hamelinck CN, Hooijdonk GV, Faaij APC (2005) Ethanol from lignocellulosic biomass: techno-economic performance in short-, middle- and long-term. Biomass Bioenergy 28(4):384–410

Hegner J, Pereira KC, Boef BD, Lucht BL (2010) Conversion of cellulose to glucose and levulinic acid via solid-supported acid catalysis. Tetrahedron Lett 51(17):2356–2358

Hu L, Li Z, Wu Z, Lin L, Zhou S (2016) Catalytic hydrolysis of microcrystalline and rice straw-derived cellulose over a chlorine-doped magnetic carbonaceous solid acid. Ind Crop Prod 84:408–417

Jönsson LJ, Alriksson B, Nilvebrant NO (2013) Bioconversion of lignocellulose: inhibitors and detoxification. Biotechnol Biofuels 6:16

Kim Y, Hendrickson R, Mosier N, Ladisch MR (2005) Plug-flow reactor for continuous hydrolysis of glucans and xylans from pretreated corn fiber. Energy Fuel 19:2189–2200

Kim I, Lee B, Park J, Choi S, Han J (2014) Effect of nitric acid on pretreatment and fermentation for enhancing ethanol production of rice straw. Carbohydr Polym 99:563–567

Kumakura M, Kaetsu I (1978) Radiation-induced degradation and subsequent hydrolysis of waste cellulose materials. Int J Appl Radiat Isot 30:139–141

Kumar P, Barrett DM, Delwiche MJ, Stroeve P (2009) Methods for pretreatment of lignocellulosic biomass for efficient hydrolysis and biofuels production. Ind Eng Chem Res 48:3713–3729

Kundu C, Lee J (2015) Optimization conditions for oxalic acid pretreatment of deacetylated yellow poplar for ethanol production. J Ind Eng Chem 32:298–304

Lai DM, Deng L, Li J, Liao B, Guo QX, Fu Y (2011) Hydrolysis of cellulose into glucose by magnetic solid acid. Chem Sus Chem 4(1):55–58

Li C, Knierim B, Manisseri C, Arora R, Scheller HV, Auer M, Vogel KP, Simmons BA, Singh S (2010) Comparison of dilute acid and ionic liquid pretreatment of switchgrass: biomass recalcitrance, delignification and enzymatic saccharification. Bioresour Technol 101:4900–4906

Li J, Zhang X, Zhang M, Xiu H, He H (2015) Ultrasonic enhance acid hydrolysis selectivity of cellulose with HCl–$FeCl_3$ as catalyst. Carbohydr Polym 117:917–922

Lynd LR, Wyman CE, Gerngross TU (1999) Biocommodity engineering. Biotechnol Program 15(5):777–793
Meena S, Navatha S, Prabhavathi Devi BLA, Prasad RBN, Pandey A, Sukumaran RK (2015) Evaluation of Amberlyst15 for hydrolysis of alkali pretreated rice straw and fermentation to ethanol. Biochem Eng J 102:49–53
Mosier N, Wyman C, Dale B, Elander R, Lee YY, Holtzapple M, Ladisch M (2005) Features of promising technologies for pretreatment of lignocellulosic biomass. Bioresour Technol 96(6):673–686
Murakami M, Kaneko Y, Kadokawa J (2007) Preparation of cellulose-polymerized ionic liquid composite by in-situ polymerization of polymerizable ionic liquid in cellulose-dissolving solution. Carbohydr Polym 69(2):378–381
Nata IF, Irawan C, Mardina P, Lee C (2015) Carbon-based strong solid acid for cornstarch hydrolysis. J Solid State Chem 230:163–168
Ninomiya K, Yamauchi T, Oginoc C, Shimizua N, Takahashi K (2014) Microwave pretreatment of lignocellulosic material in cholinium ionic liquid for efficient enzymatic saccharification. Biochem Eng J 90:90–95
Onda A, Ochi T, Yanagisawa K (2008) Selective hydrolysis of cellulose into glucose over solid acid catalysts. Green Chem 10:1033–1037
Pang J, Wang A, Zheng M, Zhang T (2010) Hydrolysis of cellulose into glucose over carbons sulfonated at elevated temperatures. Chem Commun 46:6935–6937
Perez-Pimienta JA, Flores-Gómez CA, Ruiz HA, Sathitsuksanoh N, Balan V, Sousa LC, Dale BE, Singh S, Simmons BA (2016) Evaluation of agave bagasse recalcitrance using AFEXTM, autohydrolysis, and ionic liquid pretreatments. Bioresour Technol 211:216–223
Ramadoss G, Muthukumar K (2014) Ultrasound assisted ammonia pretreatment of sugarcane bagasse for fermentable sugar production. Biochem Eng J 83:33–41
Ramli NAS, Amin NAS (2014) Catalytic hydrolysis of cellulose and oil palm biomass in ionic liquid to reducing sugar for levulinic acid production. Fuel Process Technol 128:490–498
Rick O, James RK, Joby M (2012) Hemicellulose hydrolysis using solid acid catalysts generated from biochar. Catal Today 190(1):89–97
Sakdaronnarong C, Saengsawang A, Siriyutta A, Jonglertjunya W, Nasongkla N, Laosiripojana N (2016) An integrated system for fractionation and hydrolysis of sugarcane bagasse using heterogeneous catalysts in aqueous biphasic system. Chem Eng J 285:144–156
Silva JRF, Cantelli KC, Soares MBA, Tres MV, Oliveira D, Meireles MAA, Oliveira JV, Treichel H, Mazutti MA (2015) Enzymatic hydrolysis of non-treated sugarcane bagasse using pressurized liquefied petroleum gas with and without ultrasound assistance. Renew Energy 83:674–679
Siripong P, Duangporn P, Takata E, Tsutsumi Y (2016) Phosphoric acid pretreatment of Achyranthes aspera and Sida acuta weed biomass to improve enzymatic hydrolysis. Bioresour Technol 203:303–308
Somerville C, Bauer S, Brininstool G, Facette M, Hamann T, Milne J, Osborne E, Paredez A, Persson S, Raab T, Vorwerk S, Youngs H (2004) Toward a system approach to understanding plant cell walls. Science 306(5705):2206–2211
Sousa LC, Chundawat SPS, Balan V, Dale BE (2009) 'Cradle-to-grave' assessment of existing lignocellulose pretreatment technologies. Curr Opin Biotechnol 20(3):339–347
Stocker M (2008) Biofuels and biomass-to-liquid fuels in the biorefinery: catalytic conversion of lignocellulosic biomass using porous materials. Angew Chem Int Ed 47(48):9200–9211
Suganuma S, Nakajima K, Kitano M, Yamaguchi D, Kato H, Hayashi S, Hara M (2008) Hydrolysis of cellulose by amorphous carbon bearing SO_3H, COOH, and OH groups. J Am Chem Soc 130:12787–12793
Taherzadeh MJ, Karimi K (2007) Enzyme based hydrolysis process for ethanol from lignocellulosic materials: a review. Bioresour Com 2:707–738
Takagaki A, Tagusagawa C, Domen K (2008) Glucose production from saccharides using layered transition metal oxide and exfoliated nanosheets as a water-tolerant solid acid catalyst. Chem Commun 42:5363–5365

Tan IS, Lee KT (2015) Solid acid catalysts pretreatment and enzymatic hydrolysis of macroalgae cellulosic residue for the production of bioethanol. Carbohydr Polym 124:311–321

Villière A, Lassi U, Hu T, Paquet A, Rinaldi L, Cravotto G, Molina-Boisseau S, Marais M, Lévêque J (2013) Simultaneous microwave/ultrasound-assisted hydrolysis of starch-based industrial waste into reducing sugars. ACS Sustain Chem Eng 1:995–1002

Vyver V, Peng L, Geboers J, Schepers H, Clippel F, de Gommes CJ, Goderis B, Jacobs PA, Sels BF (2010) Sulfonated silica/carbon nanocomposites as novel catalysts for hydrolysis of cellulose to glucose. Green Chem 12:1560–1563

Wang N, Zhang J, Wang H, Li Q, Wei S, Wang D (2014) Effects of metal ions on the hydrolysis of bamboo biomass in 1-butyl-3-methylimidazolium chloride with dilute acid as catalyst. Bioresour Technol 173:399–405

Werle LB, Garcia JC, Kuhn RC, Schwaab M, Foletto EL, Cancelier A, Jahn SL, Mazutti MA (2013) Ultrasound-assisted acid hydrolysis of palm leaves (Roystonea oleracea) for production of fermentable sugars. Ind Crop Prod 45:128–132

Xue J, Chen HZ, Kadar Z (2011) Optimization of microwave pretreatment on wheat straw for ethanol production. Biomass Bioenergy 35:3859–3864

Yan N, Zhao C, Gan WJ, Kou Y (2006) Polyols: a new generation of energy platform? Chin J Catal 27(12):1159–1163

Zhong C, Wang C, Huang F, Wang F, Jia H, Zhou H, Wei P (2015) Selective hydrolysis of hemicellulose from wheat straw by a nanoscale solid acid catalyst. Carbohydr Polym 131:384–391

Zu S, Li W, Zhang M, Li Z, Wang Z, Jameel H, Chang H (2014) Pretreatment of corn stover for sugar production using dilute hydrochloric acid followed by lime. Bioresour Technol 152:364–370

The Cost-Effective Stirred Tank Reactor for Cellulase Production from Alkaline-Pretreated Agriculture Waste Biomass

N.N. Deshavath, S.K. Sahoo, M.M. Panda, S. Mahanta, D.S.N. Goutham, V.V. Goud, V.V. Dasu, and Annapurna Jetty

Abstract Traditional conversion of lignocellulosic biomass into bioethanol takes place in three processes which include pretreatment, enzymatic hydrolysis, and fermentation. Among them, enzymatic hydrolysis accounts for ~40% of the total cost. Therefore, commercial production of cellulolytic enzymes by using lignocellulosic biomass as a substrate may decrease the production cost. Alkaline pretreatment of sorghum biomass was carried out with different concentrations of NaOH at 121 °C for 20–60 min for the solubilization of hemicellulose and lignin in order to increase the cellulose surface area. This pretreated biomass has comparatively rich content of cellulose than raw biomass. These pretreated substrates were subjected to cellulase production using *Phanerochaete chrysosporium* NCIM 1106 for the optimization of enzymatic activity (FPU). Therefore, maximum cellulase activity (36.8 FPU/g) was obtained during the fermentation of 0.2 M NaOH (at 60 min)-pretreated substrate. In order to enhance the cellulase production, bioreactor studies were performed under submerged fermentation. A laboratory scale stirred tank reactor (STR) was designed which was easily operational and sustainable. As a result, enzyme activity was successfully increased up to 51.3 FPU/g. In addition to this, hydrothermal pretreatment of sorghum biomass was performed for comparison study. However, cellulase activities were relatively lower than alkaline-pretreated substrates.

Keywords Tank reactor • Agriculture waste biomass • Cost • Cellulase

---

N.N. Deshavath • V.V. Goud • V.V. Dasu (✉)
Indian Institute of Technology Guwahati, Guwahati, Assam, India
e-mail: venkatadasu@yahoo.com

S.K. Sahoo • M.M. Panda
Utkal University, Bhubaneswar, Odisha, India

S. Mahanta
Ramjas College, University of Delhi, New Delhi, India

D.S.N. Goutham • A. Jetty
CSIR-Indian Institute of Chemical Technology, Hyderabad, Telangana, India

S.K. Ghosh (ed.), *Utilization and Management of Bioresources*,
DOI 10.1007/978-981-10-5349-8_3

1 Introduction

Lignocellulosic biomass is one of the most abundant renewable resources for the biosynthesis of various value-added products. Biomass consists of three types of natural polymers such as hemicellulose, cellulose, and lignin that are strongly interlinked and chemically bonded by non-covalent forces and covalent cross-linkages. Among the lignocellulosic polymers, cellulose is very strong and is resistant to break by physical and chemical treatment. Cellulose is composed of glucose units linked with β, 1–4, glycosidic bonds to form glucan polymer. The long chains of glucan polymers cross-link together and form hydrogen as well as van der Waals bonds which results in cellulose to be packed into microfibrils that is covered by lignin and hemicelluloses (Krassig et al. 2004). Several pretreatment techniques were introduced for cellulose hydrolysis, but the major problem was formation of hazardous chemicals such as furfural, 5-hydroxymethylfurfural (5-HMF), and formic acid during the pretreatment process (Palmqvist and Hahn-Hagerdal 2000). Acid pretreatment can slightly break down the lignocellulosic structure which enhances the enzymatic hydrolysis reaction as compared to raw biomass. Pretreated biomass hydrolyzed with various enzymes such as xylanase and cellulase may increase the production cost. Cellulase is a relatively expensive enzyme which affects the bioethanol production cost. During ethanol production from lignocellulosic biomass, cost of cellulase enzymes for cellulose hydrolysis is accounted to be approximately 40% of the total cost (Spano 1978). Significant cost reduction is required in order to enhance the commercial viability of cellulase production technology. Cellulase is primarily produced in nature by fungi, bacteria, plants, and even some protozoa, nematodes, and mollusks (Watanabe and Tokuda 2001). At present, the common microbial sources of industrial cellulase producers are mesophiles grown in the temperature ranging of 30–35 °C, such as *Trichoderma reesei* and *Aspergillus niger* in solid-state fermentation. Cellulase is a multienzyme complex that consists of three major components: endo-β 1–4-glucanases (EC 3.2.1.4), exo-β 1–4-glucanase or cellobiohydrolase (EC 3.2.1.91), and β-glucosidase (EC 3.2.1.21); all these enzymes collectively act on cellulose to form glucose. Endoglucanase hydrolyzes internal β, 1–4 linkages of cellulose chains and yields new reducing and nonreducing ends. Thereupon, exoglucanase cleaves disaccharide cellobiose from the nonreducing end (cellobiohydrolase) and in some cases from the reducing end (cellobiosidase), generally from the crystalline part of the cellulose chain. These cellobiose units and short-chain cello-oligosaccharides are hydrolyzed by β-glycosidase into individual monomeric units of glucose (Wilson 2009).

Development of an economical process for cellulase production is hindered because of the cost of substrate and purification process. The cost of raw material is the limiting factor in developing an economically viable process for cellulase production (Reczey et al. 1996). The cost of enzyme production could be minimized by the use of cheap cellulosic substrates (Lynd et al. 2002). Lignocellulosic biomass can be utilized as a substrate for cellulase production by various fungal strains. It is important to pretreat the raw lignocellulosic biomass, as it is an essential step for enzyme production. Therefore, in chemical pretreatment generally, alkaline pretreatment is necessary before cellulase production. Alkaline pretreatment mostly dissolves lignin and hemicellulose and can also affect the cellulose structure. Chemical pretreatment not only removes lignin but also acts as a swelling agent, which enhances the surface area of the substrate accessibility during enzyme production.

Several studies were performed to produce cellulase by solid-state fermentation of different lignocellulosic sources. In this present work, we made an attempt to produce cellulase by submerged fermentation in shake flask as well as in stirred tank reactor in a batch mode operation. Therefore, sorghum biomass was pretreated with various molar concentrations of sodium hydroxide (NaOH), at 121 °C for 20–60 min. This alkaline-pretreated biomass was used as a substrate for cellulase production by *P. chrysosporium* NCIM 1106. In addition to this, optimum cellulase-produced substrate was subjected to stirred tank reactor (STR) studies to enhance the fermentation efficiency.

2 Materials and Methods

2.1 Microorganism

Fungal strain, *Phanerochaete chrysosporium* NCIM 1106, was procured from the National Collection of Industrial Microorganisms (NCIM), Pune, India. The strain was preserved at 4 °C on potato dextrose agar (PDA) media.

2.1.1 Subculture and Spore Formation

Preserved fungal strain was subcultured on PDA media and kept for incubation for 5 days at 30 °C. This 5-day-old culture was aseptically harvested by adding the sterile distilled water containing 0.1% (w/v) of Tween 80 (Smits et al. 1996) and then transferred into Mendel's medium (Mandels and Reese 1957) for seed culture preparation which contains (g. L^{-1}) glucose, 5; urea, 0.3; peptone, 1 $(NH_4)_2SO_4$, 1.4; $MgSO_47H_2O$, 0.3; $CaCl_22H_2O$, 0.4; Tween 80, 0.2; and trace elements $ZnSO_47H_2O$, 0.0014; $FeSO_47H_2O$, 0.005; $MnSO_47H_2O$, 0.0016; and $CoCl_2$, 0.002. The initial pH of the medium was adjusted to 5 with 1 N H_2SO_4 or 1 N NaOH and incubated in aerobic condition at 30 °C for 48 h. Spore count was measured with hemocytometer and adjusted to 1×10^6 spores, mL^{-1}. This spore suspension was used as inoculum for cellulase production.

2.2 *Alkali Pretreatment*

Alkaline pretreatment was performed with various molar concentrations of NaOH, viz., 0.01 M, 0.05 M, 0.1 M, 0.15 M, and 0.2 M at 121 °C for 20–60 min under a solid-to-liquid ratio of 1:20 (w/v). In addition to this, hydrothermal pretreatment was also performed at aforementioned conditions. Reducing sugar estimation was performed with dinitrosalicylic acid (DNS) method according to Miller (1959). After the alkaline pretreatment, solid and liquid portions were filtered through 0.2 μm nylon membrane under vacuum condition. The solid portion was washed with distilled water to attain neutral pH and dried for 48 h at 45 °C. This can be used as a substrate for cellulase production experiments.

2.2.1 Chemical Composition Analysis

Chemical composition of raw biomass and pretreated biomass was determined according to the standard NREL protocol (Sluiter et al. 2011).

2.3 *Cellulase Production*

One gram of different alkali-pretreated substrate was loaded into 250 ml of Erlenmeyer flask which contains 100 ml of the Mendel's medium (Mandels and Reese 1957). The initial pH of medium was adjusted with 1 N H_2SO_4 or 1 N NaOH and sterilized by autoclaving at 121 °C for 20 min. The 1 mL of spore suspension was inoculated into cellulase production medium aseptically. At every 24-h intervals, 1 mL of sample was withdrawn from the production media and filtered through 0.2 μm syringe filter for cellulase activity measurement.

2.3.1 Cellulase Production by Stirred Tank Reactor (STR)

To enhance the enzyme activity and fermentation efficiency, cellulase production was performed in STR. The capacity of STR was 1 L with 0.6 L of working volume. As shown in Fig. 1, air inlet and outlet were guarded by HEPA–VENT ™ air filters (0.2 μm) in order to maintain aseptic conditions. The air was supplied through air pump at the rate of 1.3 vvm by a suitable sparger. The agitation was given by using magnetic stirrer with bead rotating at a speed of 150 rpm (note: high agitation or stirring speed leads to breakdown of mycelium). A sample collection port was also maintained in order to analyze the progress of the reaction. The constant temperature for reactor was maintained by keeping the reactor in glass trough containing water. pH of the media was adjusted prior to inoculation by adding acid or base according to requirement. A working volume of 600 mL of media (pH-5) contains

Fig. 1 Stirred tank reactor for cellulase production

6 g of alkaline-pretreated substrate (121 °C, 0.2 M NaOH and 60 min) which showed maximum cellulase activity during the shake flask method. This enzyme production was carried out under aerobic submerged fermentation at 30 °C for 7 days.

2.3.2 Cellulase Activity

Filter paper unit (FPU) assay was performed for the estimation of cellulase activity which is according to the method recommended by Ghose (1987) and expressed as international units (IU). One international unit of filter paper activity (IFPU) is the amount of enzyme which forms 1 μmol glucose min^{-1} (reducing sugars as glucose) during the saccharification reaction. The reducing sugar was determined using DNS method (Miller 1959).

3 Result and Discussion

3.1 *Effect of Alkaline Pretreatment on Sorghum Biomass*

The performance of the pretreatment is described by percentage of delignification and hemicellulose solubilization. NaOH pretreatment effectively solubilized and hydrolyzed the hemicellulose and lignin into alkaline hydrolysate. This black liquor contains sugars, oligosaccharides, and lignin derivatives. Percentage of biomass loss during the NaOH pretreatment was shown in Fig. 2a, and it was observed that biomass solubilization was increased with corresponding pretreatment reaction severity. The amount of hemicellulose and lignin was significantly affected during the alkali pretreatment, whereas cellulose loss was comparatively lower than other biopolymer. Previous studies indicate that hemicellulose is more susceptible when alkaline pretreatment process is used as compared to cellulose (Schmidt and Thomsen 1998). Previous works done by Hendriks and Zeeman (1998) and Taherzadeh and Karimi (2008) also agree that alkaline pretreatment could cause cellulose fibers to swell, rather than directly decompose it. The maximum biomass loss was observed at 121° with 0.2 M NaOH for 60 min.

Even though, high percentage of biomass was dissolved at high pretreatment severity levels, but the concentration of monosaccharide was comparatively less which might be due to the dissolved hemicellulose presence in the black liquor in the form of oligosaccharides which are not detected during the DNS assay. According to the DNS results, reducing sugar yield was decreased with increase in the pretreatment reaction severity (Fig. 2b). This could be due to the degradation of reducing sugars at high pretreatment severity. However, high concentration of reducing sugars yield was observed at 121 °C with 0.01 M for 20 min.

Apart from this, hydrothermal pretreatment of sorghum biomass yields comparatively more concentrations of reducing sugars than alkaline pretreatment except 0.01 M NaOH. The hydrolytic effect of water (at 121 °C) has significantly cleaved the nonstructural carbohydrates and minor portions of hemicellulose. Apart from this, alkaline pretreatment solubilized the hemicellulose and yields oligosaccharides

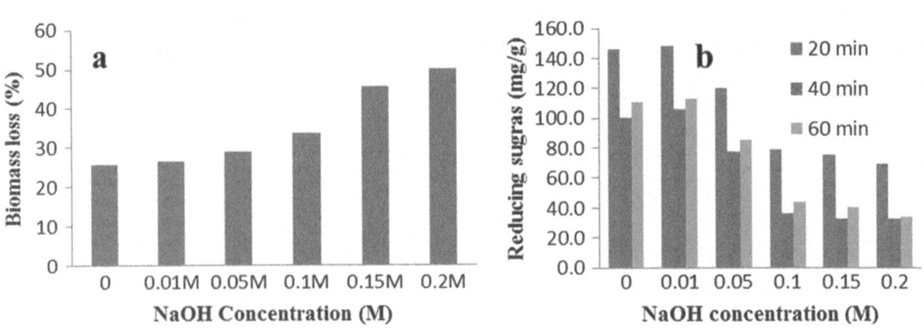

Fig. 2 (**a**) Sorghum biomass loss during the pretreatment at 60 min (**b**) Reducing sugar yield at 121 °C

Table 1 Chemical composition of raw and NaOH-pretreated sorghum biomass

Pretreatment	Cellulose (%)	Hemicellulose (%)	Lignin (%)
Raw	35.780	25.5	19.5
Water	35.780	23.2	18.46
0.01 M	36.370	21.88	20.19
0.05 M	41.370	18.34	21.22
0.1 M	45.870	12.8	25.2
0.15 M	51.670	8.5	15.3
0.2 M	60.130	3.3	11.5

rather than monosaccharide; in addition to this, reducing sugar degradation was also observed. The reducing sugar yield at 0.01 M NaOH was comparatively higher than hydrothermal treatment. This can be attributed to uniform breakage of hemicellulose to release the monomeric sugar units (xylose, arabinose). While increasing the NaOH concentration, random breakdown of hemicellulose occurred to form xylo-oligomer and concurrently monomeric sugar degradation. However, the present work is focused on maximization of hemicellulose and lignin removal for better cellulase enzyme production.

It was observed that with the increase in the NaOH concentration, cellulose content was increased after the pretreatment. This could be due to the fact that the amount of hemicellulose and lignin dissolution into alkaline black liquor was increased with corresponding NaOH concentration which eventually increased the cellulose percentage after the pretreatment. This can be validated according to the NREL protocol (Sluiter et al. 2011), and results are shown in Table 1. In addition to this, the cellulose fraction swelled and changed its configuration from crystalline form to the more reactive amorphous form (Abdul Aziz et al. 2002). This transformation significantly increases the accessibility and more amenable to the microorganism for enzyme production.

3.2 *Cellulase Enzyme Production*

After alkaline pretreatment, cellulose content increased due to the significant solubilization of hemicellulose and lignin. This could be a major advantage for the production of cellulase with different microbial strains. In this process, *P. chrysosporium* NCIM 1106 was employed to produce cellulase from alkali-pretreated substrate through submerged fermentation. Among them, it was found that sorghum biomass treated with 0.2 M for 60 min showed highest cellulose activity of 36.8 FPU/g at 144 h incubation period (Fig. 3a). The lowest cellulase activity was observed during the hydrothermal-pretreated substrate fermentation, i.e., 10.83 FPU/g. This can be observed that hydrothermal pretreatment did not show any significant effect on hemicellulose and lignin solubilization. The presence of lignin and hemicellulose in the biomass limits the cellulose accessibility to the microorganism which ultimately

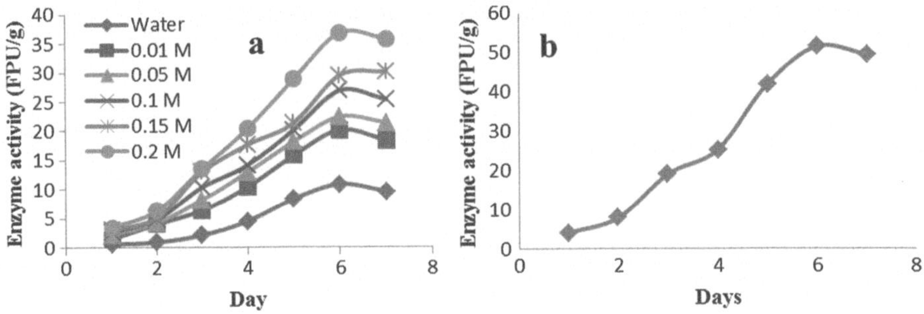

Fig. 3 (**a**) Cellulase production in shake flask (**b**) Cellulase production in stirred tank reactor

leads to poor cellulase production. In the present study, hydrothermal pretreatment at 121 °C significantly hydrolyzed the surface-bound nonstructural carbohydrates and minor portion of hemicellulose.

During the dilute NaOH pretreatment, lignocellulosic disruption occurred followed by dissolving hemicellulose, lignin, and swelling of cellulose. Lignin degradation is generally attributed to the breakdown of the a-aryl ether bonds from its polyphenolic monomers, whereas hemicellulose solubilization and cellulose swelling are consequences of hydrogen bond weakening. Sodium hydroxide (NaOH) shows the greatest decomposition and subsequent fermentation yields when compared to other alkalis, such as ammonium hydroxide, hydrogen peroxide, sodium carbonate, and calcium hydroxide (Rodríguez and Diaz 1994). Therefore, NaOH pretreatment leads to increase in the accessibility of cellulose surface area which could be a major advantage to achieve high levels of cellulase production. Rita Rani S et al. (2006) also reported that pretreatment of lignocellulosic biomass increases the cellulase yield by 33%. During the solid-state fermentation (SSF) of rice straw using *A. niger* KK2, 19.5 FPU/g.sd of enzyme activity was obtained by Kang et al. (2004). Deshpande et al. (2008) reported 27.39 FPU/g with *T. reesei* (QM 9414 mutant) by SSF using wheat straw, water hyacinth, wood straw, and their combinations as substrates. The present study closely comes to agreement with the findings of Potumarthi et al. (2012), who reported enzyme activity of 38.46 FPU/g at 28 °C in submerged fermentation using *P. chrysosporium* NCIM 1106 with pretreated corncob as a substrate.

3.2.1 Cellulase Enzyme Production in STR

Among the different alkaline-pretreated substrates, 0.2 M for 60-min pretreated biomass is found to be an optimum cellulase-producing substrate in the shake flask method. It shows comparatively high cellulase activity (36.8 FPU/g). Therefore, in order to enhance the cellulase production, reactor studies were performed using STR with 1.3vvm aeration (atmospheric air) with aforementioned pretreated

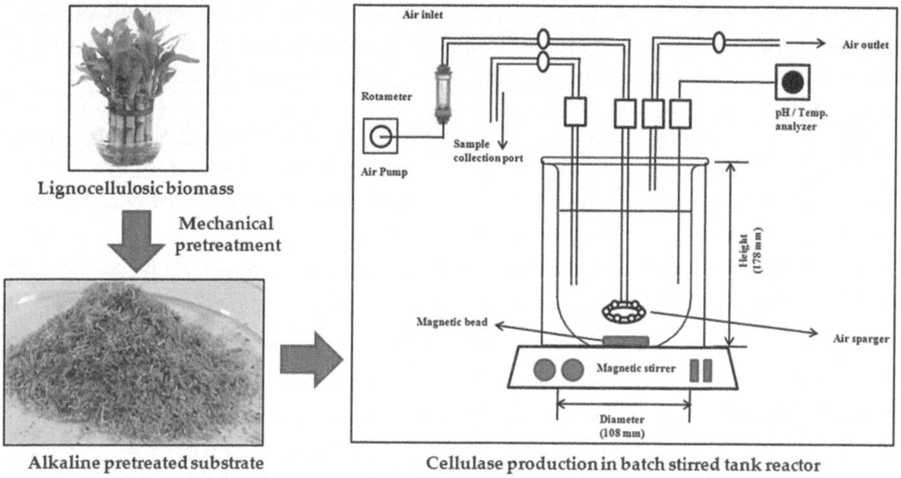

Fig. 4 Steps involved in cellulase production from lignocellulosic biomass and schematic representation of stirred tank reactor

biomass as a substrate (schematic representation is shown in Fig. 4). During the reactor studies, enzyme production was successfully increased. As compared to the shake flask method, 15 FPU/g.sd of enzyme activity increase was observed in the bioreactor studies (Fig. 3b). Also a similar result was found in Mazutti et al. (2010) studies during the production of inulinase in a packed-bed bioreactor, where the enzyme production in the bioreactor was about twofold greater than the shake flask method. The enhanced enzyme production during bioreactor studies is due to aeration and continuous stirring which provides better medium oxygenation and uniform mixing of substrate for microbial utilization, respectively. This was the main reason for increased enzyme activity during the scale-up process in STR. The bioreactor designed in the present study is cost-effective, reliable, and easy to monitor.

4 Conclusion

With increase in the NaOH concentration and reaction time, hemicellulose lignin solubilization was successfully increased. This can lead to increase the cellulose surface accessibility for cellulase-producing microbial strains. Therefore, cellulase production was increased with increase in the pretreatment severity. The maximum removal of hemicellulose and lignin was observed at 121 °C, 0.2 M NaOH for 60 min; subsequently the maximum cellulase activity was also observed during the fermentation of pretreated substrate which was obtained by the similar pretreatment condition. Less enzymatic activities was observed during the fermentation of hydrothermal-pretreated substrate. It can be observed that hydrothermal pretreatment at 121 °C may not show any significant effect on lignocellulosic complex network to solubilize hemicelluloses and lignin which ultimately leads to less accessibility of cellulose during fermentation. However, cellulase activity was significantly increased, up to 51.3 FPU/g, by laboratory scale bioreactor. Aeration and agitation play vital roles during the production of cellulase. Therefore, present study suggested that utilization of biologically produced crude cellulolytic enzymes for hydrolysis of cellulosic substrate may decrease the production cost of ethanol.

Acknowledgment The authors gratefully acknowledge Mr. Lalit Goswami for his assistance in preparation of schematic diagram.

References

Abdul Aziz A, Das K, Husin M, Mokhtar A (2002) Effect of physical and chemical pre-treatment on xylose and glucose production from oil palm press fiber. J Oil Palm Res 14:10–17

Deshpande S, Bhogmange M, Chakrabarti T, Shastri P (2008) Production of cellulose and xylanase by Trichoderma reesei (Q 9414 mutant), Aspergillus niger and mixed culture by solid state fermentation (SSF) of water hyacinth (Eichornia crassipes). Indian J Chem Technol 15:449–456

Ghose T (1987) Measurement of cellulase activities. Pure Appl Chem 59:257–268

Hendriks A, Zeeman G (1998) Pretreatments to enhance the digestibility of lignocellulosic biomass. Bioresour Technol 64:139–151

Kang S, Park Y, Lee J, Hong S, Kim S (2004) Production of cellulose and hemicellulases by *Aspergillus niger* KK2 from lignocellulosic biomass. Bioresour Technol 91:153–156

Krassig H, Schurz J, Steadman RG, Schliefer K, Albrecht W, Mohring M, Schlosser H (2004) Cellulose. Ullmann's Encyclopedia of Industrial Chemistry

Lynd L, Weimer P, Van Zyl W, IS P (2002) Microbial cellulose utilization: fundamentals and biotechnology. Microbiol Mol Biol Rev 66:506–577

Mandels M, Reese E (1957) Induction of cellulase in trichoderma viride as influenced by carbon sources and metals. J Bacteriol 73:269–278

Mazutti M, Zabot G, Boni G, Skovronski A, Oliveira D, Luccio M, Rodrigues M, Treichel H, Maugeri F (2010) Kinetics of inulinase production by solid-state fermentation in a packed-bed bioreactor. Food Chem 120:163–173

Miller GL (1959) Use of dinitrosalicylic acid reagent for determination of reducing sugar. Anal Chem 31:426–428

Palmqvist E, Hahn-Hagerdal B (2000) Fermentation of lignocellulosic hydrolysates. I: inhibition and detoxification. Bioresour Technol 74:17–24

Potumarthi R, Baadhe RR, Jetty A (2012) Mixing of acid and base pretreated corncobs for improved production of reducing sugars and reduction in water use during neutralization. Bioresour Technol 119:99–104

Reczey K, Szengyel Z, Eklund R, Zacchi G (1996) Cellulase production by *Trichoderma reesei*. Bioresour Technol 57:25–30

Rita Rani S, Sukumaran RK, Pillai A, Prema P, Szakacs G, Pandey A (2006) Solid state fermentation of lignocellulosic substrates for cellulase production by *Trichoderma reesei* NRRL 11460. Indian J Biotechnol 5:332–336

Rodríguez V, Diaz C (1994) Effect of chemical solutions sprayed on sugarcane bagasse pith to produce single cell protein physical and chemical analysis of pith. Bioresour Technol 47:159–164

Schmidt A, Thomsen A (1998) Optimization of wet oxidation pretreatment of wheat straw. Bioresour Technol 64:139–151

Sluiter A, Hames B, Ruiz R, Scarlata C, Sluiter J, Templeton D, Crocker D (2011) Determination of structural carbohydrates and lignin in biomass. National Renewable Energy Laboratory, Golden. NREL/TP-510-42618, pp 1–15

Smits J, Rinzema A, Tramper J, Van Sonsbeek H, Knol W (1996) Solid-state fermentation of wheat bran by *Trichoderma reesei* QM9414: substrate composition changes, C balance, enzyme production, growth and kinetics. Appl Microbiol Biotechnol 46:489–496

Spano L (1978) Enzymatic hydrolysis of cellulose to fermentable sugar for the production of ethanol. In: Proceedings of annual symposium on fuels from biomass, pp 671–684

Taherzadeh M, Karimi K (2008) Pretreatment of lignocellulosic wastes to improve ethanol and biogas production. Int J Mol Sci 9:1621–1651

Watanabe H, Tokuda G (2001) Animal cellulases. Cell Mol Life Sci 58:1167–1178

Wilson D (2009) Evidence for a novel mechanism of microbial cellulose degradation. Cellulose 16:723–727

Biotransformation of Municipal Solid Waste (MSW) to Bioenergy: Prospects and Potentials

Piyush Nanda, Ramkrishna Sen, and Ramalingam Dineshkumar

Abstract The rampant increase in municipal solid waste (MSW) accumulations has largely contributed to the exacerbating scenario of the environment. The lack of efficient waste processing technologies to mitigate accumulation has led to a large influx of waste into environment causing multiple damages to nature. Failure of developing efficient mechanisms to degrade polythene which constitutes a major fraction of the MSW has also been a hindrance. Scientists across the globe are trying to develop efficient methods of converting waste to energy which can help us resolve energy crisis and reduce solid waste accumulations. Although a plethora of technologies are coming up, their potential to intake unsegregated waste is limited and hence their applications are restricted. Polythene being the most stable component of MSW remains practically unaffected for decades. It is well known that polythene on combustion produces CO_2 and other gases which are responsible for air pollution. The prospects of utilizing this released CO_2 in fuel production are being investigated in this research. Microalgae has been considered as a potential candidate for biofuel production owing to their invasiveness, high CO_2 sequestering potential, and high lipid content. In this study, the growth of microalgae using polythene combustion gas as the CO_2 source was investigated.

Bioprocess optimization can enhance the CO_2 sequestration kinetics and hence cause high oil content. The whole system comprises of an aqua-separation unit to segregate MSW, an anaerobic digester for degrading organic waste, a fermentor to process lignocellulosic waste coupled with the polythene decomposition, and a waste gas bioprocess unit. This can serve the process of utilizing almost all the components of MSW for energy production and hence impart a carbon reduction advantage to the system.

Keywords Carbon dioxide sequestration • Microalgae • Municipal solid waste • Polythene combustion gas

P. Nanda • R. Sen (✉) • R. Dineshkumar
Department of Biotechnology, Indian Institute of Technology, Kharagpur, India
e-mail: rksen@hijli.iitkgp.ernet.in

S.K. Ghosh (ed.), *Utilization and Management of Bioresources*,
DOI 10.1007/978-981-10-5349-8_4

1 Introduction

Development of waste processing technologies is quintessential to a country like India where each day around 1.5 million metric tons of municipal solid waste (MSW) is produced, of which around 0.9 million metric tons is collected and only 0.158 million metric tons is processed and treated (Artuchelvi et al. 2009). Accumulation of such huge quantities of MSW causes the release of thousands of tons of obnoxious gases due to open decomposition in the environment. Approximately an average 0.400 kg per capita per day of MSW is generated, and this goes on increasing at a rate of 1.3 % per annum (Sharma and Ganguly 2014). The actual composition of MSW depends entirely on the source and location, but one of the greatest hindrances faced in developing waste processing technologies is the resistance of plastic wastes to chemical and biological treatments. Plastic waste in particular polythene occupies a considerable portion in MSW. Its inert nature is due to the long hydrocarbon chains and absence of reactive groups. Technologies to process and effectively extract value-added products from other parts of waste have been developed. In the recent past, techniques of degrading polythene using xenobiotic microbes have come to picture, but the major drawback of such a process is the unmanageably slow degradation kinetics and sensitivity of such microbes toward toxins in polythene. This is possibly due to high positive free energy of depolymerization of polythene, and the metabolism of such microbes is not exergonic enough to impart a net negative Gibbs free energy to degradation of polythene. The combustion of polythene is exergonic in contrast to one-step depolymerization of polythene, and it releases CO_2 and other combustion products under appropriate oxygen supply. Controlled combustion of polythene releases mixture of gases with CO_2 predominating in composition and varying with oxygen supply. As the process is exergonic, a very low input of heat energy is required to sustain combustion.

Microalgae are being extensively used for Bio-CSS or bio-carbon sequestration and storage because of the faster growth rate and high substrate to oil conversion ratio (Gross 2013). Microalgae such as *Chlorella minutissima* are being used in photo-bioreactors to sequester CO_2 and generate oil which can be transesterified to biodiesel or processed to bio-butanol. In this study, the growth kinetics of photosynthetic microalgae in laboratory scale photo-bioreactor using waste gas released from market polythene was investigated. Some factors which may retard the growth of microalgae such as the presence of toxins can be eliminated using ligninolytic fungi or appropriate adsorbent.

This can successfully capture the carbon from polythene and metabolize the same to triglycerides which can be transesterified to biodiesel. The mass transfer analysis and maximum theoretical yield were estimated to be 480 mg per 1000 mg of polythene burnt. Experimental analysis is being conducted to see the efficiency of such an approach in industrial scale. Toxin degradation techniques are being developed to minimize inhibitory effects of toxins.

In an attempt to couple the polythene processing unit with anaerobic digestion unit to degrade organic loading in MSW (Sharholy et al. 2007) and lignocellulosic fermentor to produce bioethanol (Mata-Alvarez et al. 2000) from lignocellulosic portion of the waste, an aqua-separation system was proposed. Such a system will separate the components of MSW according to their specific gravities. Organic loading which will mostly settle down due to relatively high density (Schmitta et al. 2011) can be fed to the anaerobic digester to generate biogas which can be further used for sustaining polythene combustion. The lignocellulosic portion can be alkali treated to separate cellulosic filtrate and can be processed to bioethanol.

The whole process can be a one-stop solution for treatment of municipal solid waste and subsequent extraction of value-added products from it. The biofuel produced can be used with gasoline or diesel in appropriate proportions as per need.

2 Materials and Methods

To execute the proposed processing system, the first requirement is polythene separated from the domestic waste, which generally comes as a mixture of polythene wastes, lignocellulosic, cellulosic, and food waste. Segregated waste is easier to process than unsegregated ones both with regard to time and money. To make the process economic with respect to segregation of wastes, this process utilizes aquaseparation method which is new to this kind.

Aqua-separation method utilizes the fact that the density of polythene and in general most of the polymers is relatively quiet less in comparison to that of food waste, cellulosic wastes, etc. (Schmitta et al. 2011). Transferring the entire waste feed, i.e., the waste collected from the localities into water tank, will cause the lighter polythene and a major portion of the lignocellulosic waste to rise up and the heavier food waste and cellulosic wastes to settle down. Hence, we separate the entire waste feed into two layers, one consisting of polythene and lignocellulosic waste and the other of food/kitchen waste and cellulosic waste. The basic advantage of this method is that generally we have soiled polythene from domestic wastes; this method will desoil the polythene waste hence making it suitable for further processes. The top layer is transferred to pretreatment chamber where it is alkaline treated to breakdown the lignin, and then the cellulose extracted is solubilized in water and transferred to fermentation unit. The polythene, which remains as the residue, is taken to the combustion unit. There it is combusted aerobically under controlled flow of air, and the heat of combustion of polythene can be used to generate electricity using steam turbines. The combustion gases having a large composition of CO_2 along with hydrocarbons and CO are first compressed into storage cylinders. The gases are passed through water with fungal inoculums, which aids in decomposition of toxins, i.e., PAH and other organic toxicants. The filtered CO_2 is then mixed with air to have a 20% CO_2 concentration and bubbled through the raceway containing algae with appropriate nutrient medium. The undissolved gases are recycled (Fig. 1).

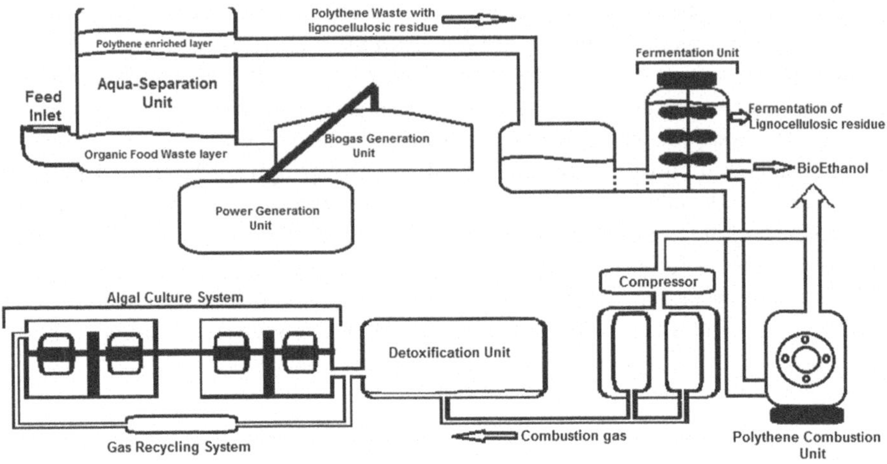

Fig. 1 The prototype design of the processing plant abstracting the whole process

Then the bottom layer of the separation unit that will be consisting of heavier kitchen/food waste and cellulosic waste will be subjected to anaerobic digestion where biogas is extracted which can be used to supply the electricity needs of the entire processing plant. Hence, make the processing system selfsustainable. The water used in separating the wastes can be reused or, in case after optimum usage, can be processed in a sewage treatment unit or using water hyacinth, whichever will be suitable. The bioethanol produced can be used in the transesterification of biooils extracted from the algae, making the system sustainable.

This makes the processing unit selfsustainable in terms of energy and chemical inputs. The further sections describe about the physical analysis of the plant.

3 Calculations and Results

3.1 Theoretical Estimation of Expected Output and Yield from the Algal Culture

It is important to estimate in advance the fuel generation per unit of the supplied waste feed so as to make a rough estimate of feasibility of the project. The entire arc of conversion of polythene to biooil takes place through various physical and biochemical transformations.

The aerobic combustion of polythene (domestically used) or LDPE yields CO_2 gas along with CO and other hydrocarbons. The percentage composition of these in the combustion products depends on oxygen supply rate (OSR), feed purity, and temperature of operation. Table 1 gives the amount of products released

Table 1 Quantity of combustion products released

Air flow, cc/min	100	100	100
Oxygen flow, cc/win	0	40	0
Heating rate, C/min	5	5	50
Carbon dioxide	88.0	1610.0	178.0
Carbon monoxide	312.0	171.0	110.0
Methane	10.0	7.0	17.0
Ethylene	40.0	33.0	70.0
Ethane	5.0	3.0	11.0
Propylene	29.0	14.0	33.0
Propane	5.0	3.0	7.0
1-butene	17.0	8.0	19.0
Butane	4.0	2.0	6.0
Trans-2-butene	6.0	4.0	9.0
Cis-2-butene	0.95	0.50	1.0
1-pentene	9.0	4.30	12.0
Pentane	2.0	1.0	3.0
1,3-pentadiene7	23.0	8.0	32.0
1-rexene	10.0	5.0	15.0
2-hexene	4.0	2.0	6.0
%plastic accounted for	32.0	61.0	34.0

Source: Office of Research and Monitoring, US Environmental Protection Agency
The quantity of each combustion product is reported in milligrams per gram of sample

(in milligram) per gram of polythene in the mentioned conditions. This helps in making a rough assumption about the amount of CO_2 released per gram polythene.

Assuming a yield of 1600 mg/g of polythene, 1.6 g of CO_2 is released when 1 g of polythene is aerobically burnt. The maximum biomass production is observed in the range of 5–10 % concentration of CO_2 (in ppm by volume). This turns around to be 387.5 mgCO_2/L.

At other conditions being optimized, algae is supposed to absorb almost twice the CO_2 its own mass (Gross 2013). Hence, assuming that after repeated recycling of the unused gases, algae absorb almost all CO_2 (theoretical maximum). This results in production of 193.75 mg algae/L of algal biomass when all 387.5 mg/L is used.

As polyculture is used, oil content in different species of algae will be different. An average content of 60–70 % will not affect rough assumptions. This turns out to be 116.25 mg oil/L.

From about 1 g of polythene, a theoretical maximum of 1.6 g of CO_2 is released. Provided this amount of CO_2 is air diluted to 1 L, extrapolating the results, this would result in production of 800 mg algae/L. About 480 mg oil/L can be extracted assuming a conversion efficiency of 60 % to oil.

3.2 Experimental Results

To determine the feasibility of such an approach to use polythene combustion gas to culture algae, a lab scale set was designed, and the growth kinetics of *Chlorella minutissima* was studied.

Locally available polythene was collected and burnt in combustion unit specially designed for collection of combustion gases. The collected gases were stored in cylinders. In the setup, in experimental photo-bioreactor (500 ml) combustion, gas diluted with air was passed, and suitable conditions were maintained for growth of algae. Bold's basal medium (BBM) (Weyer et al. 2009) was used for culture of algae, and appropriate buffer was added to suppress fluctuations in pH due to acidic gases present. The growth of algae was measured relative to a control. The growth was determined in terms of absorbance or optical density of the culture at 750 nm using UV-vis spectrophotometer. Gravimetric analysis was also conducted to quantify the final biomass (Figs. 2 and 3).

Analysis of the data made us conclude that algae was able to sequester the CO_2 present in the polythene combustion gas but presence of toxins which bioaccumulate in algae inhibits their growth. These toxins are produced due to presence of additives in polythene and their incomplete combustion. Oxygen supply can be properly regulated to ensure maximum feed mass to gas conversion and minimal carbonaceous residue production. We are looking forward to use ligninolytic fungi to decompose toxins present in the combustion gas so that algae can properly biosequester the CO_2 in combustion gas.

We are working on process optimization of this proposed model and eliminating carbon loses and inhibition.

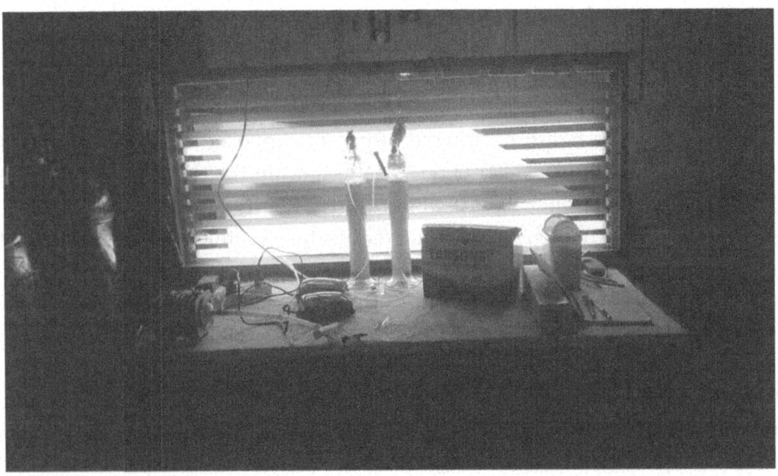

Fig. 2 Experimental setup for studying growth kinetics of algae using polythene combustion gas

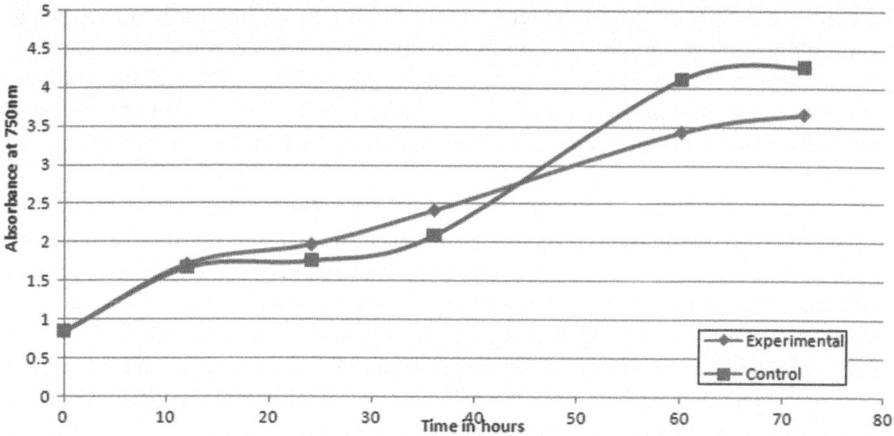

Fig. 3 Growth profile of *Chlorella minutissima* growing in medium sparged with polythene combustion gas

Table 2 Gravimetric analysis of final biomass

Sample	Final algal concentration (500 ml culture)
Control	1.18 g/L
Experimental	1.06 g/L

As it can be observed from the growth profile during mid-log phase of growth, there was a decrease in growth possibly due to the presence of toxins. After the end of the experiment, gravimetric analysis of the centrifuged biomass was conducted (Table 2).

4 Conclusion

This processing plant intakes the unsegregated domestic wastes, sustainably processes them converting them to fuel and energy. The impacts of scaling up this into industrial scale model are enormous.

4.1 Reduction in accumulation of MSW and conversion to value added products:

Around 55 million metric tons of MSW is generated each year in India. Each day around 0.4 Kg/capita of waste is generated/averaged all over the states. Due to lack of efficient and versatile technologies, a major fraction of these wastes are dumped in landfills which become the epicenter of air pollution. Open burning of wastes in certain anthropological specific region is responsible for the addition of carcinogenic dioxins in the air. There are many present-day technologies dealing with processing of MSW and production of valueadded products from it. But the ability to intake all components of waste and produce energy and fuel from it is rare to its kind.

Conversion of wastes to value-added products: Wastes are a vast reserve of chemical energy that can help us meet our energy need. Introduction of efficient, green, and cheap alternative fuel can remove our dependence on petroleum and coal-based fuel. Our country is expected to undergo acute energy shortage. The need of technologies making it possible for production of fuel and energy from cheap and readily available raw materials is required. This definition of raw materials perfectly matches with that of "waste," which is cheaply and readily available in India. Harnessing the chemical energy by the application of chemistry and biotechnology converting waste into energy is the need of the hour.

4.2 Self-Sustainable and Green Process

The whole process is designed in such way that all the energy needs within the system are met by the energy generated within it. The generation of electricity from biogas from the kitchen waste/organic waste and the energy generated by polythene combustion are thoroughly utilized in meeting the energy needs within the plant, hence making the process selfsustainable. The biochemical and biological feed must be externally supplied that is cheap and selfreplicating.

4.3 Decomposition of Waste Polythene

There are a number of technologies coming up for degradation of polythene including that of pyrolysis of polythene, gasification, and microbial degradation using fungi, waxworms, bacteria, etc. But most of them either require high energy and chemical input which does not give them edge on ecofriendly and sustainable ground or in case of microbial degradation are ecofriendly and sustainable but highly inefficient and slow. So, polythene degradation and extraction of value-added products from it is the need of the hour. Though a large fraction of polythene can be recycled, again this recycling process is carbon positive. Also, we are looking for extraction of bioplastics from the biooils generated from algae.

The growth kinetics of microalgae is being studied using polythene combustion gas as source of carbon dioxide. The process is being optimized for enhanced oil productivity. The enhancement in amount of oil obtained per gram of algae by supplying external stimuli is also being studied.

Acknowledgment I would extend my sincere gratitude and thanksgiving to Prof. Ramkrishna Sen, Department of Biotechnology, IIT Kharagpur, and Mr. Dinesh Kumar Ramalingam for their immense support in conducting the experiments.

References

Artuchelvi J, Sudhakar M, Arkatkar A, Doble M, Bhaduri S, Uppara PV (2009) Biodegradation of polyethylene and polypropylene

Density of Waste Materials (n.d.) http://www.epa.vic.gov.au/business-and-industry/lower-your-impact/~/media/Files/bus/EREP/docs/wastematerials-densities-data.pdf

Gross M (2013) Development and optimisation of algal cultivation systems

Mata-Alvarez J, Mace S, Llabres P (2000) Anaerobic digestion of organic solid wastes. An overview of research achievements and perspectives

Schmitta E, Buraa R, Gustafsona R, Cooperb J, Vajzovica A (2011) Converting lignocellulosic solid waste into ethanol for the State of Washington: an investigation of treatment technologies and environmental impacts

Sharholy M, Ahmad K, Mahmood G, Trivedi RC (2007) Municipal solid waste management in Indian cities – a review

Sharma D, Ganguly R (2014) Management of urban solid waste in Solan

Weyer KM, Bush DR, Al Darzins, Willson BD (2009) Theoretical maximum algal oil production

Preparation and Characterisation of Solid Catalysts for Saccharification of Biomass

A. Mallick and M. Mukhopadhyay

Abstract Biomass saccharification has assumed a significant importance in the context of the modern-day energy crisis and climate change scenarios. Waste biomass, which can constitute a significant portion of solid wastes, can be converted into value-added chemicals like ethanol by this process. The present investigation deals with the development of a solid acid catalyst for biomass saccharification using coconut shell, a cheap and abundant raw material, which has not been explored previously in this field. Coconut shell has been carbonised with zinc chloride at 723 K for 1 h to produce activated carbon which has been sulphonated with conc. H_2SO_4 (98%) at 403 K for 16 h to develop the solid acid catalyst. The catalyst has been characterised by scanning electron micrography, X-ray diffraction, FTIR spectroscopy and nitrogen adsorption. The X-ray diffraction studies have shown a graphene sheet content of 39% in the catalyst, while the FTIR spectra show the presence of SO_3H, phenolic OH and COOH groups. The specific surface area measured by nitrogen adsorption was 10.162 m^2/g. The catalyst has been used to hydrolyse pretreated sawdust from *Acacia nilotica* heartwood as well as microcrystalline cellulose under experimental conditions specified by central composite design. The yields of total reducing sugars in the hydrolysates have been analysed by UV spectrophotometry, and the produced sugars were identified by HPLC. Glucose constituted almost all of the produced sugars with negligible amounts of galactose being formed. The maximum sugar yield was 91% for pretreated sawdust and 93% for microcrystalline cellulose, indicating the excellent catalytic property of the catalyst. The results indicate the suitability of coconut shell as a source for developing biomass saccharification catalysts, as well as the efficacy of such a catalyst in the saccharification process.

Keywords Lignocellulosic biomass • Biomass saccharification • Sulphonated activated carbon

---

A. Mallick (✉)
Heritage Institute of Technology, Kolkata, India
e-mail: abhyuday.mallick@heritageit.edu

M. Mukhopadhyay
University of Calcutta, Kolkata, India

S.K. Ghosh (ed.), *Utilization and Management of Bioresources*,
DOI 10.1007/978-981-10-5349-8_5

1 Introduction

Biomass constitutes a significant portion of solid wastes that are generated in human settlements, especially urban areas. The most common method of disposing of biomass is by landfilling or incineration. But these useful energy sources can also be disposed off in a novel way, i.e. converting them to value-added chemicals. As the main component of lignocellulosic biomass, cellulose is a biopolymer consisting of many glucose units connected through *β*-1,4-glycosidic bonds. Breakage of *β*-1,4-glycosidic bonds by acids leads to the hydrolysis of cellulose polymers, resulting in the formation of glucose or other oligosaccharides. These sugars can be further fermented to produce ethanol, which can be used as transportation fuel in blends with gasoline. Thus these novel methods of biomass waste disposal have the potential to reduce fossil fuel consumption and greenhouse gas emissions, which have a great implication in the modern industrial society. Mineral acids like HCL and H_2SO_4 have been used in the hydrolysis of cellulose. Since they suffer from problems of product separation, reactor corrosion and poor recyclability, emphasis is being placed on developing more efficient catalysts for these reactions. Research works are being carried out for the possibility of replacing mineral acids with solid acidic catalysts for these processes. Solid catalysts can be easily recycled for further use and pose no environmental hazards.

2 Literature Survey

Suganuma et al. (2008) developed a solid acid catalyst by sulphonation of activated carbon. The solid acid catalyst prepared by them had nanosized graphene sheets along with SO_3H, phenolic OH and COOH groups. This catalyst reduced the activation energy of lignocellulose hydrolysis to 110 kJ mol^{-1}.

A sulphonated porous catalyst was prepared from wood powder with high specific surface area (Kitano et al. 2009). Carbonisations at different temperatures were done, and 723 K was found as the optimum temperature for carbonisation with the carbon catalysts containing high densities of micro- and mesopores.

Huang et al. (Huang and Yao 2013) studied the hydrolysis of cellulose to glucose by solid acid catalysts. Advances in the hydrolysis of cellulose by different types of solid acids, such as sulphonated carbonaceous-based acids, polymer-based acids and magnetic solid acids, were summarised.

A sulphonated carbonaceous material was prepared from lignosulphonate, and its usefulness was studied as an esterification catalyst (Lee 2013).

Salmi et al. (2014) developed kinetic models for the hydrolysis of *O*-acetyl-galactoglucomannan (GGM), a hemicellulose appearing in coniferous trees. Mineral acid catalysts and heterogeneous solid catalysts were used to hydrolyse the hemicelluloses.

Various solid acid catalysts were prepared which were further used for hemicellulose hydrolysis (Cara et al. 2013). The reactions were conducted under neutral ph and relatively mild temperature and pressure conditions in water as the reaction medium.

Sulphonated silica/carbon nanocomposites were successfully developed as reusable, solid acid catalysts for the hydrolytic degradation of cellulose into high yields of glucose (Van de Vyver et al. 2010).

Onda et al. (2008) for the first time performed solid acid catalysis for the hydrolysis of cellulose. The sulphated and the sulphonated catalysts showed a remarkably high yield of glucose.

The objective of the present work is to develop a solid acid catalyst based on activated carbon prepared from coconut shell and study the effectiveness of the developed catalyst in the hydrolysis of pretreated sawdust from *Acacia nilotica* heartwood. Coconut shell, which is a highly abundant and cheap raw material, has not been explored so far for the development of solid acid catalysts for biomass hydrolysis. Literature on hydrolysis of actual lignocellulosic biomass by solid acid catalysts is also scant, with most researchers focussing on hydrolysis of pure cellulose and starch. Hence on this front, the present work is a novel research in this domain. The study of the recyclability of the developed catalyst has also been incorporated.

3 Materials and Methods

Chemicals used were:

- $ZnCl_2$ (Merck, >95%)
- NaOH (Merck, 97%)
- H_2SO_4 (Merck, 98%)
- HCl (Merck, 35%)
- NaCl (Merck, 99%)

Sawdust from *Acacia nilotica* heartwood was the substrate for hydrolysis. The composition of *Acacia nilotica* wood is detailed in Table 1. After pretreatment (Mallick et al. 2016), the initial lignin content was reduced by 60%.

Table 1 Composition of *Acacia nilotica* heartwood

Components	Percentage
Cellulose	40
Hemicellulose	30
Lignin	27

3.1 Experimental Procedure

Coconut shell was used for the manufacture of activated carbon. Its composition on dry basis is given below:

Cellulose – 33.61%	Lignin – 36.51%
Pentosans – 29.27%	Ash – 0.61%

The high cellulose and pentosan content and low ash content make coconut shell an excellent substrate for preparing activated carbon. The high lignin content makes it suitable for preparation of the solid acid catalyst since lignin being an aromatic polymer, the produced activated carbon should contain a large proportion of aromatic rings, which would facilitate easy sulphonation.

3.1.1 Carbonisation

The coconut shells were crushed to obtain a definite particle size passing through 100 mesh. The shells were then impregnated with 10% $ZnCl_2$ solution for 24 h. The shells were filtered and dried to room temperature. The coconut shells were then taken in silica crucibles and carbonised at 723 K in a muffle furnace for 1 h. Nitrogen purging was done to maintain an inert atmosphere. The carbonised mass so obtained was washed with deionised water to remove traces of chlorides. It was then air-dried for a day.

3.1.2 Sulphonation

In borosilicate glass beaker, 400 ml of sulphuric acid was taken. The carbonised mass was then slowly added to it and heated at 403 K for 16 h in a castor oil bath for digestion with continuous stirring by a magnetic stirrer. Following the digestion process, the mixture was cooled to room temperature and was diluted by adding 1000 ml of deionised water to the reaction mixture. A black precipitate was formed which was then filtered and washed repeatedly with deionised water to remove SO_4^{2-} ions. The particle size of the sulphonated activated carbon was around 100 mesh.

3.1.3 Catalyst Characterisation

The developed catalyst was analysed for estimating the acidity in terms of SO_3H group content and total acidity in terms of SO_3H, phenolic OH and COOH group content (Lee 2013). The specific surface area of the catalyst was measured by nitrogen adsorption in a BET apparatus. Scanning electron micrography of the catalyst was carried out to determine its surface structure. X-ray diffraction studies of the activated carbon prior to sulphonation were conducted to determine the graphene sheet content. The functional groups in the catalyst were determined by FTIR spectroscopy.

3.1.4 Estimation of Catalyst Performance in Hydrolysis of *Acacia nilotica* Heartwood

The performance of the developed catalyst was estimated in the hydrolysis of pretreated *Acacia nilotica* heartwood. For design of the hydrolysis experiments, central composite design was followed. Accordingly the hydrolysis reactions were carried out between 372 and 413 K. The time of reaction was varied between 30 min and 2 h. The catalyst/biomass ratio was varied from 1:1 to 2:1. Appropriate amounts of biomass and catalyst were taken in a 500 ml conical flask. Two hundred millilitres of deionised water was added. The flask was placed in an autoclave and hydrolysis was carried out. The pH of the reaction mixture with the catalyst present was 1. A conical flask containing only the pretreated sawdust without the catalyst was also placed in the autoclave. After reaction, hydrolysates were collected and analysed for total reducing sugars (TRS) by UV spectrophotometry (Miller 1959), and identification of the reducing sugars was carried out by HPLC. For studying the reusability of the catalyst, the residue (catalyst + biomass) remaining in the conical flasks was filtered out and weighed. Then the residues were transferred in other conical flasks, and hydrolysis was again carried out under same experimental conditions. The hydrolysates from this second batch of hydrolysis were also analysed for total reducing sugars by UV spectrophotometry. The catalyst was also used for hydrolysis of microcrystalline cellulose to compare the catalytic performance.

4 Results and Discussions

The SO_3H content in the catalyst was found to be 1.27 mmol/g, and the total acidity was determined to be 10 mmol/g. The specific surface area of the developed catalyst was 10.162 m^2/g. The pore volume was 0.36 $cm^3\ g^{-1}$, and the average pore diameter was 1 μ. The XRD pattern of the activated carbon has been shown in Fig. 1.

The graphene sheet content in the prepared activated carbon was determined according to Liou and Huang (2013). It was found to be 39%. The FTIR spectrum of the sulphonated activated carbon catalyst is shown in Fig. 2.

The peaks at 1027 cm^{-1} and 889 cm^{-1} are due to the presence of SO_3H group. The peaks corresponding to 1703 and 1383 cm^{-1} can be attributed to the presence of C=O stretching and carboxylic OH bending vibrations, respectively, indicating the presence of COOH group. The peak at 3530 cm^{-1} is due to phenolic OH stretching vibration, indicating the presence of phenolic OH. The peak at 812 cm^{-1} indicates aromatic C–H bending, and peaks at 1515 and 1643 cm^{-1} indicate aromatic C=C

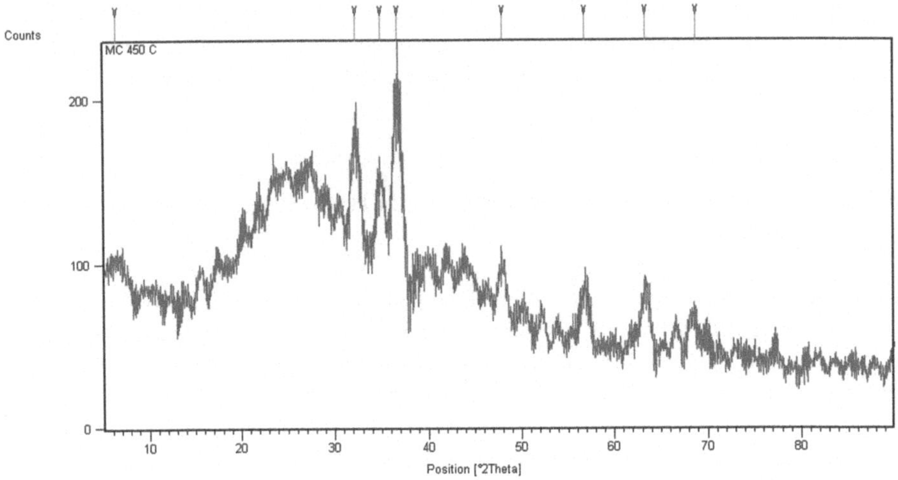

Fig. 1 X-ray diffraction pattern of activated carbon from coconut shell, prior to sulphonation

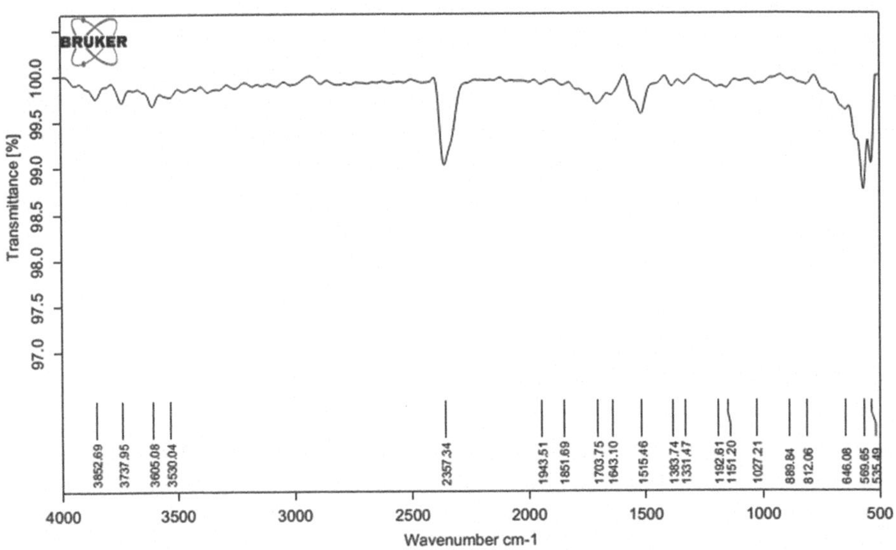

Fig. 2 FTIR spectra of produced sulphonated activated carbon catalyst

bending. This indicates the highly aromatic structure of the developed catalyst. Figure 3a, b shows the ultrastructure of the catalyst. As can be seen from the figures, the porosity of the catalyst is quite high, with average pore diameter ranging from 5 to 10 μ.

The principal hydrolysis product was glucose with negligible amounts of galactose. The glucose selectivity of the catalyst was more than 99%. The maximum

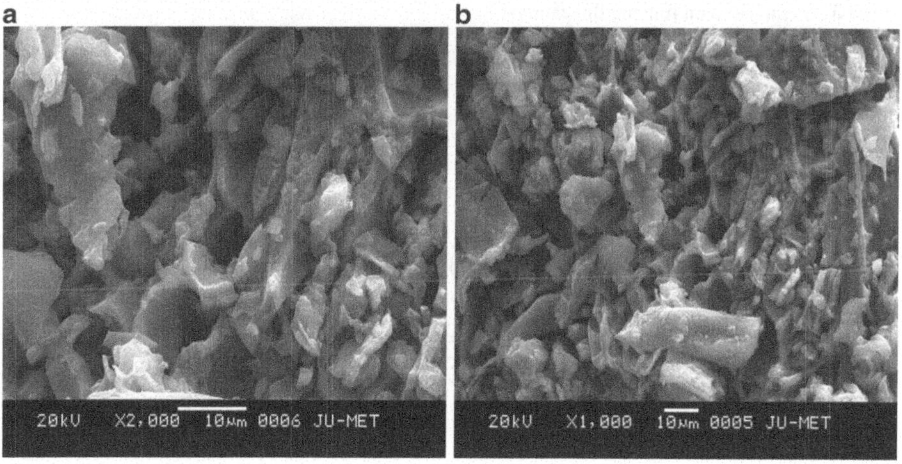

Fig. 3 (**a**) SEM of developed catalyst, (**b**) SEM of developed catalyst

Table 2 Hydrolysis experimental results

Maximum sugar yield (%)		t (°C)	c/b	t (h)
Sawdust	Microcrystalline cellulose			
83.3	87	140	1:1	1
87.1	90	140	1.5:1	1
91	93	140	2:1	1

yields of reducing sugars for hydrolysis of pretreated sawdust and microcrystalline cellulose have been detailed in Table 2.

The optimum time of reaction was found to be 1 h. For reaction times more than 1 h, the yield of reducing sugars decreased significantly at higher temperatures. This is due to the degradation of sugars to products like furfural, 5-hydroxymethylfurfural, acetic acid, etc. (Wei et al. 2008). Hence the studies on reusability of the catalyst were carried out for 1 h only. The yields of reducing sugars during recyclability studies are detailed in Table 3.

As evident from Table 2, the yield of reducing sugars is very high compared to earlier works in this field (Saganuma et al. 2008; Cara et al. 2013; Van de Vyver et al. 2010). The yield of sugars without the catalyst was negligible under the same experimental conditions, which indicate the excellent catalytic activity of the developed catalyst. The decrease in the sugar yield when the catalyst was reused is due to the leaching out of loosely bonded SO_3H groups into the aqueous medium. The rate of this leaching increases at higher temperatures; hence, the percentage decrease in sugar yield is more at higher temperatures. The surface area of the developed catalyst is low in comparison with that developed by Kitano et al. (2009). It can be inferred from this that large hydrophobic molecules will not be adsorbed properly on the catalyst; hence, the efficacy of this catalyst in hydrophobic acid-catalysed

Table 3 Sugar yield on reusing the catalyst

Maximum sugaryield (%)	t (°C)	c/b	t (h)
59.52	140	1:1	1
54.42	120	1:1	1
47.61	99	1:1	1

reactions, e.g. esterification, will be low. But for a hydrophilic reaction like biomass saccharification, relatively smaller cellulose units are adsorbed and stabilised effectively onto the catalyst surface, which then makes them amenable to attack by H^+ ions coming from adjacent SO_3H groups on the catalyst surface.

5 Conclusions

The present investigation has been successful in developing a solid acid catalyst from coconut shell, a cheap and abundant raw material, for biomass saccharification. Heterogeneous catalytic hydrolysis of actual lignocellulosic biomass (viz. *Acacia nilotica* heartwood) has been explored, which has been, up to now, a relatively unexplored area. A high yield of reducing sugar and high glucose selectivity has been obtained at moderate temperatures (413 K) and relatively short reaction time (1 h). The catalyst can also be reused though with a moderate drop in sugar yields. All these indicate the applicability of the catalyst in industrial biomass saccharification units which will lead to increased glucose production and consequently increased production of bioethanol. This will also reduce the problem of accumulation of solid biomass wastes since every type of biomass waste can be hydrolysed to value-added chemicals using this catalyst.

6 Future Scope of Work

In the future, the kinetics of the heterogeneous catalytic hydrolysis needs to be explored which can lead to further improvement in the development of saccharification catalysts.

Acknowledgements The authors wish to thank the laboratory staff members of the Department of Chemical Engineering, University of Calcutta, for their cooperation in the experimental work. The authors also acknowledge the SEM section of the Department of Metallurgical Engineering and Material Science, Jadavpur University, India, for their cooperation. Finally the authors would like to gratefully acknowledge the authorities of University of Calcutta, for providing the financial support for this research work.

References

Cara PD, Pagliaro M, Bmekawy A, Brown DR, Verschuren P, Raveendran Shiju N, Rothenberg G (2013) Hemicellulose hydrolysis catalysed by solid acids. Cat Sci Technol 3:2057–2061

Huang Y-B, Yao F (2013) Hydrolysis of cellulose to glucose by solid acid catalysts. Green Chem 15:1095–1111

Kitano M, Arai K, Kodama A, Kousaka T, Nakajima K, Hayashi S, Hara M (2009) Preparation of a sulfonated porous carbon catalyst with high specific surface area. Catal Lett 131:242–249

Lee D (2013) Preparation of a sulfonated carbonaceous material from lignosulfonate and its usefulness as an esterification catalyst. Molecules 18:8168–8180

Liou Y-J, Huang W-J (2013) Quantitative analysis of graphene sheet content in wood char powders during catalytic pyrolysis. Mater Sci Technol 29(5):406–410

Mallick A, Ash SN, Mahapatra DK (2016) Pretreatment of *Acacia nilotica* sawdust by catalytic delignification and its fractal kinetic modeling. J Inst Eng India Ser E 97(1):39–45

Miller GI (1959) Use of dinitrosalicylic acid reagent for determination of reducing sugar. Anal Chem 31(3):426–428

Onda A, Ochi T, Yanagisawa K (2008) Selective hydrolysis of cellulose into glucose over solid acid catalysts. Green Chem 10:1033–1037

Saganuma S, Nakajima K, Kitano M, Yamaguchi D, Kato H, Hayashi S, Hara M (2008) Hydrolysis of cellulose by amorphous carbon bearing SO3H, COOH and OH groups. J A Chem Soc 130:12787–12793

Salmi T, Murzin DYU, Maki-Arvela P, Kusema B, Holmbom B, Willfor S, Warna J (2014) Kinetic modeling of hemicellulose hydrolysis in the presence of homogeneous and heterogeneous catalysts'. AICHE J 60:1066–1077

Van de Vyver S, peng L, Geboers J, Schepers H, de Clippel F, Gommes CJ, Goderis B, Jacobs PA, Sels BF (2010) Sulfonated silica/carbon nanocomposites as novel catalysts for hydrolysis of cellulose to glucose. Green Chem 12:1560–1563

Wei Q, Su-ping Z, Qing-li X, Zheng-wei R, Yong-jie Y (2008) Degradation kinetics of xylose and glucose in hydrolysate containing dilute sulfuric acid. Chin J Process Eng 8(6):1132–1137

Biodegradation Behaviour of Cellulose-Reinforced PMMA Composites in Pond Water

Shubhalakshmi Sengupta, Papita Das, Siddhartha Datta, Sunanda Sain, Aniruddha Mukhopadhyay, and Dipa Ray

Abstract Acrylics and polyolefins are widely used synthetic plastics in daily consumer products which are non-biodegradable in nature. An accumulation of these solid wastes in the environment poses ecological threats and requires novel management techniques. Researchers have now focussed their work on developing novel biodegradable polymer materials and isolating and identifying microorganisms which have the potential to degrade these polymeric materials. Isolating and identifying these microorganisms having potential to degrade polymers and polymer composites are required for developing newer biotechnological techniques for management of these solid wastes in the environment. The present work studies the biodegradation behaviour of PMMA and micro-/nano-cellulose-reinforced PMMA (polymethyl methacrylate) composites in pond water. The weight loss data revealed improved biodegradability in cellulose-reinforced PMMA composites in comparison to the synthetic PMMA. Scanning electron microscopy (SEM) images revealed effective biodegradability of the composites in pond water. The microorganism (fungus) was isolated, and its biodegradation behaviour was studied.

Keywords Biodegradation • SEM • Polymer • Microorganisms

1 Introduction

Plastic materials are used widely in our lives. These materials have their sources in petroleum products and are thus environmentally hazardous. Plastic wastes like carrier bags, refused sacks and other packaging materials are mainly buried in soil.

S. Sengupta (✉) • P. Das • S. Datta
Department of Chemical Engineering, Jadavpur University, Kolkata, India
e-mail: sengupta.shubha@gmail.com; papitasaha@gmail.com; sdatta_che@rediffmail.com

S. Sain • A. Mukhopadhyay
University of Calcutta, Kolkata, India
e-mail: sunanda.sain@gmail.com; am_cuenvs@yahoo.co.in

D. Ray
University of Limerick, Limerick, Ireland
e-mail: roy.dipa@gmail.com

S.K. Ghosh (ed.), *Utilization and Management of Bioresources*,
DOI 10.1007/978-981-10-5349-8_6

They are resistant to biodegradation and result in serious environmental pollution. In order to solve this environmental problem, development of biodegradable polymers is the need of the hour. Plant fibres, agricultural and forest products are often used as alternative resources for product development in industries, in order to achieve environmentally sustainable biomaterials (Tudorachi et al. 2000). These cellulosic materials are widely used in the development of various products (e.g. twines, ropes, insulating materials, felts, fleece, non-woven materials, geotextile, fillers in polymer composites, etc.). Cellulose in plant fibres is often used as reinforcing filler in polymers. Cellulose (α-cellulose) is a major component of plant fibres. It is one of the most important natural polymer, a renewable material and a source for development of sustainable materials in industries. Incorporation of natural renewable fillers like cellulose in synthetic polymers is expected to enhance its properties and also improve its biodegradability. As cellulose is susceptible to degradation by microorganisms, its incorporation in synthetic polymers also renders biodegradability to it. Few research works have been carried out on biodegradability of synthetic polymers on incorporation of cellulose as fillers. Among the synthetic polymers, polymethyl methacrylate (PMMA) is widely used for manufacturing of various products, automotive parts, patio roofs, aircraft windscreens, etc (Sain and Khatua 2011). There are also reports on the development of cellulose-reinforced polymer composites having various application potentials. However, studies on biodegradability of the cellulose-reinforced PMMA composites are very few. Isolation of microorganisms is important for more effective degradation of the polymer composites. Specific microorganisms with ability to degrade polymers can be isolated from soil, compost, sewage, sludge, etc., through comprehensive screening and can thereafter be utilised for bioaugmentation in the context of solid disposal. Therefore, disposal of a specific polymer into a field enriched with microorganisms specific for the degradation of that polymer can significantly improve the rate of biodegradation.

2 Literature Review

There are few research works on the biodegradability of PMMA. Biodegradation studies were reported to have been performed in enzymatic route for PMMA grafted onto sago starch (Qudeseih et al. 2007). In this study, α-amylase, which was specific for starch degradation, was used. So, little degradation was achieved for the PMMA component. Considerable degradation was also achieved in the soil environment by the incorporation of starch cinnamate in PMMA (Thakore et al. 2001) and also with cellulose fibres in aerobic compost environment (Maity et al. 2013). Bhat et al. (Maity et al. 2013; Sain et al. 2014) had studied blend miscibility and biodegradability of polymer blends (PMMA with cellulose acetate, CA, and cellulose acetate phthalate, CAP). In this study along with the blends, films of the blends were fabricated through solution casting method using the solvents acetone and dimethylformamide (DMF), respectively. They were subjected to four different biodegradability methods (soil burial test, degradation in activated sludge, enzymatic degradation and degradation in phosphate) followed by water absorption tests. It was observed that blends were degraded effectively as was revealed by the weight loss data. It was further analysed that increase in CA and CAP content in the blend compositions increased the degradability. Maity et al. (2013) had also previously studied biodegradation of PMMA/cellulose nanocomposites by degrading the composites in an aerobic compost. In situ suspension polymerisation technique was deployed along with ex situ solution dispersion technique. The degradation was studied for a period of 60 days. Higher weight loss was observed in the in situ prepared composite films. Gel permeation chromatography (GPC) and nuclear magnetic resonance (NMR) studies also showed significant changes in the chemical structure of the biodegraded films.

Again, isolation and identification of polymer-degrading microorganisms are important as this knowledge provides valuable information for their use in bioaugmentation processes. The isolation, identification and study of metabolism of polymers by fungi have been well reported by some researchers (Cook et al. 1981). Fungi, due to their wide availability, faster growth rate in soil, robust nature and being a source of diverse enzymes, are widely used in bioremediation (Artham and Doble 2010). Sahoo et al. (Sahoo and Samal 2007) investigated the biodegradability of PMMA/montmorillonite nanocomposites. Unreinforced PMMA showed poor biodegradation behaviour due to its hydrophobic nature. The addition of MMT as a nanofiller in the PMMA polymer matrix increased the rate of degradation due to their hydrophilic nature of MMT. The specific microorganism, *B. cereus*, helped to increase the rate of biodegradation of PMMA/MMT nanocomposite than unreinforced PMMA. Prasantha et al. (2005) studied the biodegradability of chitosan-grafted PMMA films. They found that 50% weight loss was due to biodegradation evidenced by *A. flavus*. This microorganism helped in biodegradation by consuming chitosan, but it could not break the PMMA chains even after 25 days of degradation. *Aspergillus niger* was found to degrade starch grafted with PMMA (Moreno-Chulim et al. 2003).

In this study for the first time, biodegradation of PMMA composites was carried out in pond water, and their biodegradation behaviour was subsequently studied.

3 Materials and Methods

3.1 Materials

The raw material jute was procured from the local market for extracting cellulose nanofibre. Tung oil was bought from the local market. The monomer methyl methacrylate (purity of ≥99%), benzoyl peroxide (BPO, 98% purity), polyvinyl alcohol (PVOH, 96–99% hydrolysed) and sodium chlorite ($NaClO_2$), NaOH, acetone, maleic anhydride (MA), sulphuric acid and chloroform (>99% purity) were purchased from Merck, Germany.

3.2 Extraction of the CNF from Jute

One gram of jute was treated with 0.7% of sodium chlorite ($NaClO_2$) to remove the lignin fraction. The lignin removed fraction was treated with 17.5% NaOH solution for 15 min and then macerated. Then alkali was removed, and the jute was subjected to acid hydrolysis in 47% sulphuric acid for 3 h under constant stirring. The prepared cellulose nanofibre was washed repeatedly, and the concentrated mass was freeze-dried at −20 °C at a pressure of 15 Pa using Eyela Freeze Dryer FD-5N, Japan (Sain et al. 2012).

3.3 Fabrication of PMMA/Cellulose Composite Films

To prepare the PMMA/cellulose composite films, PMMA granules were first dissolved in chloroform, and a measured amount of cellulose nanofibre (10 wt% with respect to polymer weight) was dispersed and sonicated for 2–3 h and then solution casted in petri dishes and dried at 50 °C for 30 min in hot air oven.

3.4 Collection of Pond Water

The pond water was collected from three sites of a pond in Ballygunge Science College, Kolkata. This pond was subjected to a lot of solid waste dumping including plastic wastes.

3.5 Weight Loss

The collected pond water was poured in equal amounts in glass vials and PMMA film, and the composites were kept in it, and weight loss was recorded after 10, 20 and 30 min respectively. The loss in weight of the degraded samples was obtained following the equation, $\%\text{Weight loss} = \{(W_o - W_t)/W_o\} \times 100$, where W_o is the weight of the films before biodegradation and W_t is the weight of the films after biodegradation at time 't'.

3.6 Scanning Electron Microscopy (SEM)

The composite film surface morphology before and after 30 days of biodegradation was examined under SEM (Zeiss EVO 18, Carl Zeiss, Germany) at an accelerating voltage of 5 kV. The 2.5% glutaraldehyde fixed fungus on the composite film surface was also observed under SEM (accelerating voltage of 15 kV).

3.7 Isolation of the Fungal Strain

3.7.1 Media Composition and Preparation

Potassium di-hydrogen phosphate (KH_2PO4), ferrous sulphite ($FeSO_2.7H_2O$), sodium nitrate ($NaNO_3$), potassium chloride (KCl), magnesium sulphate ($MgSO_4.7H2O$) and ammonium chloride (NH_4Cl) were obtained from Merck, Germany, in order to prepare the minimal media in the composition (Sain et al. 2014). Agar powder (Merck, Germany) was used in the minimal broth for agar media. Nutrient agar media was also procured from Merck, Germany, for preparing the nutrient agar media.

3.7.2 Isolation of Fungal Strain

The fungal strain from pond water was isolated. The film samples were taken out from the pond water after 30 days and immersed in 10 ml of physiological saline. This was shaken for 1 h at room temperature and then several dilutions were made to isolate distinct fungal colonies. The distinct fungal colonies were replica plated on the minimal agar media where the sole source of carbon was the composite sample which was cut in even squares of 15 × 15 mm dimensions and embedded in the media. A control with only the minimal media was also kept. The colony which grew encircling the composite sample was isolated for further studies.

3.8 Weight Loss Study of the Composites in Liquid Culture Shaking Method by the Fungal Strain

The isolated pure fungal culture's biodegradation behaviour was evaluated, where the composites were added to a conical flask containing previously sterilised minimal media. The fungal isolate spore suspension was added (2 μl in 50 ml) to the test medium, properly stoppered and incubated at 37 °C with reciprocal shaking for 7, 14 and 21 days. The degradation was monitored by determining the weight loss of the composite films at the stated intervals.

4 Results and Discussion

4.1 Weight Loss

The extent of the biodegradation of the samples at three sites of the pond was evaluated by measuring their weight loss as a function of the time of biodegradation, as shown in Fig. 1. All the samples showed weight loss especially at 30 days. Spot 3

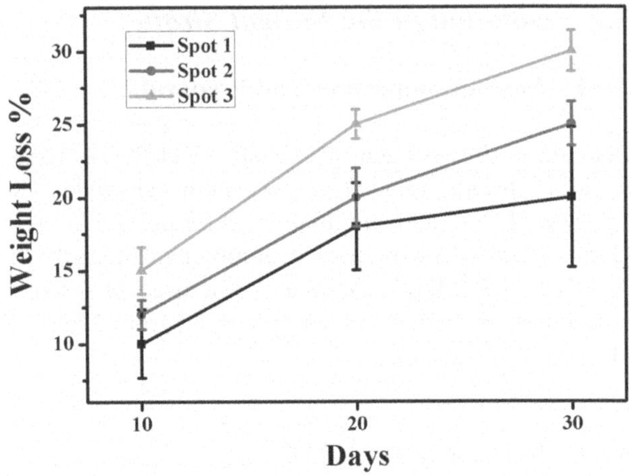

Fig. 1 Weight loss of the composite films

recorded the highest weight loss percentage. Therefore, in the presence of the pond water microorganisms, the cross linking between the polymers was broken. The PMMA films did not show any weight loss. The cellulose contained in the composites thus facilitated the route effective biodegradation by breaking of the polymer chains and their subsequent uptake as the sole source of carbon. As the composite showed the best result in spot 3, further studies were conducted with water collected from that spot.

4.2 Scanning Electron Microscopy (SEM)

The PMMA films and the composite kept in pond water collected from spot 3 for 30 days were evaluated under SEM (Fig. 2). The PMMA film showed roughness of surface, but the SEM micrograph also revealed great extent of degradation of the polymer composite surface in the presence of cellulose filled. Its presence rendered pronounced biodegradability of the composites. Growth of fungal hypha was also seen, which indicated degradation of polymer composites by the fungal microbiota.

4.3 Weight Loss Study of the Composites in Liquid Culture Shaking Method by the Fungal Strain

In Fig. 3, the inoculation of the pure culture of the fungal strain in liquid minimal media revealed higher percentage of weight loss. Thus this result indicates that the isolated fungal strain isolated from spot 3 of the pond water was responsible for the biodegradation of the composite film.

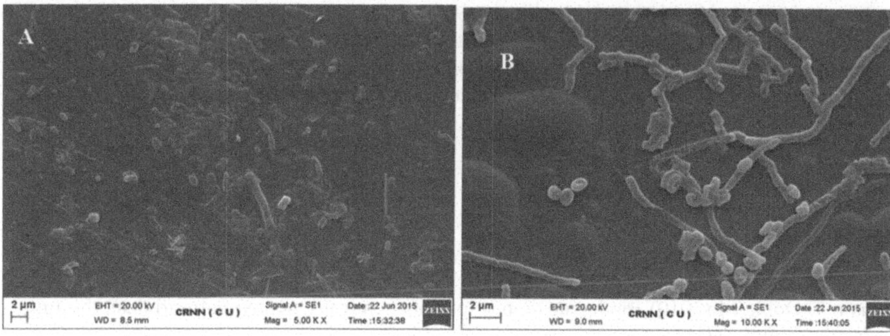

Fig. 2 SEM micrographs of (**a**) PMMA film and (**b**) composite of spot 3 after 30 days of degradation

Fig. 3 Weight loss of the composite films

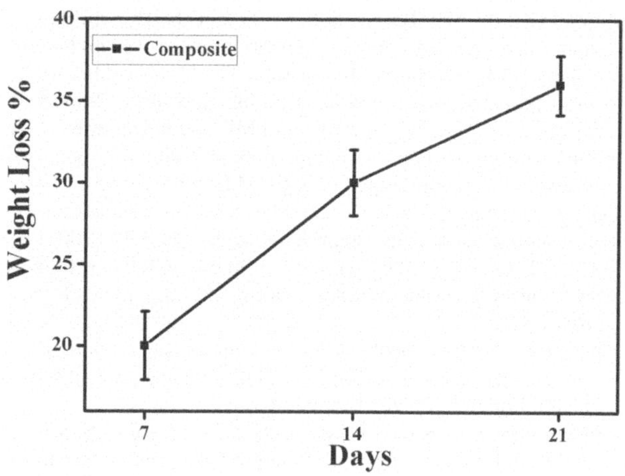

5 Conclusion

The biodegradation of the ex situ prepared PMMA composite was studied. Being close to a solid waste dumping ground, it was found to contain microorganisms especially capable of degrading a PMMA reinforced with cellulose fillers. The presence of cellulose fillers facilitated the biodegradability of the composites. The SEM micrographs also revealed fungal growth on the surface revealing the microorganism responsible for the degradation. This study has provided a platform for identifying fungi in pond water which are capable of degrading polymer wastes. As a lot of polymer wastes are dumped in freshwater bodies, identifying microorganisms adapted to degrade them provides an interesting research idea towards effective bioaugmentation of polymeric materials.

Acknowledgement S. Sengupta is thankful to UGC for granting her the UGC D. S. Kothari Postdoctoral Fellowship [Sanction No.: CH/15-16/0163 (S-63)]. Authors are also thankful to Ms. Priya Banerjee, Research Scholar, University of Calcutta, for her help in conducting the SEM studies and also to Mrs. Rimpa Majumder, student, Rabindra Bharati University, for her help in the work.

References

Artham T, Doble M (2010) Biodegradation of physiochemically treated polycarbonate by fungi. Biomacromolecules 6:612–626

Cook WJ, Cameron JA, Bell JP, Huang SJ (1981) Scanning electron microscopic visualization of biodegradation of polycaprolactones by fungi. J Polym Sci B Polym Phys 19:159–165

Maity S, Sain S, Ray D, Mitra D (2013) Biodegradation behaviour of PMMA/cellulose nanocomposites prepared by in-situ polymerization and ex-situ dispersion methods. Polym Degrad Stab 98:635–642

Moreno-Chulim MV, Barahona-Perez F, Canche-Escamillia G (2003) Biodegradation of starch and acrylic-grafted starch by Aspergillus niger. J Appl Polym Sci 89:2764–2770

Prasantha KVH, Lakshman K, Shamlal TR, Tharananthan RN (2005) Biodegradation chitosan-graft-polymethylmethacrylate film. Int Biodeterior Biodegrad 56:115–120

Qudeseih IY, Kabbashi NA, Al-Khatib MF, Alam MZ, Ateih MA, Fandi KG, Mamun A, Rahman MZ, Muyibi SA (2007) Enzymatic biodegradability behaviour of poly methyl methacrylated grafted sago starch biopolymer. IIUM Eng J 8:37–45

Sahoo PK, Samal R (2007) Fire retardancy and biodegradability of polymethylmethacrylate/montmorillonite nanocomposite. Polym Degrad Stab 92:1700–1707

Sain S, Khatua BB (2011) Synthesis of highly exfoliated PS/Na+-MMT nanocomposites by suspension polymerization using Na+-MMT clay platelets as suspension stabilizer. Macromol Res 19(1):44–52

Sain S, Ray D, Mukhopadhyay A, Sengupta S, Kar T, Ennis JC, Rahman KSMP (2012) Synthesis and characterization of PMMA-cellulose nanocomposites by in situ polymerization technique. J Appl Polym Sci 126:E127–E134

Sain S, Sengupta S, Kar A, Mukhopadhyay A, Sengupta S, Kar T, Ray D (2014) Effect of modified cellulose fibres on the biodegradation behaviour of in-situ formed PMMA/cellulose composites in soil environment: isolation and identification of the composite degrading fungus. Polym Degrad Stab 99:156–165

Thakore IM, Desai S, Devi S (2001) Compatibility and biodegradability of PMMA/starch cinnamate blends in various solvents. J Appl Polym Sci 79:488–496

Tudorachi N, Cascaval CN, Rusu M (2000) Biodegradable polymer blends based on polyethylene and natural polymers: degradation in soil. J Polym Eng 20:287–304

Stubble Decomposition (*In Situ*) of Two Rice Varieties Through Microbial Inoculation

Nilay Borah, Prassana Kumar Pathak, Rajen Barua, Kailash Hazarika, Amrita Phukon, and Kashyap Porag Bezbaruah

Abstract Stubbles of rice varieties *Mahsuri* (*Taichung 65/Mayang Ebos 6080/2*) and *Ranjit* (*Pankaj* x *Mahsuri*) were treated after harvest of the crop by spraying microbial inoculums – laboratory culture of cellulose degrading microorganism (CDM) or *yogurt* (commercial) or mixture of CDM and *yogurt* with glyphosate (0.205% solution in water) either with sugar or without sugar for in situ decomposition. Microbial inoculums were also sprayed with glyphosate and urea, and the treatments were compared with control plots, viz., untreated, water spray, and glyphosate spray. At the end of the fourth month, *Mahsuri* showed faster decomposition than *Ranjit* with corresponding decreases in dry biomass and percent organic carbon of the stubbles following treatment with CDM culture or *yogurt* with glyphosate solution. Addition of sugar to the spray mixture was not significant. The reduction in dry biomass and organic carbon in stubble was up to 61.1 and 45.3% in *Mahsuri* and up to 47.1 and 46.4% in *Ranjit*, respectively, after 4 months of the treatments. Significant increase in diversity of weeds in terms of number of species was observed in plots with higher stubble decomposition.

Keywords CDM • *Yogurt* • *In situ* decomposition • Rice variety

1 Introduction

About one third of the world's total rice-growing area is grown as rainfed under lowland situation (IRRI 2016), of which more than 90% occurs in Asia (Singh et al. 2016). East India, with 13.9 million ha, accounts for more than half and more than one third of the total lowland rice area of the region and the country, respectively, under rainfed situation. Assam occupies about 1.4 million ha rice area with up to

N. Borah (✉) • R. Barua • A. Phukon • K.P. Bezbaruah
Department of Soil Science, Assam Agricultural University, Jorhat, Assam, India
e-mail: nilayborah@rediffmail.com

P.K. Pathak
Directorate of Research (Agri), Assam Agricultural University, Jorhat, Assam, India

K. Hazarika
Department of Nematology, Assam Agricultural University, Jorhat, Assam, India

S.K. Ghosh (ed.), *Utilization and Management of Bioresources*,
DOI 10.1007/978-981-10-5349-8_7

50 cm submergence (Ahmed 2013), and these areas are usually kept fallow after winter rice, leaving huge quantities of biodegradable crop residues in the field after harvest. The decomposition of the rice stubbles mostly takes place under anaerobic environment as it extends up to the next crop season. Anaerobic decomposition of rice stubbles had been implicated for high levels of methane emission (Liou et al. 2003) and adverse effects on the succeeding crop (Cannell and Lynch 1984) possibly through production of organic acids (Shan et al. 2008) toxic to rice roots and their breakdown into harmful gases (Brouder and Hill 1995) and loss of potential plant nutrients. The rice crop residues undergo very little decomposition during first 3 months after harvest until the pre-monsoon rain due to low temperature and rainfall (Borah et al. 2016b). Addition of nitrogen fertilizer was shown to enhance rice straw decomposition under low temperature and humidity (Arlauskienė et al. 2016). Spraying of liquid microbial culture containing cellulose degrading organisms including yeast caused faster decomposition of wheat straw in summer with or without soil cover but had no effect in rice straw during winter due to inhibition of microbial activity by low temperature (Fang et al. 2015). Similarly, treating rice stubbles with CDM or *yogurt* cultures as spray mixture of glyphosate reduced dry biomass and C/N ratio of crop residues during both winter (Borah et al. 2016b) and summer (Borah et al. 2016a) seasons. The area under modern variety in India has increased from little more than 40% in 1978 (Herdt and Capule 1983; Dalrymple 1986) to 53.6% in 2000 (Janaiah et al. 2006). In Assam, the proportion of area under modern variety to total rice acreage increased from 0.24 in 1978 (Roy and Bezbaruah 2002) to 0.76 in 2010, and *Ranjit* and *Mahsuri* are grown in about two third of the state's rice area under modern variety (Deka and Gauchan 2012). The straw composition significantly varies among the rice varieties (Vadiveloo and Phang 1996; Shen et al. 1998). Besides, due to differential preferences for surface hydrology class (Tuong et al. 2000) *vis-à-vis* rainfall uncertainty, the sowing time of the varieties differs and the crop is harvested from November to December. Thus, the efficiency of any decomposition method would vary greatly as the temperature falls sharply with declining or no rainfall during this period. Variation in the caobon dioxide release rate and percent carbon release from straw was observed among thiry rice varieties during six week long study using perfusing ssystem (Villegas-Pangga et al. 2000). Accordingly, this study was undertaken to evaluate the effect of spraying CDM or *yogurt*, with glyphosate on in situ stubble decomposition of rice varieties *Ranjit* and *Mahsuri*.

2 Materials and Methods

2.1 Location and Climate

The location of the experiment, conducted during November 2015 to April 2016, was situated at the Instructional cum Research (ICR) Farm (26°44′N, 94°10′E, and 91 m above MSL) of Assam Agricultural University, Jorhat, India. The soil of the experimental site was clay loam in texture with pH 5.5 and organic carbon content 7.2 g/kg. The mean daily temperature of the experimental site decreases from November to January and then increases from February to April with a mean maximum temperature of 28 °C in November to 23 °C in January and then 24 °C in February to 28 °C in April and with a mean minimum temperature of 16 °C in November to 8 °C in January and thereafter 13 °C in February to 19 °C in April. Similar to daily temperature, the monthly rainfall also decreases from a mean of 24 mm in November to 12 mm precipitation in January, which then increases from 70 mm in February to 217 mm with 15 rainy days in April. The relative humidity fluctuates from 75% in November to 71% in January to 75% in April, while the mean total duration of bright sunshine hours remains at 8–10 h/day during November to April. Except for an average of 7 days in January, the other months from November to April are not affected by fog.

2.2 *Rice Variety*

The rice varieties *Mahsuri* and *Ranjit* were evaluated for in situ decomposition through microbial spray. *Mahsuri* (*Taichung 65/Mayang Ebos 6080/2*) is a tall variety with average plant height of 130 cm and duration of 135 days. *Ranjit* was developed between Pankaj × *Mahsuri*, while Pankaj was developed as a cross between Peta and Tangkai Rotan. *Ranjit* is a semidwarf variety of 150–155 days duration and average plant height of about 110 cm.

2.3 *Microbe Culture*

The cellulose degrading microbe (CDM) strains were screened by growing the isolates in a minimal agar plate consisting of yeast extract (0.2%), KH_2PO_4 (0.1%), $MgSO_4$ (0.5%), and a soluble form of cellulose, carboxymethyl cellulose (0.5%). The appearance of clear zone around colonies indicated dissolution of carboxymethyl cellulose (Apun et al. 2000) as reported elsewhere (Gogoi et al. 2015). Commercial *yogurt*, available in local market for consumption, was used as microbe culture for spray. *Lactobacillus* species present in yogurt are facultative anaerobe or microaerophilic in nature and are well competent with cellulose degrading microbe (CDM). It was hypothesized that the isolated CDM strain and *Lactobacillus* in *yogurt* would perform well as aerobic decomposer because facultative anaerobes have a dual nature, i.e., generating ATP in presence of oxygen (aerobic growth), and possess fermentative properties in absence of oxygen.

Rice stubbles after harvest of transplanted winter rice in individual plot of size 6 × 5 m were treated with CDM culture (10^6–10^7 effective colony-forming unit per ml) and *yogurt* obtained from local market. CDM culture (1 ml) or *yogurt* (5 g) was mixed with each liter of water or of 0.205% glyphosate solution (5 ml of 41% EC commercial formulation per liter of water) or of 0.205% glyphosate and 0.5% table sugar (sucrose) and used for spraying on the rice stubbles. A plot of identical area was also sprayed with water using the same sprayer with a spray volume of 600 l/ha. The CDM culture (1 ml) or *yogurt* (5 g) was also sprayed separately after mixing with solution containing 0.5% urea (fertilizer grade) in water. The spray was done on 5 November 2015 for *Mahsuri* and on 16 November 2015 for *Ranjit* with hollow cone nozzle fitted manual knapsack sprayer. The treatments were assigned to plots with three replications in randomized block design. Pertinent observations were taken and parameters were analyzed, and the data were subjected to statistical analysis for interpretation.

2.4 Rice Stubble Dry Weight

Stubbles were collected from 1 × 1 m area in each plot before imposing the treatments and at 30, 60, 90, and 120 days after spray. The stubbles were cleaned, processed, and weighed as per procedure followed by Borah et al. (2016a).

2.5 Carbon Content of Rice Stubble

The samples of rice stubbles were prepared (Borah et al. 2016a), and the organic carbon content was determined following the procedure outlined by Walkley and Black (1934).

2.6 Diversity of Weed Species

The diversity of weed species was assessed at 4 months after the treatments by counting the number of species present in each plot under each category (grasses, broad leaved, and sedges).

2.7 Statistical Analysis

The treatment means of individual parameter were compared after analyzing the data through a one-way ANOVA. The difference between two means was ascertained as significant or nonsignificant by testing the values through least significant difference (LSD) test.

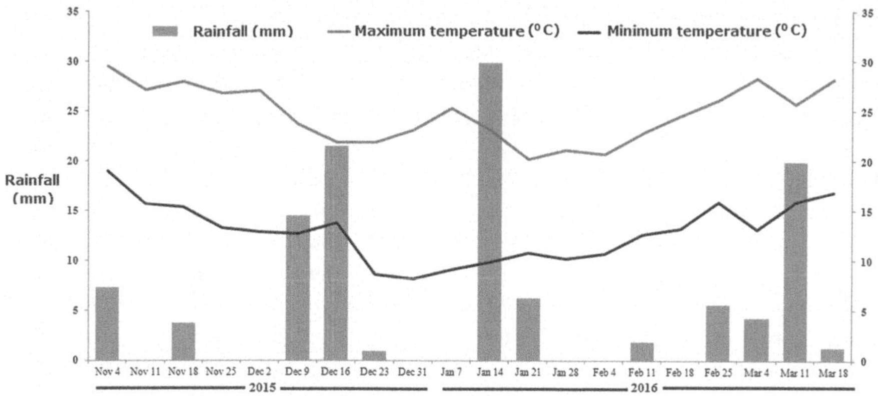

Fig. 1 Mean weekly maximum and minimum temperature and rainfall for standard meteorological weeks with ending date during the experiment

3 Results and Discussion

3.1 Rainfall and Temperature

The mean maximum and minimum temperature and rainfall of each standard meteorological week during the period of study are depicted in Fig. 1. The average maximum temperature during the period (5 November 2015 to 18 March 2016) fluctuated between 20.2 °C (15–21 January 2016) and 28.3 °C (26 February to 4 March 2016), and the minimum temperature was 8.2 °C (24–31 December 2015) to 16.9 °C (12–18 March 2016). Both the mean weekly maximum and minimum temperatures decreased after imposition of the treatments (11 November 2015) up to 21 January 2016 (tenth week) and thereafter increased till the completion of the experiment. The highest mean weekly rainfall received during the period was 29.4 mm (8–14 January 2016), and eight rainfall-free weeks were recorded during the study period.

3.2 Rice Stubble Dry Weight

The rice stubble dry weight, irrespective of the varieties, was not affected by the treatments up to 60 days of spray, but differed significantly at 90 and 120 days after the treatments (Table 1). Application of CDM or *yogurt* with herbicide significantly decreased stubble dry weight in both the varieties at 90 and 120 days after treatments. The mixing of sugar with herbicide-microbial culture did not produce significant effect on stubble decomposition. The percent decrease in stubble dry biomass at each month after treatment is shown in Fig. 1a, b. The comparative

Table 1 Rice stubble dry weight (g/m^2) at different days after treatment

	Variety – Mahsuri					Variety – Ranjit				
Treatments	0	30	60	90	120	0	30	60	90	120
Untreated	383	414	423	390	307	562	600	582	542	508
Water spray	374	411	421	381	315	560	590	572	537	492
Glyphosate (2.05 g a.i./l) spray	389	381	371	359	270	535	523	513	477	391
Glyphosate + CDM	381	364	344	301	151	550	519	490	396	321
Glyphosate + CDM + sugar	387	362	340	287	158	571	531	486	374	305
Glyphosate + *yogurt*	390	373	356	305	169	539	508	481	383	319
Glyphosate + *yogurt* + sugar	397	378	361	307	175	562	528	490	379	312
Glyphosate + CDM + *yogurt*	381	363	346	294	163	574	545	486	369	304
Urea 5% + CDM	376	418	428	389	307	539	575	568	518	459
Urea 5% + *yogurt*	379	423	435	401	318	555	592	572	532	469
$LSD_{P=0.05}$	NS	NS	48	47	45	NS	NS	74	70	63
CV (%)	9	8	7	8	11	11	8	8	9	9

decrease in stubble dry weight over initial was significant at 90 and 120 days after spraying of CDM or *yogurt* with glyphosate, compared to other treatments. Similar decrease in biomass weight of rice straw was earlier reported (Eusufzai et al. 2013; Borah et al. 2016a). There was gain in stubble dry weight in both the varieties without herbicide application which was attributed to the ratooning ability of the varieties (Sanni et al. 2009; Ringera et al. 2011) (Fig. 2).

3.3 Organic Carbon Content of Rice Stubble

The carbon content of the stubble was not affected by the treatments up to 60 days after the treatments (Table 2), but thereafter significant decrease was observed for treatment with glyphosate and CDM culture or glyphosate and *yogurt* in both the varieties. A decrease in organic carbon content of rice stubble with application of glyphosate-microbial culture was reported (Borah et al. 2016a) and was ascribed to glyphosate that facilitated effective inoculation of microorganisms (Madsen et al. 1978; Thelen et al. 1995; Hetherington et al. 1998) followed by decomposition of rice stubble through acidification (Borah et al. 2016a). The relative decrease in organic carbon content of stubble at different stages after the treatments is shown in Fig. 3a, b. Similar to rice stubble dry weight, the organic carbon content was not affected by the treatments up to 60 days after spray. Application of CDM culture or *yogurt* decreased organic carbon content of rice stubble significantly at 90 and 120 days after treatment. The results are in conformity to those reported by Borah et al. (2016b) for rice variety *Ranjit*.

The relatively slower decomposition of rice stubble and mineralization of organic carbon during first 60 days after spray may be attributed to decreasing temperature and low rainfall. During this period (5 November 2015 to 6 January 2016), about

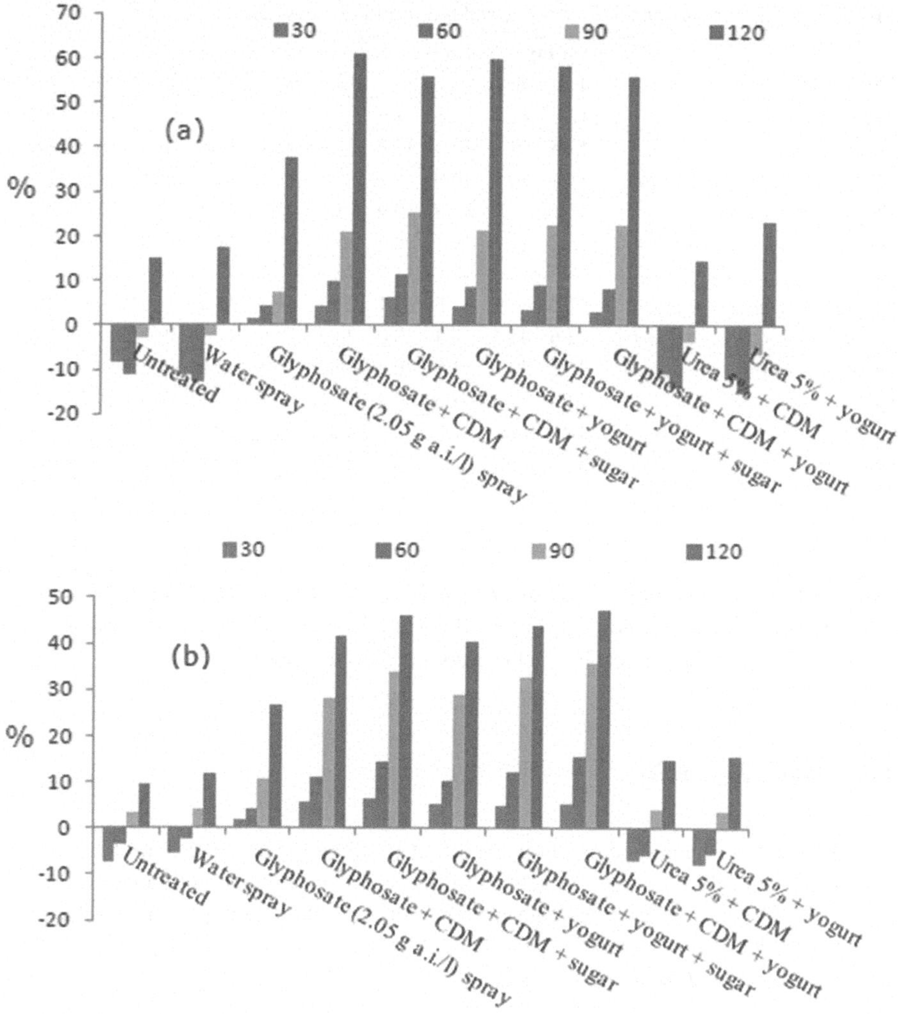

Fig. 2 Decrease in stubble dry weight (%) over initial at different days after treatment for (**a**) Mahsuri and (**b**) Ranjit rice varieties

39.0 mm rainfall was received with decreasing mean weekly temperature as against 65.9 mm rainfall with increasing mean weekly temperature in the latter part (7 January to 18 March 2016). This might have accounted for the enhanced decomposition of the rice stubble after 60 days of spray. Between the varieties, *Mahsuri* showed relatively higher reduction in dry weight of the stubble over *Ranjit*. This may be ascribed to mainly two reasons – first, due to the difference in initial substrate volume and mass. *Ranjit* had higher residue left in the field than *Mahsuri*, which might have accounted for its slower decomposition compared to *Mahsuri*.

Table 2 Organic carbon content (%) of rice stubble at different days after treatment

	Variety – Mahsuri					Variety – Ranjit				
Treatments	0	30	60	90	120	0	30	60	90	120
Untreated	42.8	41.3	39.8	38.6	34.8	44.5	43.5	41.8	40.7	35.1
Water spray	43.5	42.3	40.6	38.9	34.9	42.2	41.0	40.1	38.9	34.0
Glyphosate (2.05 g a.i./l) spray	41.4	40.2	39.0	37.3	31.9	41.3	40.3	39.1	37.6	31.8
Glyphosate + CDM	43.1	41.2	37.9	30.9	24.6	43.8	42.2	39.4	32.7	23.3
Glyphosate + CDM + sugar	43.2	41.8	38.3	28.2	23.9	45.5	44.0	41.0	32.4	25.5
Glyphosate + *yogurt*	46.2	45.0	40.8	31.0	26.9	43.5	41.7	39.3	31.2	24.5
Glyphosate + *yogurt* + sugar	41.4	40.2	37.6	29.1	23.4	41.4	39.8	37.6	30.6	22.6
Glyphosate + CDM + *yogurt*	42.6	41.6	38.4	28.0	23.2	42.6	41.2	39.2	31.9	24.9
Urea 5% + CDM	46.2	44.8	43.9	40.2	35.2	43.5	42.6	41.7	39.3	32.7
Urea 5% + *yogurt*	44.1	42.1	41.3	39.2	35.1	42.2	41.1	40.3	38.5	32.5
$LSD_{P=0.05}$	NS	NS	NS	3.8	4.0	NS	NS	NS	3.8	3.4
CV (%)	5.1	6.3	6.8	6.6	7.9	7.0	6.4	5.7	6.3	6.9

Second, the faster decomposition of *Mahsuri* over *Ranjit* might be due to difference in brittleness of the stubbles. Higher total carbon loss as CO_2 or CH_4 in brittle than non-brittle rice straw had earlier been reported (Johnson et al. 2006).

3.4 *Weed Diversity*

The diversity of weed species under different treatments was recorded at 120 days after spray, and the data are presented in Table 3. Significantly higher number of broad-leaved weed species was observed with spraying of glyphosate and CDM culture or *yogurt* compared to the others. Similar observations were recorded in earlier studies also (Borah et al. 2016a, b).

4 Conclusion

The stubbles of rice varieties *Mahsuri* and *Ranjit* could be decomposed faster by spraying of glyphosate with CDM culture or with commercial *yogurt*. Between the varieties, *Mahsuri* showed higher reduction in stubble dry weight over *Ranjit*, but did not differ in organic carbon content. Addition of sugar or urea to the spray mixture was not effective, irrespective of microbe cultures. The pace of

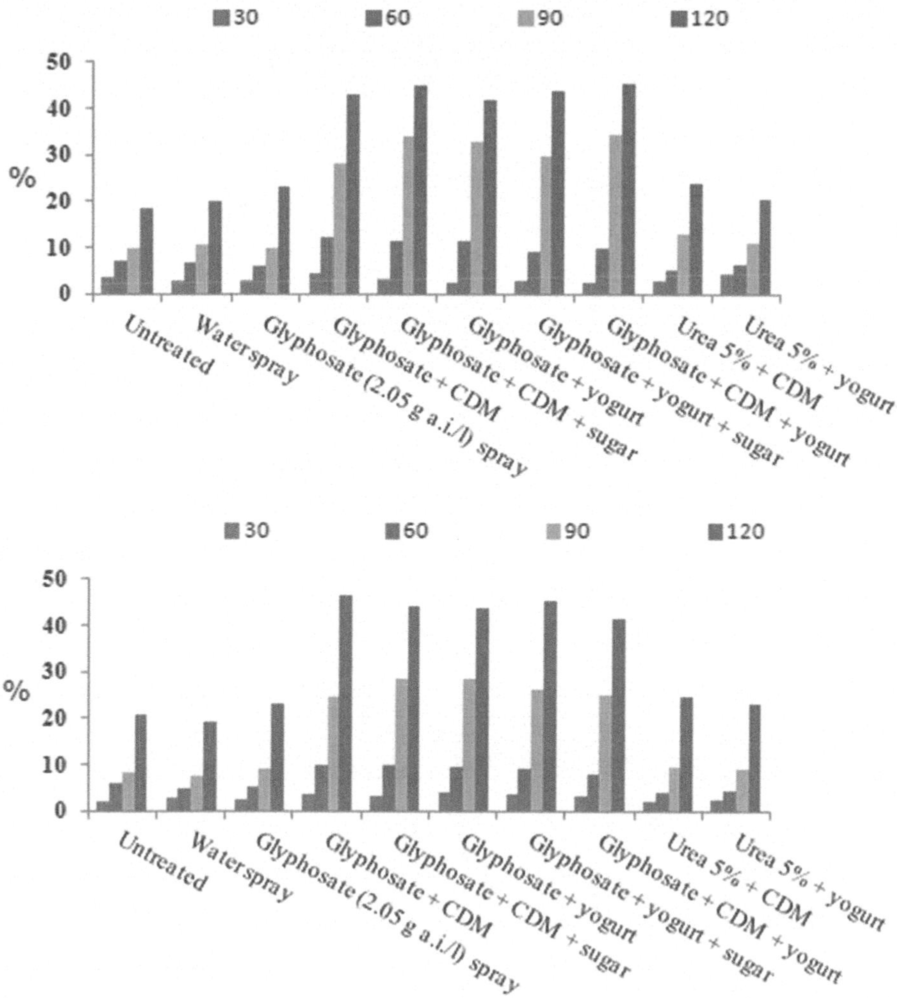

Fig. 3 Decrease in total organic carbon content of stubble (%) over initial at different days after spray for (**a**) Mahsuri and (**b**) Ranjit varieties

decomposition of rice stubbles was faster with increase in temperature and rainfall. Nonselective herbicide glyphosate played a major role in desiccation of rice stubbles and subsequent decomposition by microbial strains. Stubble decomposition significantly enhanced emergence of broad-leaved weed species after 4 months of rice harvest.

Table 3 Species diversity of weeds at 120 days after treatment of rice stubbles

Treatments	Variety – Mahsuri				Variety – Ranjit			
	[a]Number of species/plot (30 m^2)							
	[b]G	[b]BL	[b]S	Total	[b]G	[b]BL	[b]S	Total
Untreated	1.34 (1.3)	1.93 (3.3)	0.88 (0.3)	2.34 (5.0)	1.34 (1.3)	1.46 (1.7)	1.1 (0.7)	2.04 (3.7)
Water spray	1.46 (1.7)	1.84 (3.0)	1.05 (0.7)	2.40 (5.3)	1.34 (1.3)	1.17 (1.0)	1.1 (0.7)	1.86 (3.0)
Glyphosate (2.05 g a.i./l) spray	1.66 (2.3)	1.68 (2.3)	0.88 (0.3)	2.34 (5.0)	1.66 (2.3)	1.05 (0.7)	1.2 (1.0)	2.11 (4.0)
Glyphosate + CDM	1.46 (1.7)	2.61 (6.3)	1.05 (0.7)	3.02 (8.7)	1.34 (1.3)	1.95 (3.3)	1.2 (1.0)	2.48 (5.7)
Glyphosate + CDM + sugar	1.58 (2.0)	2.73 (7.0)	0.88 (0.3)	3.12 (9.3)	1.22 (1.0)	2.20 (4.3)	1.2 (1.0)	2.61 (6.3)
Glyphosate + *yogurt*	1.56 (2.0)	2.40 (5.3)	1.05 (0.7)	2.89 (8.0)	1.46 (1.7)	2.04 (3.7)	1.2 (1.0)	2.61 (6.3)
Glyphosate + *yogurt* + sugar	1.46 (1.7)	2.61 (6.3)	0.88 (0.3)	2.97 (8.3)	1.22 (1.0)	2.41 (5.3)	1.2 (1.0)	2.80 (7.3)
Glyphosate + CDM + *yogurt*	1.44 (1.7)	2.48 (5.7)	1.05 (0.7)	2.91 (8.0)	1.34 (1.3)	2.02 (3.7)	1.2 (1.0)	2.54 (6.0)
Urea 5% + CDM	1.46 (1.7)	1.58 (2.0)	1.05 (0.7)	2.20 (4.3)	1.46 (1.7)	1.34 (1.3)	1.2 (1.0)	2.11 (4.0)
Urea 5% + *yogurt*	1.29 (1.3)	1.46 (1.7)	1.05 (0.7)	2.04 (3.7)	1.22 (1.0)	1.46 (1.7)	1.1 (0.7)	1.95 (3.3)
$LSD_{P=0.05}$	NS	0.48	NS	0.50	NS	0.44	NS	0.30
CV (%)	20.7	13.2	31.0	10.1	16.2	14.8	14.7	8.1

[a]Square root transformed $\left[\sqrt{(x+0.5)}\right]$ value, the observed data (x) in parentheses

[b]*G* grasses, *BLW* broad leaved, *S* sedges weed species

References

Ahmed T (2013) Twenty years of achievements of the EIRLSBN: accomplishments of the Regional Agricultural Research Station, Titabar, Assam. In: Collard BCY, Ismail AM, Hardy B (eds) EIRLSBN: twenty years of achievements of rice breeding. International Rice Research Institute, Los Banos. ISBN-978-971-22-0295-7

Apun K, Jong BC, Salleh MA (2000) Screening and isolation of a cellulolytic and amylolytic *Bacillus* from sago pith waste. J Gen Appl Microbiol 46:263–267

Arlauskienė A, Velykis A, Šlepetienė A, Janušauskaitė D (2016) Comparison of postharvest practices used for cereal straw decomposition in a clay loam soil. Acta Agric Scand Sect B Soil Plant Sci 66(6):523–533

Borah N, Barua R, Pathak PK, Barua IC, Hazarika K, Phukan A (2016a) Rice stubble decomposition by cellulose degrading microbe and *yogurt* with glyphosate under rainfed upland ecosystem. Int J Agri Sci 8(20):1350–1353

Borah N, Barua R, Nath D, Hazarika K, Phukon A, Goswami K, Barua DC (2016b) Low energy rice stubble management through in situ decomposition. Procedia Environ Sci 35:771–780

Brouder SM, Hill JE (1995) Conjunctive use of farmland adds value ..... Winter flooding of ricelands provides waterfowl habitat. Calif Agric 49(6):58–64

Cannell RQ, Lynch JM (1984) Possible adverse effects of decomposing crop residues on plant growth. In: Organic matter and rice. Inter-national Rice Research Institute, Los Banos. ISBN 971-104-104-9, pp 455–476

Dalrymple DG (1986) Development and spread of high-yielding rice varieties in developing countries. Bureau for Science and Technology, Agency for International Development, Washington, DC, p 46

Deka N, Gauchan D (2012) Patterns of adoption of improved rice varieties and farm-level impacts in stress-prone rainfed areas of Assam, India. In: Pandey S, Gauchan D, Malabayabas M, Bool-Emerick M, Hardy B (eds) Patterns of adoption of improved rice varieties and farm-level impacts in stress-prone rainfed areas in South Asia. International Rice Research Institute, Los Baños. ISBN 978-971-22-0287-2

Eusufzai MK, Deb SK, Maeda T, Fujii K (2013) Mass loss and C and N release from decomposing fresh and composted residues as affected by cold climate conditions. Environ Nat Resour Res 3:116–127

Fang Z, Liu Y, Ding W, Chen Y, Qin K (2015) Effect of microbial spraying on wheat and rice straw returning to ditch. Trans Chin Soc Agric Eng (Trans CSAE) 31(23):187–194. (in Chinese with English abstract). doi:10.11975/j.issn.1002-6819.2015.23.025

Gogoi G, Sharma P, Buragohain P, Phukan A, Baruah R (2015) Isolation and characterization of compost accelerating microbes. Trends Biosci 8(21):5854–5859

Herdt RW, Capule C (1983) Adoption, spread, and production impact of modern rice varieties in Asia. International Rice Research Institute (IRRI), Los Baños, p 59

Hetherington PR, Marshall G, Kirkwood RC, Warner JM (1998) Absorption and efflux of glyphosate by cell suspensions. J Exp Bot 49:527–533

IRRI (2016) http://www.knowledgebank.irri.org/ericeproduction/0.3._Rice_environments.htm. Accessed on 14 July 2016

Janaiah A, Hossain M, Otsuka K (2006) Productivity impact of the modern varieties of rice in India. Dev Econ 44(2):190–207

Johnson SE, Angeles OR, Brar DS, Buresh RJ (2006) Faster anaerobic decomposition of a brittle straw rice mutant: implications for residue management. Soil Biol Biochem 38:1880–1892

Liou RN, Huang SN, Lin CW, Chen SH (2003) Methane emission from fields with three various rice straw treatments in Taiwan paddy soils. J Environ Sci Health Part B Pestic Food Contam Agric Wastes B38(4):511–527

Madsen HEL, Christensen HH, Gonlieb-Petersen C (1978) Stability constants of copper (II), zinc, manganese (II), calcium, and magnesium complexes of N-(phosphonomethyl) glycine (glyphosate). Acta Chem Scand A32:79–83

Ringera TJ, Kinyamario IJ, Amugune ON, Kanya IJ (2011) Ratooning ability and its effect on grain yield of upland Nerica rice varieties in Central Kenya. Int J Agron Agric Res 4:44–55

Roy N, Bezbaruah MP (2002) Agricultural growth and regional economic development (a study of Barak Valley). Mittal Publucations, New Delhi. ISBN 81-7099-845-X, p 65

Sanni KA, Ojo DK, Adebisi MA, Somado EA, Ariyo OJ, Sie M, Akintayo I, Tia DD, Ogunbayo SA, Cisse B, Sikirou M, Adekoya MA (2009) Ratooning potential of interspecific Nerica rice varieties (*Oryza glaberima* x *Oryza sativa*). Int J Bot 5:112–115

Shan Y, Cai Z, Han Y, Johnson SE, Buresh RJ (2008) Organic acid accumulation under flooded soil conditions in relation to the incorporation of wheat and rice straws with different C:N ratios. Soil Sci Plant Nutr 54:46–56. doi:10.1111/j.1747-0765.2007.00218.x

Shen HS, Ni DB, Sundstùl F (1998) Studies on untreated and urea-treated rice straw from three cultivation seasons: 1. Physical and chemical measurements in straw and straw fractions. Anim Feed Sci Technol 73:243–261

Singh RB, Woodhead T, Papademetriou MK (2016) Strategies to sustain and enhance Asia-Pacific rice production. FAO Corporate Document repository. http://www.fao.org/docrep/006/y4751e/y4751e0s.htm. Accessed on 14 July 2016

Thelen KD, Jackson EP, Penner D (1995) The basis for the hard water antagonism of glyphosate activity. Weed Sci 43:541–548

Tuong TP, Kam SP, Wade L, Pandey S, Bouman BAM, Hardy B (2000) Characterizing and understanding rainfed environments, In: Proceedings of the international workshop on characterizing and understanding rainfed environments, 5–9 December 1999, Bali, Indonesia. International Rice Research Institute, Los Baños, p 488. ISBN 971-22-0152-X

Vadiveloo J, Phang OC (1996) Differences in the nutritive value of two rice straw varieties as influenced by season and location. Anim Feed Sci Technol 61(1–4):247–285

Villegas-Pangga G, Blair G, Lefroy R (2000) Measurement of decomposition and associated nutrient release from straw (*Oryza sativa* L.) of different rice varieties using a perfusion system. Plant Soil 223(1):1–11

Walkley A, Black IA (1934) An examination of the Degtjareff method for determining soil organic matter, and proposed modification of the chromic acid titration method. Soil Sci 37:29–38

Cassava Processing Wastes: Options and Potentials for Resource Recovery in Nigeria

C.G. Achi, A.O. Coker, and M.K.C. Sridhar

Abstract Agro-food processing industries are major contributors of wastes in most developing countries. With Nigeria leading in cassava food production, little attention has been paid to provide a sustainable and profit-oriented solution to the problem of solid waste resulting from cassava processing industries.

Considering the global effort to promote sustainability in the areas of food production, processing and waste management and also the need for resource recovery and utilisation to enhance cassava food value chain, this study assessed the quality and amount of waste in selected cassava industries with regard to cassava production rates and current waste management practices (from six randomly selected cassava industries) in Ibadan Nigeria. The potentials and various options for cassava waste utilisation were explored.

Six cassava production sites were randomly selected for the purpose of this study. Information gathered through personal field observations and key informant interviews showed that between 1.5 and 3 tons of solid (peels and pulp) waste and between 3 and 6 m^3 of liquid wastes were generated daily during cassava processing from a daily supply of between 6 and 8 tons of cassava tubers. Between 25 and 37% of solid wastes usually result from production of cassava tubers with only 25% of the total available waste being utilised as livestock feed.

The recovery of this huge amount of waste resource in terms of animal feed, biomass for energy production and biosolids from spent slurry has the potential to increase the cassava food value chain significantly.

Keywords Cassava food value • Resource recovery • Waste management • Biosolids • Clean energy

C.G. Achi (✉) • A.O. Coker
Department of Civil Engineering, Faculty of Technology, University of Ibadan, Ibadan, Nigeria
e-mail: achicgjr@gmail.com

M.K.C. Sridhar
Department of Environmental Health Sciences, Faculty of Public Health, University of Ibadan, Ibadan, Nigeria

S.K. Ghosh (ed.), *Utilization and Management of Bioresources*,
DOI 10.1007/978-981-10-5349-8_8

1 Introduction

In recent times, there has been a global movement towards adopting sustainable practices in the use of natural resources and management of waste materials in many commercial and industrial sectors. The eleventh goal of the Sustainable Development Goals (SDGs) "sustainable cities and communities" clearly emphasises the importance of this global movement. The sustainability of a practice depends on the amount of resource that can be recovered and reused from the raw waste materials and also the sustained availability of the waste material stream. These resources recovered from waste can be translated to a substantial amount of fiscal savings resulting from the conservation of fresh material resources and utilisation of recovered resources.

The need to adopt sustainable best practices has been extended to agro-food processing industries. Agro-food processing industries constitute a significant part of every nation's economy. A typical example of an agro-food processing industry is cassava processing industry.

Cassava, a starchy staple food crop, is regarded as a primary food crop in Africa; this is due to its resistance to drought and diseases and its capacity to provide a reliable and an inexpensive source of carbohydrate for human consumption. Cassava tuber is processed into myriads of food items which vary sometimes based on cultural and customary differences and preferences (FAO 2006). Most cassava processing activities in Nigeria are predominantly done by subsistence farmers (Sangodoyin and Amori 2013). During cassava production and processing, large amounts of wastes (solid and wastewater) are generated. Most of these waste materials are discharged indiscriminately around the processing environment, and this usually leads to environmental pollution and sometimes obstruction of waterways and drains.

The perennial problem of food processing waste such as cassava can never be overemphasised in sub-Saharan Africa, especially in Nigeria, where cassava productions rank highest globally. With the growing production of cassava in Nigeria came the emergence of waste materials waiting to be utilised. These waste materials are currently not being utilised optimally. About 25% of peels are recovered through sun-drying during dry season and less than 20% during wet season. This is due to constraints associated with drying and concerns about safety of use, particularly hydrocyanide- and mycotoxin-related food poisoning. Drying peels in the open—practically impossible during the rainy season—takes 2–3 days. Consequently, peels are left to rot in heaps or set on fire—polluting the nearby air, soil and groundwater and wasting a potential feed resource.

The dried peels are used as livestock feed, leaving a significant amount of unutilised solid residues to go to waste. All of the waste effluents are currently being discharged indiscriminately into nearby streams and rivers thereby polluting them.

The waste effluents from cassava industries have been reported by many researchers to be toxic, containing cyanide (Siller and Winter 1998; Kaewkannetra et al. 2009), low in pH and high in BOD and COD (Hien et al. 1999; Luo et al. 2010; Sun

Table 1 Physico-chemical composition of cassava bagasse/pulp (g/100 g dry weight)

Composition	Soccol (1994)	Cereda (1994)	Sterz (1997)	Vandenberghe et al. (1998)
Moisture	5.02	9.52	10.70	11.20
Protein	1.57	0.32	1.60	1.61
Lipids	1.06	0.83	0.53	0.54
Fibres	50.55	14.88	22.20	21.10
Ash	1.10	0.66	1.50	1.44
Carbohydrates	40.50	63.85	63.40	63.00

Source: Pandey et al. (2000)

et al. 2012; Intanoo et al. 2014) and hence if not properly treated would have a negative impact on the environment and water resources (Oparaku et al. 2013).

The solid component of cassava wastes (peels and bagasse), due to their physico-chemical composition (see Table 1), has potentials to be utilised as feedstock for livestock (Ubalua 2007), as compost material (Sangodoyin and Amori 2013) and substrate for methane (energy) recovery (Cuzin et al. 1992; Ubalua 2007; Adelekan and Bamgboye 2009; Eze 2010; Panichnumsin et al. 2010; Oparaku et al. 2013) and biosolids (Ghimire et al. 2015). Studies are still being conducted in major cassava-producing regions in Africa and Asia focusing on exploring the resource recovery potential of cassava processing wastes and other potentials for value addition.

Beyond the problem of pollution identified in Nigeria due to poor management of cassava processing residues is the need to provide an alternative energy source for the rural cassava processors who are usually exposed to smoke and soot from burning firewood. Most of these farmers, mostly women, have limited resources to embark on large-scale production using modern heating technologies.

Considering these challenges, and the need to add value to the cassava food processing chain through waste utilisation and resource recovery, this study assessed the rate of cassava production and amount of unutilised waste, current waste management practices and options available for resource recovery in the Nigerian context.

2 Methodology

2.1 Selection of Study Locations

Six cassava processing industries were randomly selected for the purpose of this study. All the industries are located within Ibadan, southwestern Nigeria. The cassava industries were visited with the aim of assessing the current practices in terms of cassava processing operations, products and wastes management measures in place. The cassava industries are located in the following areas of Ibadan: Moniya, Eleyele 1, Eleyele 2, Ojoo, Agbowo and Egbeda.

3 Study Design

In order to achieve the aim and objectives of this study, the following methods were adopted:

- Extensive literature review on the subject matter
- Field visits in Ibadan for collecting primary and secondary data from small- and medium-sized cassava industries
- Face-to-face meetings (in-depth interviews) with cassava farmers and processors and assessment of current practices
- Assessment of daily cassava delivery and production rates
- Characterisation of cassava processing wastes

Investigations on the current practices at the selected cassava industries were carried out using the methods of key informant interview and direct personal observation.

The following data were collected and used to make comparison as regards management and waste disposal practices.

1. Name/location of the cassava processing site
2. Operations and major cassava products
3. Number of the workers and scale of the processing system
4. Cassava processing rate
5. The volume of cassava wastewater effluent produced per unit Kg of cassava
6. The weight of solid waste produced per unit Kg of cassava
7. The amount and volume of solid and liquid waste produced, respectively, in a cassava processing site
8. Solid waste disposal method employed

Based on the observations made at the cassava industries, the food products predominantly processed from cassava processing at the selected site are garri (Yoruba and Igbo customary types), fufu and lafun.

4 Cassava Processing Steps

Various steps involved during the processing of cassava tubers into the desired food product are as represented below (Figs. 1, 2 and 3):

In order to quantify the amount of waste that can be generated from a unit weight of cassava tuber processed, a basket full of cassava tubers was weighted before and after peeling and the amount of waste (peels, bagasse and effluent) recorded. This was used to estimate the total amount of waste generated daily based on amount of cassava tubers supplied daily.

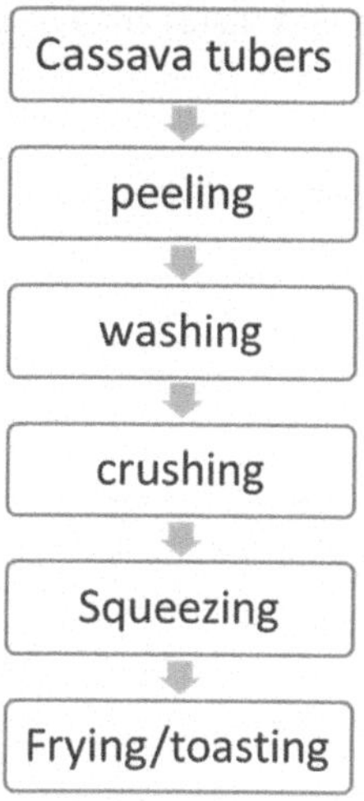

Fig. 1 Garri processing

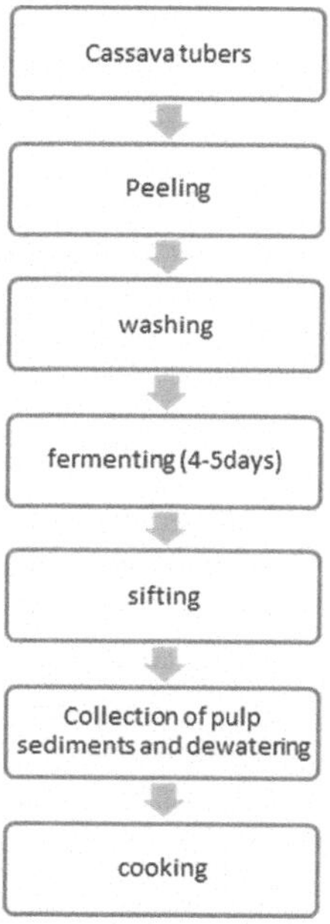

Fig. 2 Fufu processing

5 Results and Discussion

The information collected from each industry is summarised in Table 2:

These common features were recorded in all the sites visited.

1. All the sites were located close to a stream or flowing river which served as a final disposal point for the waste effluents (Plates 1, 2, 3 and 4).

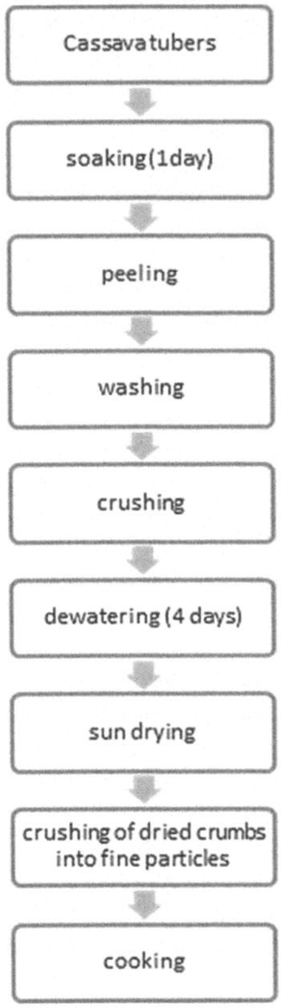

Fig. 3 Lafun processing

2. No strict measures have been put in place for the management of solid and waste effluent and as such are disposed indiscriminately or dumped without check; the vast majority of cassava peels resulting from the processing of this root is either abandoned nearby the processing sites, used as landfill or burnt. Between 20 and 25% of cassava peels and fibrous waste (spent pulp) are sun-dried (after having being washed, to remove dirt, and drained) and fed to pigs and goats as can be seen in Plate 5.
3. The only energy source available for cassava processing is firewood for cooking and roasting cassava products (Plate 6)

With regard to the current energy option for cooking and frying cassava into finished products, it was observed that firewood is the only energy option available to the processors. Certain challenges have been reported and observed in the use of firewood. During frying operations, there is a challenge of maintaining a uniform heating using firewood. Smouldering of firewood exposes workers (mostly women) to toxic fumes besides exposure to cyanide during peeling and fermentation processes (Plate 7).

Table 2 Cassava processing and waste disposal practices in six different sites in Ibadan

Parameters/locations	Eleyele 1 (Temidire)	Eleyele 2 (Ologuneru road)	Ojoo Barracks	Moniya	Agbowo	Egbeda
Daily cassava root supply (tonnes)	6–8	1–2	4–5	3–5	4–5	7–9
No. of workers	200–250	20–30	70–80	40–50	80–100	500–550
Final product	Garri, fufu and starch	Garri and fufu	Garri and fufu	Garri, fufu, starch and lafun	Fufu only	Garri, fufu, lafun and starch
Solid waste (peels) disposal practice	10–25% of peels are used as livestock feed	Utilised as livestock feed	Utilised as livestock feed	Sun-dried and used as livestock feed	Utilised as livestock feed	10–25% of peels are used as livestock feed
Fibrous waste (bagasse) disposal practice	Disposed of in a nearby bush	Disposed in a nearby bush	10–25 used as pig feed, the rest disposed off	10–25 used as pig feed, the rest disposed off	10–25 used as pig feed, the rest dumped alongside other municipal wastes	Thrown away in the bushes
Cassava effluent disposal practice	Discharged in nearby drains leading to nearby water body	Discharged indiscriminately	Discharged into nearby water body	Discharged in nearby drains leading to nearby water body	Discharged into nearby water body	Discharges into nearby river
Distance to nearest water body (metres)	Approximately 200	100	150	200	50	100
Weight of peels produced daily (kg)	1464–2049	244–512	976–1281	732–1281	976–1281	1708–2305
Volume of effluent produced daily (litres)	>5712	>1428	>3570	>3570	>3570	>6426

Plate 1 Women peeling cassava in a typical cassava processing industry

Plate 2 Heap of unutilised cassava bagasse/pulp

Beyond the health of the processors, there is also the environmental impact with regard to the current practice of using firewood. There is a major concern with the pressure put on forest trees amidst the growing cassava production rates, due to sole dependence on forest woods and also the emission of greenhouse gases from rotting cassava wastes in waste dumps leading to ozone layer depletion. Hence the need to embrace a resource-oriented approach towards addressing these challenges.

Plate 3 Cassava peels dumped beside a cassava industry in Ibadan

Plate 4 Effluent produced from *fufu* making

6 Current Trends and Future Perspective in Cassava Waste Utilisation

In view of the enormous amount of waste generated daily from cassava processing centres in Nigeria (>1 tonne) which is currently not being utilised fully, except for the little amount being utilised as livestock feed, there is an obvious need to channel these waste materials into a more profitable use. Although the International Livestock Research Institute (ILRI) with its CGIAR research partners has initiated an innovative technique of processing cassava peels into high-quality cassava peel

Plate 5 Traditional method of drying cassava peels in the sun

Plate 6 Piles of firewood for cooking and frying cassava products

(HQCP) mash for use as livestock feed, the feed produced from HQCP has the potential to replace 10–20% of maize in poultry diet and 30–40% diet for cattle, sheep, goat and pigs. Based on their assessment, approximately 40 million ton of cassava waste is generated in Africa and if harnessed properly can provide up to 150 million jobs, provide about 13 million tonne of feed per annum and result to about 3,900 million USD per annum.

There is still the need to provide alternative uses for these waste materials in the rural community, so as to meet other local needs—energy and organic fertiliser.

Plate 7 Women frying of garri using firewood and mud stove

Research is currently ongoing in the areas of energy recovery from cassava waste products co-digested with livestock wastes and other domestic wastes. Through these research efforts, sufficient energy would be generated to meet the local energy demand in cassava processing industries.

Furthermore, through composting, waste products and biosolids resulting from cassava processing are converted to useful organic manure for improving soil fertility. Stabilised digestates from aerobic digestion of cassava waste materials are also being explored for use in organic farming and growing of mushrooms.

7 Conclusion

In Nigeria, the problem of poor infrastructural facilities, including lack of sustainable energy options, has been identified among many other challenges currently facing subsistent cassava farmers. In order to prevent environmental impacts arising from the huge waste streams generated during cassava processing, various agricultural wastes should be gathered and converted to useful products.

Based on the current assessment of cassava production rate and waste generation in Ibadan City, there is a significant potential for utilising cassava waste as a sustainable energy source for meeting energy demands in rural cassava processing industries. The daily supply of cassava waste (>1 tonne) is sufficient to maintain a steady supply of substrates for biogas production to meet up with basic energy demands.

According to study carried out by Cuzin et al. (1992), 5 tons of cassava roots produced 1 ton of cassava meal and 1.5 tons of cassava peel; 1.5 tons of cassava peel is needed to produce 121 m^3 CH_4, 1200 kWh or 121 m^3 CH_4 (calorific value of methane, 9.95 kWh/m^3 CH,) which are required for drying 1 ton of cassava meal. Furthermore, use of cassava wastes as substrate for producing biogas would decrease the potential damages to forests as it reduces the overall amount of firewood employed to produce heat (for cooking or other purposes).

Apart from the already stated energy need for heating, the bioenergy recovered could also be used to power generators, which will be used to light up the processing environment during dark working hours. It would also provide a more convenient means of supplying water through pumping to the cassava processors from wells. Most cassava processing industries usually suffer from scarcity of water and energy supplies. In essence, this will reduce the stress of the women involved in cassava processing, who would have to go through the drudgery of drawing water from alternative sources and deep wells.

A sustainable utilisation of cassava wastes for various purposes mentioned above would go a long way in making life a lot easier for the rural farmers and ultimately adding significant financial value to cassava production business.

References

Adelekan BA, Bamgboye AI (2009) Comparison of biogas productivity of cassava peels mixed in selected ratios with major livestock waste types. Afr J Agric Res 4(7):571–577

Cereda MP (1994) Caratecrizacßão de resõduos da industrializacßãoda mandioca. In: Cereda MP (ed) Industrializacßão da mandioca no. Pauliceia, SãoPaulo, Brazil, pp 1150

Cuzin N, Farinet JL, Segretain C, Labat M (1992) Methanogenic fermentation of cassava peel using a pilot plug flow digester. Bioresour Technol 41:259–264

Eze JI (2010) Converting cassava (*Manihot* spp) waste from Garri processing industry to energy and bio-fertilizer. Glob J Res Eng 10(4):113–117

FAO (2006) Food and Agriculture Organization of the United Nations web page accessed at: http://faostat.fao.org/site/567/DesktopDefault.aspx?PageID=567#ancor

Ghimire A, Sen R, Annachhatre AP (2015) Biosolid management options in cassava starch Industries of Thailand: present practice and future possibilities. Procedia Chem 14(2015):66–75

Hien PG, Oanh LTK, Viet NT, Lettinga G (1999) Closed wastewater system in the tapioca industry in Vietnam. Wat Sci Technol 39(5):89–96

Intanoo P, Rangsanvigit P, Malakul P, Chavadej S (2014) Optimization of separate hydrogen and methane production from cassava wastewater using two-stage upflow anaerobic sludge blanket reactor (UASB) system under thermophilic operation. Bioresour Technol 173(2014):256–265

Luo G, Xie L, Zou Z, Wang W, Zhou Q, Shim H (2010) Anaerobic treatment of cassava stillage for hydrogen and methane production in continuously stirred tank reactor (CSTR) under high organic loading rate (OLR). Int J Hydrog Energy 35:11733–11737

Oparaku NF, Ofomatah AC, Okoroigwe EC (2013) Biodigestion of cassava peels blended with pig dung for methane generation. Afr J Biotechnol 12(40):5956–5961

Pandey A, Soccol CR, Nigam P, Soccol VT, Vandenberghe LPS, Mohan R (2000) Biotechnological potential of agro-industrial residues. II: cassava bagasse. Bioresour Technol 74(2000):81–87

Panichnumsin P, Nopharatana A, Ahring B, Chaiprasert P (2010) Production of methane by co-digestion of cassava pulp with various concentrations of pig manure. Biomass and Energy 34:1117–1124

Raewkannetra P, Imai T, Garcia-Garcia FJ, Chiu TY (2009) Cyanide removal from cassava mill waste water using Azotobacter vinelandii TISTR 1094 with mixed microorganisms in activated sludge treatment system. J Hazard Mater 172(2009):224–228

Sangodoyin AY, Amori AA (2013) Aerobic composting of cassava peels using Cowdung, sewage sludge and poultry manure as supplements. Eur Int J Sci Technol 2(8):22–34

Siller H, Winter J (1998) Degradation of cyanide in agro-industrial or industrial wastewater in an acidification reactor or in a single-step methane reactor by bacteria enriched from soil and peels of cassava. Appl Microbiol Biotechnol 50:384–389

Soccol CR (1994) Contribuiçoão Estudo da Fermentação no Estado Sólido em Relacç ão com a Producç ão de Aácido Fumárico, Biotransformac¸ ão de Resíduo Sólido de Mandioca por Rhizopuse Basidiomacromicetos do Género Pleurotus. Curitiba, Tese (Professor Titular), Universidade Federal do Parana

Stertz SC (1997) Bioconversão da Farinha de Mandioca Crua (Manihot Esculenta, Crantz) por Fungos do Gĕnero Rhizopus em Fermentacß~ao no Estado Solido. Tese Mestrado, Universidade Federal do Parana, Curitiba

Sun L, Wanb S, Yu Z, Wanga Y, Wanga S (2012) Anaerobic biological treatment of high strength cassava starch wastewater in a new type up-flow multistage anaerobic reactor. Bioresour Technol 104(2012):280–288

Ubalua (2007) Cassava wastes: treatment options and value addition alternatives. Afr J Biotechnol 6(18):2065–2073

Vandenberghe LPS, Soccol CR, Carta FS, Lebeault J-M, Milcent PF, Machado L (1998) Enzymatic hydrolysis of liquid and solid wastes of cassava root industry for production of metabolites by fermentation. COBEQ 98, Porto Alegre, Brazil

Vandenberghe LPS, Soccol CR, Lebeault JM, Krieger N (1998) Cassava wastes hydrolysate an alternative carbon source for citric acid production by Candida lipolytica. Paper presented in Internatl. Congr. Biotech'98, Portugal

Sustainable Organic Waste Management in Neighbourhoods Through Productive Urban Landscapes

P.K. Amritha and P.P. Anilkumar

Abstract In most developing countries, rapid economic growth and industrialization have transformed its landscape with significant effects on the environment. One major effect of this urbanization is the increase in waste generated and is more evident in developing nations since the capacity of these cities to collect, process or reuse and dispose solid waste is limited and is not sustainable. Disposing waste on land has been the most common and cheapest way of disposing it. But due to the uncontrolled dumping, the land has reached its carrying capacity which eventually affects the environment and its visual quality. Moreover, the quality of the wastes which has the potential for reuse in productive purposes is not considered.

In this paper, land (vacant land, open dumps/landfills) is considered as specific example case of urban landscapes which can be conserved and transformed to a productive space where organic fraction of the urban waste is processed and used for landscape development simultaneously. In addition to organic waste management the productive urban landscapes also contribute to a range of functions like food or non-food/ornamental crops for personal consumption or marketing purposes, nutrient recycling, biodiversity and visual quality that benefits the community. It also emphasizes the role of urban planners and landscape architects in ensuring that, when neighbourhood level plans are made and zoning is done, appropriate area is earmarked for such productive task, based on the number of household and the quantity of waste generated within an urban area. Thus, the paper explores the possibility of integrating the proposed facility into the urban fabric so that it is a multifunctional and a sustainable component which can be applied flexibly providing benefits to the people/community.

Keywords Productive urban landscape • Organic/biodegradable waste • Sustainability • Waste management

P.K. Amritha (✉) • P.P. Anilkumar
National Institute of Technology, Calicut, India
e-mail: amritha.pk@gmail.com

S.K. Ghosh (ed.), *Utilization and Management of Bioresources*,
DOI 10.1007/978-981-10-5349-8_9

1 Introduction

It is realized that there is a significant increase of solid waste generated the world over due to the increase in population, improved living standards, rapid economic growth and urbanization. In most cases, no segregation of waste at its source happens, or if segregation happens, then both biodegradable and non-biodegradable are mixed up at collection or at the disposal units. These units are usually open dumps/landfills, any vacant land or any other neglected area. This leads to deterioration of land and its resources and eventually gives rise to environmental degradation and health impairment. Moreover, such land areas which were originally located on the outskirts of any city has now shifted and become a part of the urban landscape as cities sprawl. Thus, they have largely manipulated the landscape of a city and have been converted to an unsightly land use which is often ignored.

In an Indian context, biodegradable waste constitutes major portion of the total waste. But when it is disposed along with non-biodegradable waste, their potential to provide nutrients for the plant growth which is an essential element of the landscape is less tapped or not tapped at all. A conceptual framework for identifying the waste category which has the potential for developing landscapes and to use the same for developing open spaces (as productive landscapes) in an urban area within a neighbourhood is the aim of this paper. The paper tries to integrate this aspect with a waste management policy which steadily facilitates a series of landscape development options and thereby redefining open spaces within an urban area using such nutrients from a sustainable/aesthetic perspective.

Nutrient part of the biodegradable waste can support plant life and thus become the base for developing productive landscape. For this, a clear and conceptually sound strategy needs to be evolved. This should also be based on the background of waste generation, its composition as well as the characteristics of the waste and the socio-economic context of the region. Further, the work proposes to do this based on existing literature-based inputs, field studies that explore deeper into the nutrient potential of the biodegradable waste to support plant life, identify feasible plant varieties and explore climatic congeniality to do this.

The current paper discusses the scope and feasibility aspects of implementing the concept of productive landscapes while planning and designing open spaces within an urban area. Through productive landscapes, the work simultaneously aims to develop green spaces which are used for growing food/ornamental gardens along with spaces within the city which manages its organic waste efficiently. Eventually the biodegradable fraction of urban waste adds to the aesthetic and visual quality of the urban landscape.

2 Review of Literature

There are many studies conducted connecting productive urban landscape development from the agriculture perspective. However extending this concept to urban landscape for managing the biodegradable fraction of the waste, there exist only limited studies. A pivotal/seminal work in this regard is the theoretical concept of CPUL (continuous productive urban landscapes) coined by architect Andrea Viljoen, to develop a network of productive open spaces which offers a space for growing food (Viljoen et al. 2005). This idea has inspired landscape interventions like the introduction of urban agriculture into the city fabric. Globally urban agriculture has evolved to meet the needs of the residents in the city. Most of the related examples were focused more on improving the livelihoods of the poor in the urban areas by providing food at lesser rates (Redwood 2009) and also giving job opportunities for the locals. 'Organoponics' is yet another attempt developed in Havana were raised beds were mixed with soil and organic matter which helped in the growth of food products. Such urban organic gardens first came up as a solution for the lack of food security. Many cities in developed countries have also recognized the extensive benefits of urban agriculture and planning or policy strategies which have been developed to support food production. Though the importance of food system through urban agriculture in planning has been recognized, very little research has been done in integrating functions like waste management by using the same space/land for both the purposes in a sustainable manner. The current paper is in this direction.

In India, 51 % of the total waste is organic in nature (Asnani 2006; Ranjith 2012). But most of these wastes (about 90%) are disposed without any proper segregation on to open dumps/landfills along with other wastes (Sharholy et al. 2008). At such sites, compaction, levelling of the waste and the cover of the waste, is rarely observed and is devoid of leachate collection, landfill gas monitoring and collection equipment (Bhide et al. 1998), unlike sanitary landfills. In a sustainable waste management, reduction, reuse and recycling of waste are most preferred options considering its environmental benefits (Asnani 2006). With regard to organic wastes, its segregation, decomposition and stabilization through different natural cycles form the basis of recycling. Treatment for organic waste recycling adopted in India is composting and anaerobic digestion. Composting involves biodegradation of organic waste into water, CO_2, energy and composted matter where the process is predominantly aerobic, and the waste volume is reduced to 50–85 % (Sharholy et al. 2008). Full-scale composting technologies are already demonstrated in many towns and cities. But their applications on landscape development are limited on account of poor marketing (Asnani 2006). It is said that only 9% of MSW is treated by composting (Gupta et al. 2007). Anaerobic digestion works out in the absence of air, where, anaerobic microorganisms act on the waste to release methane (which can be used as biogas for cooking), CO_2 and an organic residue which is a good manure. In fact, this process can occur naturally in a landfill/opendumps if the waste in it is not turned and aerated regularly. This condition if not controlled can cause serious environmental issues. Researches in the field show that if there is enough space available, then aerobic composting is a better option (Sharholy et al. 2008). But one of the key issues that we face today especially in cities is the limited site availability for processing waste or for setting up new disposal sites (TERI 1998). Moreover, the capacity of existing landfills to hold more waste exceeds its limit calling for an upgradation/rehabilitation. Many rehabilitation projects have been reviewed where these dumpsites are closed and restored for some facilities or remediated to a controlled sanitary landfill (Sharma et al. 2004); not many studies have been explored in light of restoring a site so that the site continues as a disposal site and at the same time used for a different facility thus making it multifunctional.

The current paper looks into a framework where organic waste dumping on landfills/open dumps acts as a source for landscape development while conserving the space for further disposal and processing the waste and thus reusing it. Any upgrading measures must ensure that at every stage of the process, environmental issues and aesthetics are not overlooked.

3 Methodology

Organic solid waste in most cities is a significant and essential input for developing landscape since the widespread use of compost as fertilizer for plant growth has always been an accepted practice. Among the organic wastes, food wastes are always on the top of the list and are one of the major components of the waste generated in a household. The solid waste of organic nature generated in a household are usually collected as a larger system of solid waste management and are transported to some major dump locations within and outside the city, or it may be processed individually by each household in a home composting system. But all these processes seem to be not very conducive to maximize the utilization of the organic waste in developing productive urban landscapes. Along with utilizing the organic waste as a resource, the paper tries to look at the urban setting and its open areas/ landfills as a resource to be tapped for developing productive landscape.

Urban landscapes are shaped by natural and social process that take place in an urban fabric. This makes it significant from all its perspectives, right from parks, private open spaces and recreational areas to, land used for dumping waste. As such, their locations and maintenance influence the built environment. Landfills which were located in the city outskirts earlier have now become part of the cityscape. But in most cases, improper treatment and maintenance of landfills have affected its visual stability making it an ignored and isolated part of the urban fabric which needs upgradation. This issue is more enhanced when there is a limitation in the site availability for processing waste or for setting up new landfill sites (TERI 1998) calling for an upgradation/rehabilitation. A thoughtful process of managing the potential organic waste of a city thereby recycling a major portion of MSW within the landfill site to create a distinctive urban landscape is challenging and promising at the same time. Any kind of upgradation measures must ensure that at every stage of the process, environmental issues and aesthetics are not overlooked. In an attempt to conserve and develop urban open spaces and to simultaneously come up with a solution for disposal of urban organic waste from a landscape development perspective, the model (Fig. 1) explores alternatives such as upgrading existing dilapidated or unused open spaces to productive landscapes or creating new spaces for productive landscapes. Figure 2 shows the methodology for upgrading existing disposal sites for developing productive landscapes.

The municipal government is responsible for the waste collection, transport and disposal in most cities. It is observed that most often the authorities are unable to cope with the increasing demands in a formal waste management system, and this is specifically the case in India. This demands a decentralized yet innovative system for waste management. The conceptual model explained in this paper initiates a waste management strategy at a neighbourhood/community level. A neighbourhood level was mainly selected as a base since the scale of a neighbourhood can be very effective for land use planning and design to incorporate sustainability principles (Van der Ryn et al. 1986).

In short, to develop a productive urban landscape in a new site, a suitable patch of land needs to be identified within a neighbourhood considering the quantity of organic waste generated and the site conditions. The site needs to be divided into beds/pits of depth considering the ease of maintenance. The size of the pits is to be

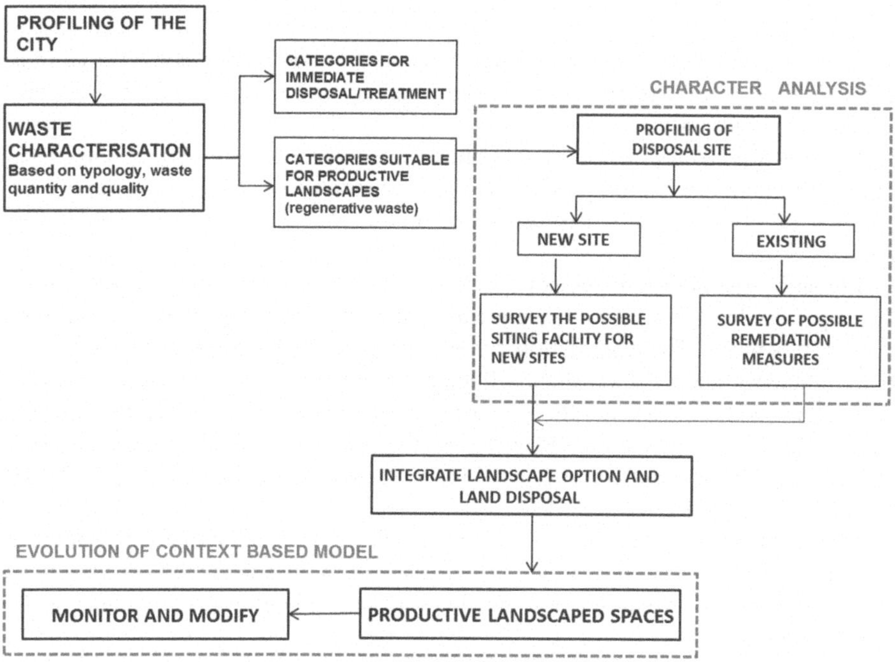

Fig. 1 Productive landscape – a conceptual model

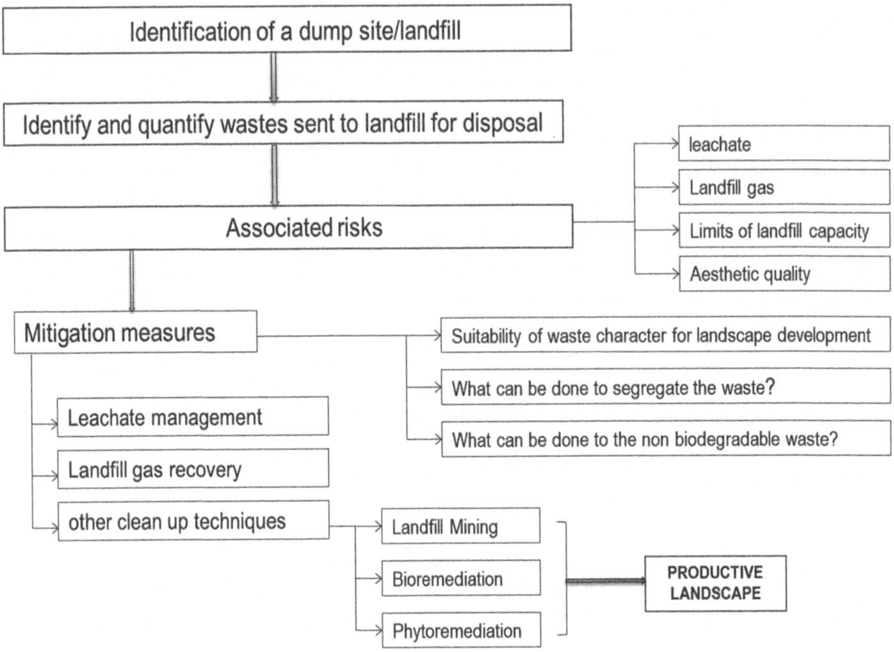

Fig. 2 Methodology for upgrading an existing disposal site

divided with respect to the plot size selected and the amount of waste diverted to it. Typically the whole process will undergo three different phases. It is briefly explained as follows.

Phase I (disposal phase) – in this phase, everyday kitchen waste was disposed in the designated beds/pits prepared on the land identified. Daily soil cover of soil, compost or any additives which can speed up the degradation can be added at this phase.

Phase II (degradation phase) – in this phase the contents in the beds/pits are kept open to degrade naturally. Frequent turning and breaking clods if any can be done in this phase. The degradation rate of the waste can be accelerated using additives. Once degraded, the substrate will be in the form of compost mixed with additives (if any added). Considering the waste settlement (which almost settles down to 50 %) which has happened in the beds/pits, the contents from one pit is transferred to another, and the emptied pit is used for disposing waste thus utilizing it for the second cycle of process.

Phase III (planting phase) – in this phase plants with suitable growth rate were selected and planted to utilize the nutrient content in the waste for their growth. It includes planting, irrigating, pruning, harvesting, replanting and monitoring the whole process.

As an output of the work, a database can be prepared that connects/correlates possible depths of the waste deposited, degradation time permitted and a range of plant growth period (which helps in fixing the plant species) to the number of households and their expected quantity of waste generation within the neighbourhood premises. At the neighbourhood level, if these variables as listed are fixed/decided, the authorities can determine the area required for developing productive landscapes in a particular neighbourhood, and also it will help in determining the kind of plant species that can be selected depending on the area available and their growth patterns. The plants selected can be food crops or non-food/ornamental crops such as flowering plants for personal consumption or marketing purpose.

4 Discussions

To summarize, this paper explores the possibility of treating organic waste generated in a neighbourhood and its simultaneous use in urban landscape development by developing a concept called productive landscape. The waste settlement which can happen (about 40–50%) in the process of using only biodegradable kitchen waste can contribute in ensuring enough space to run the process in cyclic fashion where disposing waste and planting alternate. This in turn enhances the chance of development of a productive landscape that can be implemented as a continuous and sustainable process. Certain operational and maintenance practices (sorting, turning and compaction) helps to maintain an aerobic condition through which in turn reduces the particle size and thus making the degradation faster. This also helps to evaporate excess moisture which prevents leachate formation in most phases of the experiment. However during rainy season (high humidity), measures need to be taken to control leachate formation. A minimum slope should be provided to assist surface runoff and thereby to promote reasonable surface drainage. It is observed that this is very crucial as undrained water can increase moisture content thereby promoting leachate formation, and in most cases, this can adversely affect the plant growth. Moreover, the substrate needs to be fully decomposed, and there is a time required for this, and only after this, planting can normally be done. In a productive landscaped space, this period taken for the substrate to decompose fully can be utilized positively for abating/mitigating environmental impacts caused by excess leachate, odour or the stability of the substrate if any.

At the end of phase II, the refuse is in the form of stable compost. As per the research concept, the plant uses the fertilizer potential of this compost for its growth at the same location. It should be noted that additives like daily soil cover, compost (which can be externally added) or other additives such as coir pith, chemical inoculants, earthworms (vermicomposting), etc. could reduce the foul odour generated from waste. The framework proposed has to incorporate this measure in a sustainable way. Also it has been reported that the odour from an aerobic site is less pungent than compared to an anaerobic site. When the foul smell is reduced/avoided, other nuisances such as the presence of crows and dogs feeding on the waste also get reduced. To ensure avoidance of order problem, suitable cover provision also needs to be streamlined and systematically enforced. Also, the covering can help the site during rains which otherwise increases the moisture content and eventually slows the degradation rate.

In phase III, fast-growing plants which are easy to plant and maintain and conducive to a given climate are only used. Care should be taken so that such plants do not over-dominate and restrict the growth of other plants on the landfill. In a productive urban space, to attain a cyclic process of disposing and developing vegetation, agricultural plants which give maximum yield in a shorter period (aiding crop rotation) can be selected. The remaining planting medium (soil/substrate) after harvesting can also be used as a daily cover during further processing or can be marketed. The proposed framework anticipates to utilize the service of a maintenance person periodically on site. Most of the points discussed here can be implemented through suitable instruction/training given to him.

As a result, a database can be prepared which can provide useful inputs to prospective local bodies/neighbourhood managers to determine the area required for developing the proposed productive landscape patch given other required essential parameters like depth degradation time and the plant growth period. It also explains the type of plants with respect to its growth period that can be chosen for the available space. To run the process in a cyclic fashion, you need to determine the number of days the waste is dumped in the subsequent beds till the first bed is used for the next cycle of dumping. The area requirement is calculated with respect to the amount of organic waste per person per day, the density of the waste as applicable to the study area and the number of days a bed requires to complete a cycle.

A cyclic process explained as a part of productive landscapes depends on the time required for the plant growth. So plant species which has a longer growth period is not suitable. A systematic selection of plants and understanding its growth regime will be part of the urban scale development of landscaped landfill. Selecting agricultural plants which give maximum yield in a shorter period of time thus aiding crop rotation can be selected. Climate is yet another determining factor of the decomposition process. But how exactly the seasonal changes influence this process in different soil conditions can be an area for further research to be conducted. Another limitation is the scarcity of land in urban areas which are suitable for this process based on its location, size and proximity to other resources (like water, solar access and a well-drained soil). Moreover, the issue of scarcity of land aggravates further when the proposed productive function competes with other commercial developments which provide more profit out of the same land. This aspect of the problem can be resolved by educating the public about the wide range of benefits from productive landscapes and their integration with waste management in the urban areas. While the proposed function is best implemented through community-based approach, there is a need for continuous monitoring of such spaces, and that requires a caretaker/maintenance system with an operational cost. The knowledge and skills necessary to manage these systems for multiple functions need expertise on environmental, social and economic aspects of management of the waste mentioned.

5 Conclusions

In many cities, the unsegregated waste that is disposed on land has become a serious threat causing health and environmental risks and landfills/open dumps or any vacant lands, which are used for disposing of waste, hitherto are turning out to be more of an environmental hazard. From the perspective of the concept discussed in this paper, such spaces need to be decommissioned and need to be transformed and upgraded to a productive land parcel of greater utility for the urban population. The paper proposes a methodology for such spaces and other new open spaces available to be made productive by implementing environmentally sound and sustainable strategies for disposing the major portion of municipal solid waste (which is biodegradable in nature) and utilizing the same for landscape development. Such productive spaces within an urban fabric can add aesthetic value and support/supplement economy (through landscape/food/manure production) and will eventually take the city's people closer to a self-sustained community.

References

Asnani PU (2006) Chapter 8. Solid waste management, India Infrastructure Report, 160–189

Bhide AD et al (1998) Solid waste management in Indian urban centers. Int Solid Waste Assoc Times ISWA 1:26–28

Gupta PK et al (2007) Methane and nitrous oxide emission from bovine manure management practices in India. J Environ Pollut 146(1):219–224

Ranjith KA (2012) Sustainable solid waste management in India. M S thesis, Earth Engineering Centre, Columbia University, New York

Redwood M (2009) Agriculture in urban planning – generating livelihoods and food security. Earthscan, London

Sharholy M, Ahmad K, Mahmood G, Trived RC (2008) Municipal solid waste management in Indian cities – a review. J Waste Manag 28:459–467

Sharma HD et al (2004) Geo environmental engineering: site remediation waste containment, and emerging waste management technologies, Chapter 24: End uses of closed landfills, Wiley

TERI(earlier Tata energy research institute) (1998) Solid waste management in India: options and opportunities, Report

Van der Ryn S et al (1986) In sustainable communities- a new design synthesis for cities, suburbs and towns. New Catalyst Books, Gabriola Island

Viljoen A et al (2005) CPULS: continuous productive urban landscapes – designing urban agriculture for sustainable cities. Elsevier, Amsterdam

Part II
Biogas

Feasibility Study on Implementing Kitchen Waste-Based Biogas Plant at Tezpur University, Assam

Sachankar Buragohain, Dipam Patowary, Sampriti Kataki, Barkhang Brahma, Gunajit Dev Sarma, Rupam Patowary, Helen West, Michèle Clarke, and D.C. Baruah

Abstract The growing demand for cooking fuel and non-reliable supply of commercial cooking fuel has become a major concern in recent years. In this context, biogas, a clean and renewable energy source, can be a supplement to non-renewable conventional cooking fuel. In India, small-sized biogas plants have been prevalent in domestic sectors with mixed degree of success. However, the application of community-sized biogas plants is very limited. In case of residential institute like Tezpur University, Assam, India, considering the huge amount of food waste generated, conversion of kitchen waste into useful cooking gas (biogas) through anaerobic digestion can be a better option to supplement the elevated requirement of LPG. In this work, feasibility study of renewable energy-based cooking system (biogas plant, size 50 m^3), implemented in one of the hostels of Tezpur University, is thoroughly examined from commissioning to operational stage, in order to assess the barriers and carriers of such renewable energy technology. A study investigated the performance of the installed plant, feedstock characteristics, composition and economic assessment of biogas-fuelled cooking at Tezpur University campus. Performance analysis and economic assessment of the 50 m^3 biomethanation plant showed that it can be a viable option for utilization of the food waste generated in educational institutions through production of clean cooking fuel. However, proper monitoring of feeding rate and quality is critical for smooth performance of the biogas system.

Keywords Energy • Kitchen waste • Biogas • Digester

S. Buragohain • D. Patowary • S. Kataki • B. Brahma • G.D. Sarma
R. Patowary • D.C. Baruah (✉)
Department of Energy, Tezpur University, Napaam 784028, Assam, India
e-mail: baruahd@tezu.ernet.in

H. West
School of Biosciences, University of Nottingham, Sutton Bonington Campus, Leicestershire LE12 5RD, UK

M. Clarke
School of Geography, University of Nottingham, University Park, Nottingham NG7 2RD, UK

S.K. Ghosh (ed.), *Utilization and Management of Bioresources*,
DOI 10.1007/978-981-10-5349-8_10

1 Introduction

Sustainable supply of energy is one of the main challenges that mankind is facing, particularly because of imbalance between energy supply and demand and the need to tackle the rapid climate change. In this context, use of renewable energy sources is becoming increasingly essential. With the growing energy demands, due to depletion of fossil fuel resources and population explosion, the need for renewable energy source has drawn the attention for all sectors of development. The theoretical potential for renewable energy exceeds current and projected global energy demand by a large extend. However, the main challenge still remains: to capture and utilize a sizable share of that potential to provide the desired energy demand in a cost-effective and environmentally sound way (IPCC Report 2013).

One of the key technologies for the sustainable use of biomass-based renewable energy technology is biogas technology. Anaerobic digestion (AD) process that generates biogas has been considered as one of the appropriate waste-to-energy conversion technologies and is widely used in the treatment of different organic wastes (Ten Braummeler 1993), municipal solid waste, sewage sludge, food waste, animal manure, plant biomass, crop residues, energy crops, etc. The process of anaerobic digestion converts biodegradable organics into biogas which comprises methane (55–75%) and carbon dioxide (25–45%) with a calorific value of 20 MJ/m^3 (Myles 1987; Steffan et al. 2000). Biogas, a clean energy, is considered as the appropriate technology that meets the basic need for cooking fuel in rural India. So far, biogas production has been quite dominant at household and community levels rather than large scale in India. On the one hand, biogas remains the primary energy obtained, whereas, on the other hand, the digested slurry obtained from anaerobic digestion can be efficiently utilized as fertilizer for agricultural applications as it is superior in terms of its nutrient content and can be used as a soil conditioner and plant nutrient.

Considering the suitability of wide range of feedstock for anaerobic digestion, food waste could also be considered as a potential AD feedstock because of its biodegradable nature and abundance. Almost one third of the total food produced for human consumption each year is wasted accounting to about 1.3 billion tonnes as per the study of global food waste published in 2011 by The Food and Agriculture Organization of the UN. This major portion of food waste is mostly distributed in the developing countries like India (FAO 2011). The food waste mostly includes unconsumed food and food preparation leftovers. With growing concerns over waste disposal, increasing cost of energy supply, a judicious and scientific approach of conversion of food waste to energy has become an inevitable option. Anaerobic digestion of food waste has been an economically viable energy conversion route through production of biogas. Many factors related to feedstock characteristics, digester design and operation conditions affect the performance of anaerobic digestion processes (Hawkes 1980; Fischer et al. 1986). Therefore, physical and chemical characteristics of the organic waste are important for designing and operating anaerobic digesters. In general, characteristics of food wastes are highly variable depending on their sources with a moisture content in the range of 74–90%, volatile solid to total solid ratio of 80–97% and carbon to nitrogen ratio of 14.7–36.4% (Zhang et al. 2007). Anaerobic digestion of considerable amount of food and vegetable waste can be a viable option with reference to educational institute for the production of biogas. Further, in case of residential institutions, cooking is one of the major energy-consuming sectors. With alarming concerns such as fuel price hike and LPG shortage, it has become difficult to meet the growing energy demand most of the time. Thus, in order to meet the cooking energy needs, biogas can be considered as a better alternative if it can be economically utilized as a clean source of energy. Keeping in view of the above discussion, this study was initiated to examine the feasibility of converting the food waste into biogas energy through a 50 m^3 community-sized biogas plant present in one of the hostels (*Patkai Men's Hostel*) of Tezpur University, Assam, India. The different activities involved since the commissioning to the operational stage of the project were examined in order to assess the barriers and carriers of such renewable energy technology. The study also investigated the performance of the installed plant, feedstock characteristics, composition and economic assessment of biogas-fuelled cooking at Tezpur University campus and was also carried out to examine the viability of the system.

2 Materials and Methods

2.1 Background Data Collection and Monitoring of Construction and Operation of Biogas Plant (50 m^3)

Before commissioning of the plant, background information was collected from the hostels of Tezpur University in order to understand the cooking energy consumption. A questionnaire survey was carried out comprising questions on LPG usage, food waste generation, etc. From the date of construction to the date of commissioning, up-to-date details of the progress of work, material procurement, human labour engaged, etc. were properly recorded. The biogas plant was a Shakti Surabhi-type biogas digester developed for kitchen waste-based biogas production by Vivekananda Kendra – Natural Resources Development Project (VK-NARDEP), Kanyakumari, India.

2.2 Feedstock Characteristics

Following parameters of the food waste were determined, viz. total solid (TS), total suspended solid (TSS), ash, total organic carbon (TOC), total nitrogen (TN) and pH. The TS was determined according to the standard procedure (ASTM Test No. D-271-48). A total suspended solid (TSS) (amount of filterable solids in a sample) was measured by passing the sample through a glass fibre filter. Ash content of the feedstock sample was measured with the standard procedure of ASTM Test No. D-271-48. The total organic carbon was determined using the TOC analyser (Liqui TOC II of make Elementer). The food sample was first grinded properly and then filtered using a filter bag of size 150 microns before analysis. The liquid sample that was obtained was tested for total organic carbon. Total inorganic carbon (TIC) and total carbon (TC) were also determined from the TOC analyser. The total nitrogen of the food sample was measured by the standard Kjeldahl method using the Kjeldahl apparatus. Volatile suspended solids (VSS) are the amount of volatile solids in a dried filtered sample. It is measured by igniting the dried filtered sample in a muffle furnace at 550 °C.

2.3 Determination of Biogas Yield

Biogas plant was fed with an amount of 85–90 kg of food waste daily. The biogas yield of the plant was measured at an interval of 24 h. The holder height was measured regularly by taking the difference of the height of the holder of the day with that of the previous day; the volume of the gas holder was measured which corresponds to the amount of biogas yield per day. In case of usage, the usage hours were noted, and correspondingly approximate amount of biogas used was calculated by considering the hourly fall of the holder height. Thus by considering the difference in height as well as the drop during usage, the amount of biogas formed was calculated in m^3. The amount of biogas produced per kg of total solid was also calculated.

2.4 Determination of Composition of Biogas

Biogas composition was measured at different intervals regularly in terms of methane content. It was measured using a biogas analyser (Ambetronics Engineers Private. Ltd., Model No: MS panel 830213). The biogas analyser is equipped with IR 5000 infrared gas detector which is a microcontroller-based gas detector that continuously monitors the CH_4 and CO_2 in %V/V and H2S in ppm levels. For the present study, only methane content was recorded.

2.5 Determination of Equivalent Amount of LPG

For the determination of equivalent amount of LPG, a water boil test was done. An amount of 6 L of water was taken in a utensil and was allowed to boil using biogas until it reaches 100 °C. During the operation, the initial height, the final height of the holder and the time duration for its boiling were noted. The same was repeated by using LPG and the initial and final weight of the LPG cylinder were noted. Thus, by comparing the volume of biogas to that of the amount of LPG, equivalent amount of LPG was calculated.

2.6 *Economic Analysis of the Biogas Plant*

The economic analysis of the 50 m^3 biogas plant was carried out by considering all the cost, i.e. capital cost, maintenance cost, electrical cost, etc., for a period of 10 years considering an increase of 4.5% price rise every year. The payback period was calculated by calculating the cumulative savings that could have been incurred in case of LPG. As per the number of LPG savings incurred during a month of experimentation and the prevailing price of LPG, the payback period was calculated. The payback period was computed as the smallest value of n that satisfies the Eq. 1. where *Bn and Cn* represent the benefits and costs, respectively, associated with the investment at each period of *n*.

$$\sum_{n=0}^{\mathbf{nsp}} \left(Bn - Cn \right) \geq 0 \tag{1}$$

3 Results and Discussions

The university comprises a total of 12 hostels, which generate a huge amount of food waste per day and are either drained or taken by the pig bearers of the locality. The present energy (LPG) consumption pattern of waste generation of the hostels is shown in Table 1. The minimum monthly requirement in case of men's hostel is 1 cylinder/day, whereas that of women's hostel is 0.7 cylinders/day. The per capita consumption of LPG was found to be in the range of 3.03–4.96 MJ per day which corresponds to Pobitora Madam Curie women's hostel and Saraighat CV Raman men's hostel, respectively. The average per capita consumption of LPG was found to be 3.92 MJ per day. The daily requirement in the hostels estimates at an average of 1.5–2 cylinders aggregating an average of around 36 to 40 cylinders per month.

3.1 *Installation of the Plant*

The construction of the plant started in September 2013 under the supervision of technician from VK-NARDEP, Kanyakumari. The various phases of construction are shown in Fig. 1 below. The construction of the gas holder was done initially by fabricating the structure with MS rod and tin plates and then reinforcing it with fibreglass-reinforced plastic (FRP). The entire process took 2 months to complete.

Table 1 LPG consumption pattern in different hostels in Tezpur University

Hostel	No. of boarders	Approximate number of cylinders required/day	MJ equivalent of LPG	Energy density (MJ/capita)	Food waste available (kg/day)
Saraighat CV Raman men's Hostel	350	1.06	693.90	4.96	95
Patkai Men's Hostel	386	2.16	1413.98	3.66	92
Charaideo Men's Hostel	270	1.55	1014.66	3.76	112
Nilachal Men's Hostel	411	2.16	1413.98	3.44	74
Kanchenjunga Men's Hostel	390	2.16	1413.98	3.63	139
Dhansiri Women's Hostel	139	0.91	595.70	4.29	32
Pragjyotika Women's Hostel	136	1.00	654.62	4.81	30
Bordoichila Women's Hostel	104	0.70	458.23	4.41	24
Suwansiri Women's Hostel	183	1.00	654.62	3.58	49
Kopili Women's Hostel	186	1.00	654.62	3.52	44
Pobitora Madam Curie Women's Hostel	316	1.00	654.62	3.03	88
New Women's Hostel	126	0.78	510.60	4.05	25

3.2 Feedstock Characterization

The feedstock mainly food waste that were available are characterized as shown in Table 2.

3.3 Determination of Biogas Yield

The biogas production of the installed plant was monitored for a period of 2 months. The gas production was also measured in m^3/kg TS of food waste. During this period, the gas production was in the range of 0.2–0.5 m^3/kg TS of food waste. The highest production achieved was 0.63 m^3/kg TS of food waste, whereas the lowest production was 0.5 m^3/kg TS. These values directly related to the maximum and minimum feed fed to the plant on the previous day of gas measurement during the period of study.

Fig. 1 (**a**–**d**) Construction phase of the biogas plant

Table 2 Characteristics of the feedstock

Parameter	
Total solid (TS), %	27.17
Ash, % TS	3.6
Total suspended solid (TSS), g/l	250
Total carbon(TC), mg/l	5128.82
Total organic carbon (TOC), mg/l	5073.78
Inorganic carbon	64.06
Total nitrogen (TN), %	3.36
C:N	15

At times when gas production was low, pH of the digestate (residue that remains after digestion) was measured. It was observed that the pH had a direct impact on the production of biogas. The production of biogas was low when the pH of the digestate was in the range of 5–6, and it was comparatively higher when its pH was in the range of 6–7.5. In case of food waste, because of the heterogeneous nature of food, the amount of volatile fatty acid formation is higher during anaerobic digestion process, which hinders the growth of methanogenic bacteria leading to decrease in biogas production. Therefore, when the gas production was found to drop, to

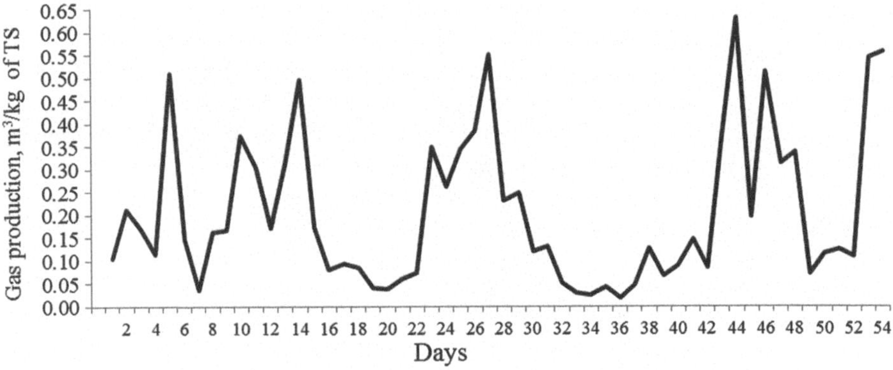

Fig. 2 Gas production in m^3/kg total solid of food waste or a period of 54 days

raise the pH, cow dung was fed to the digester until the pH of the digestate rose to 7. Figure 2 shows the gas production in m^3/kg total solid of food waste for the period of study.

3.4 Variation of Methane Content of Biogas

The composition of biogas in terms of methane content (%V/V) was measured using a biogas analyser. The calibration was done by analysing the biogas sample with GC, and the % of CH_4 obtained was set in the biogas analyser using the same sample of biogas.

The CH_4 was studied for a period of 36 days. It was seen that initially the CH_4 was in the range of 45.8%, and gradually it decreased to 30.1% at which the burner emitted yellow flame and even stopped burning due to high concentration of CO_2 in the biogas. Therefore, feeding was done with cow dung to make the system stabilized. Cow dung feeding increased the % of CH_4. Once the system was stabilized again, food waste feeding was started. During the period of the study, the highest % of CH_4 obtained was 55.20%. The overall % changes of CH_4 are shown in Fig. 3.

3.5 Determination of Equivalent Amount of LPG

For the determination of equivalent amount of LPG, the test was carried out on four different days. And due to the variation of CH_4 (%), there was slight variation in the amount of equivalent LPG obtained. It was found that 1 m^3 of biogas is equivalent to 0.25–0.35 kg of LPG w.r.t. prevailing conditions of the system.

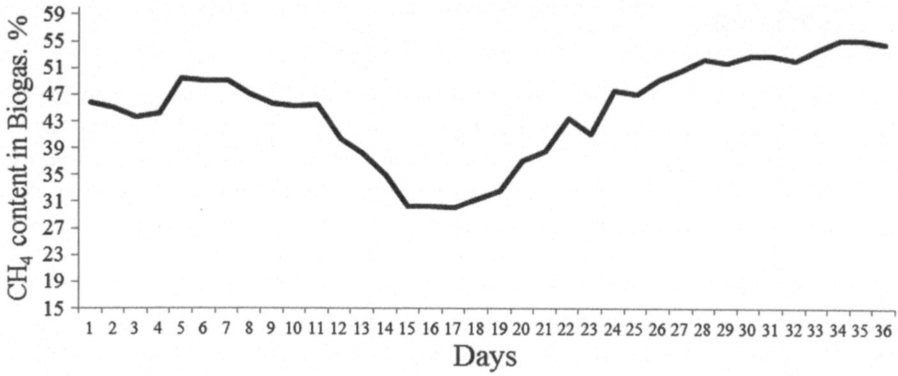

Fig. 3 Variation of CH_4 (%) content of biogas

Table 3 Overall benefits and costs for 10 years

Year	Capital cost (Rs)	Overall maintenance cost Electricity cost (Rs/annum)	Labour wage (Rs/annum)	Maintenance Cost (Rs/annum)	LPG saving (Rs/annum)	Gross expenses (Rs/annum)	Cumulative saving (Rs/annum)
1	1,000,000	5395	18,000	10,000	148,920	1,033,395	−884,475
2		5638	18,810	10,450	155,621	34,898	−763,752
3		5892	19,656	10,920	162,624	36,468	−637,597
4		6157	20,540	11,411	169,942	38,109	−505,764
5		6434	21,465	11,925	177,589	39,824	−367,999
6		6723	22,431	12,461	185,581	41,617	−224,035
7		7026	23,440	13,022	193,932	43,489	−73,592
8		7342	24,495	13,608	202,659	45,446	83,620
9		7673	25,597	14,221	211,779	47,491	247,907
10		8018	26,749	14,860	221,309	49,629	419,587

3.6 Economic Analysis of the Biogas Plant

As per the version of Vivekananda Kendra – Natural Resources Development Project (VK-NARDEP), the daily amount of biogas generated would be equivalent to approximately 30 cylinders per month. But during the experimentation stage, the maximum amount of LPG cylinder that could be saved during a month was found to be 14. This may be due to the acidic nature of heterogeneous food waste fed to the plant, which eventually hindered methanogenic bacteria growth, and biogas generation was low. As the plant was in the experimentation stage, the exact amount of LPG saving cannot be estimated. As per the calculations made, an average of ten LPG cylinders can be saved in a month considering the proper condition of the biogas plant. The overall cost analysis is shown in the Table 3. It can be seen that the payback period of the biogas plant is 7.36 years (Table 3).

4 Conclusions

Cooking energy consumption is expected to increase with continuous increase in the population of Tezpur University. Considering this, renewable energy-based cooking energy (biogas) could be a viable option to supplement the daily energy needs in the future. The present study was based on the feasibility study of implementing biogas-based cooking from kitchen waste in one of the hostels at Tezpur University. Performance analysis and economic assessment of the 50 m^3 biomethanation plant showed that it can be a workable option for utilization of the food waste generated in educational institutions through production of clean cooking fuel 'biogas'. However, proper monitoring of feeding rate and quality is critical for smooth performance of the biogas system.

References

Fischer JR, Iannotti EL, Durand J (1986) In: Goswami I, Yogi D (eds) Anaerobic animal manure. Agriculture and energy, alternative energy in Agriculture. CRC Press, Inc, Florida

Food and Agriculture Organization of the United Nations (2011) Global food losses and food waste. Interpack 2011, Dusseldorf

Hawkes DL (1980) Factors affecting net energy production from mesophilic anaerobic digestion. In: Stafford DA, Wheatley BI, Hughes DE (eds) Anaerobic digestion: [proceedings of the first international symposium on anaerobic digestion, held at University College, Cardiff, Wales, September 1979]. Applied Science Publishers, London

IPCC (2013) Summary for policymakers. In: IPCC special report on renewable energy sources and climate change mitigation

Myles RM (1987) Practical guide to Janata biogas plant technology. In Practical guide to Janata biogas plant technology, AFPRO

Steffen R, Szolar O, Braun R (2000) Feed stock for anaerobic digestion. Making energy and solving modern waste problem. Available from www.adnett.org/dl_feedstocks.pdf

Ten Braummeler E (1993) Dry anaerobic digestion of the organic fraction of municipal solid wastes. Ph.D. thesis, Wageningen University, The Netherlands

Zhang R, El-Mashad HM, Hartman K, Wang F, Liu G, Choate C, Gamble P (2007) Characterization of food waste as feedstock for anaerobic digestion. Bioresour Technol 98(4):929–935

The Potential of Biogas Recovery from Anaerobic Co-digestion of Fecal Sludge and Organic Waste

Hoang Le Phuong and Nguyen Thi Kim Thai

Abstract According to some reports, the amount of fecal sludge (FS) that is also known as septic tank sludge or septage in urban areas of Vietnam is relatively high. It can be up to hundreds of tons per year in some big cities (Hanoi Urban Environment. The situation and solutions of management, collection and treatment of fecal sludge, International Conference on Management of sludge from the drainage system and sanitation (FS3 – 2015) – Proceedings sludge management in Vietnam opportunity to improve, pp 25–27, 2015). Therefore, the management and treatment of septic tank are an urgent problem currently. Ingredients of fecal sludge such as total nitrogen (TN), total phosphorus (TP), and alkalinity are high, but the ratio of C/N is often lower (Montangero A, Strauss M, Fecal sludge treatment, Eawag/Sandec, 2004; Klingel F et al, Fecal Sludge Management in Developing Countries, Eawag/Sandec, 2002; Thai et al, Fecal sludge management from the sanitation, Science and Technology Publishing, Hanoi, 2013). So, the anaerobic co-digestion fecal sludge with other organic wastes with C/N higher can recover biogas. This study was conducted with experiment model in mesophilic fermentation conditions during 40 days in the solid waste laboratory of National University of Civil Engineering. Fecal sludge (FS) and organic waste (OW) have mixed with the ratios of 4:1, 3:1, and 2:1 by weight in three parallel models. Experimental results showed that at ratio of 3:1, it yielded higher biogas production with 514.33 Nl/kgVS of feed. Also the ratio and parameters of process such as the change of its height, temperature, pH, and COD are consistent with anaerobic digestion process, and the amount of fecal sludge treated is high relatively.

Keywords Fecal sludge • Anaerobic digestion • Biogas

H. Le Phuong (✉)
Thai Nguyen University of Technology, Thai Nguyen Province, Vietnam
e-mail: phuongqtn@tnut.edu.vn

N.T.K. Thai
National University of Civil Engineering, Hanoi City, Vietnam

S.K. Ghosh (ed.), *Utilization and Management of Bioresources*,
DOI 10.1007/978-981-10-5349-8_11

1 Introduction

In urban areas of Vietnam, the majority of houses use septic tank as one-site sanitation because of its simplicity and stability. Besides, it also has public toilets in some cities. There are about 77% of households in urban, 40% in suburban, and 19% in rural which use septic tank (World Bank 2012). The efficiency of BOD and suspended solids treatment of the septic tank usually ranges from 0 to 50% depending on the design of septic tank, water discharge, and habit of emptying tanks so the organic matter in fecal sludge is usually high (World Bank 2012). Meanwhile, the amount of fecal sludge generated each year is quite large, specifically in Hanoi and Ho Chi Minh cities with about 500 tons per day (Ho Chi Minh City Urban Drainage Co., Ltd. 2015). Moreover, the quantity of this sludge will be increasing with the development of urbanization. According to the strategy of integrated management of solid waste in Vietnam up to 2025 and the vision to 2050 assigned by the Prime Minister until 2025, Vietnam must ensure that 100% of solid wastes are collected and treated, of which 90% are recycled, reused, recovered energy or produced organic fertilizer, and 100% of fecal sludge in urbans and 50% in suburban are collected and treated (The Prime Minister's Office 2009). However, in the most of urban, the majority of fecal sludge is not managed and treated properly; it is dumped directly into the environment in various forms. The private collectors often pour sludge into sewage system, fishponds, and lakes illegally. The public collectors usually buried fecal sludge in landfills. As the result, fecal sludge remains an unregulated environmental pollutant in Vietnam.

Vietnam has some urban cities such as Hanoi and Ho Chi Minh City, using co-composting method fecal sludge with organic waste. This method is simple but has some drawbacks such as needing dewatering of sludge before composting, supplying gas, and generating odors, leachate, and CO_2 during the composting (Strauss et al. 2003). To overcome these drawbacks, anaerobic digestion recovering biogas is a solution. According to some studies, anaerobic digestion of fecal sludge at a temperature of 15–30 °C can generate 15–90 ml biogas/g dry fecal sludge, and the biogas will be higher if conditions of the process are optimal (Strande et al. 2014). However, it has not yet been proven for fecal sludge itself in semi-centralized to centralized treatment in urban areas; because the composition of nutrients in fecal sludge is not optimal, the rate of C/N is often low (Sandec Training Tool 1.0 – Module 5, 2008; Strauss and Montangero 2002). Besides, market waste (MW) with high carbon content is also generated in a large amount from the markets. Anaerobic co-digestion of different wastes such as municipal waste solid, sewage sludge, animal dung and other biowaste can produce high volume of biogas with high metan content (Babel et al. 2009; Khai 2002; Völgeli et al. 2014). Therefore, anaerobic co-digestion of FS and MW can create biogas higher. This study was done to evaluate the potential of biogas recovery from fecal sludge when it is mixed with market waste at different ratios in laboratory, as a basis for further research on optimal operating conditions.

2 Materials and Methods

2.1 Raw Materials and Reactor Preparation

Collect Fecal Sludge (FS) Fecal sludge is drawn at the public toilets and the septic tanks in Hanoi and is gathered in the sludge tank at Cau Dien plant in Hanoi. This tank is brushed scum and stirred and then takes about 80 l and transferred to the laboratory.

Organic Waste (OW) Organic waste was collected at Dong Xuan and Long Bien markets in Hanoi. It included the fruits and vegetable removed. The organic waste was collected at 6–8am and was transferred to the experimental reactor immediately. The organic waste was chopped into sizes from 1 to 3 cm and mixed well.

Experiment Model The experiment models are prepared according to Fig. 1.

2.2 Determination Characteristics of Raw Materials and Co-digestion

FS and OW are analyzed for various parameters such as moisture content (MC), total solids (TS), volatile solids (VS), total nitrogen (TN), and total organic carbon (TOC).

FS and OW are mixed at different ratios of 4:1, 3:1, and 2:1 by weight, respectively. The reactors are operated under the temperature 30 °C, which represents mesospheric condition. The retention time is 40 days. The amount of biogas, temperature, pH, and COD are monitored every day.

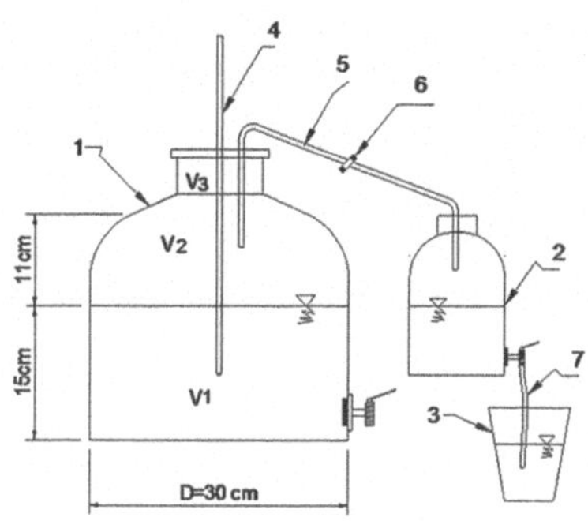

Note for figure 1:

Bottle A: the reactor by glass with volume is 20 liters

Bottle B: collecting biogas from bottle A with volume is 5 liters

Pot C: collecting water from bottle B

1. Thermometer to measure temperature in bottle A
2. Soft plastic pipe to collect biogas from bottle A
3. Clamp valve
4. Valve to take sampling
5. Valve always opening
6. Plastic tube to collect water into a basin C

Fig. 1 The experiment model

3 Results and Discussion

3.1 Characteristic of Fecal Sludge and Organic Waste

The characteristics of fecal sludge and organic waste are shown in Table 1.

%MC and TN in FS is high, but the rate of C/N is 11.95, which is relatively low compared with conditions of anaerobic digestion; the optimum C/N ratio is in the range of 20–30 (Khai 2002; Parkin and Owen 1986), whereas, in OW the C/N rate is 47.41. Consequently, mixing FS to OW can have suitable nutrients appropriately for anaerobic digestion.

3.2 Results of Anaerobic Co-digestion Process

3.2.1 The Change of Temperature in the Reactors

The changes of temperature in reactors are shown in Fig. 2.

From Fig. 2, it can see that for FS to OW ratio at 3:1, the highest temperature was 36–37 °C on day 14–22 and then decreased to 31 °C on day 40. For FS to OW ratio at 4:1, the fastest increase in the temperature is observed during the first days; it could reach up to 35.5–36.5 °C on day 9–16 and then decreased to 30 °C on day 40. For FS to OW ratio at 2:1, the increase of temperature is slowest, it only could reach up to 36–37 °C on the day 17–23, and then it also dropped to 32 °C on day 40. The anaerobic digestion has made increase the temperature of the reactors; however, this temperature is also less than 40 °C that is suitable for mesospheric condition. This is similar to some researches that agree that energy released as heating during anaerobic digestion is enough to maintaining optimal temperature in mesopheric condition (Pohland 1968). The more quickly the digestion took place, the faster the temperature is raised and then falls rapidly. Therefore, it can see that at ratio 4:1, anaerobic digestion happened fastest and at ratio 2:1 was slowest in three ratios of this experiment.

3.2.2 The Change of pH in the Reactors

The changes of pH in reactors are shown in Fig. 3.

Table 1 Characteristics of FS and OW before digestion

Raw material	MC (%)	TS (%)	VS (%TS)	TN (mg/gTS)	TOC (mg/gTS)	C/N
FS	94.73	5.27	71.08	30.54	365.07	11.95
OW	48.3	51.7	78.2	9.2	436.14	47.41

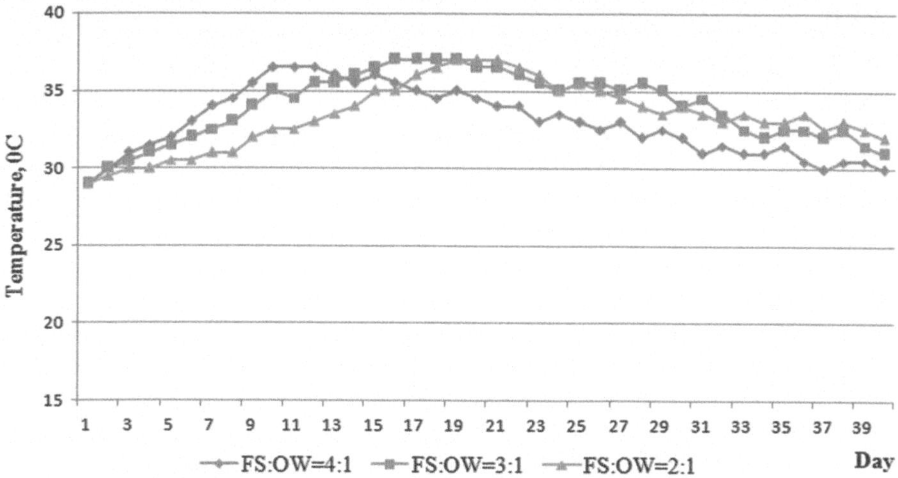

Fig. 2 The change of temperature in the reactors

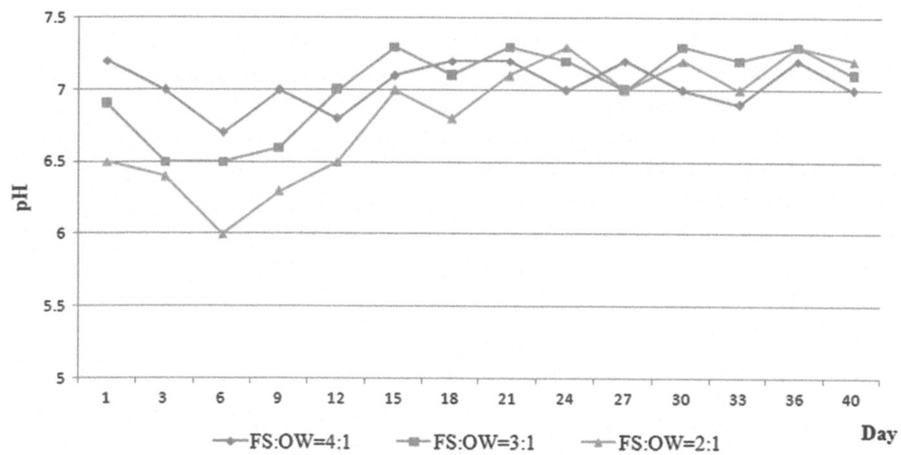

Fig. 3 The change of pH in the reactors

In the anaerobic digestion, pH is the most sensitive parameter. Most anaerobic bacteria operate at the suitable pH of 6.5–7.5, optimally from 6.8 to 7.2. The methane production can reduce if the pH is less than 6.3 or greater than 7.8 (Stronach et al. 1986; Jiunn-Jyi Lay et al. 1997). From Fig. 3, it can be seen that for FS to OW ratio at 4:1 and 3:1, pH dropped in the first days and then increased. However, it only ranged from 6.5 to 7.5 relatively suitable for anaerobic digestion process. Only for FS to OW ratio at 2:1 when pH was 6.0 on day 6, and then it rose and ranged from 6.3 to 7.3. For FS to OW ratio at 4:1, the lowest of pH was 6.7 and it fluctuated from 6.7 to 7.2; for ratio at 3:1, the lowest of pH was 6.5 on day 3 and day 6, and then it increased and fluctuated from 6.6 to 7.3. According to the result of some

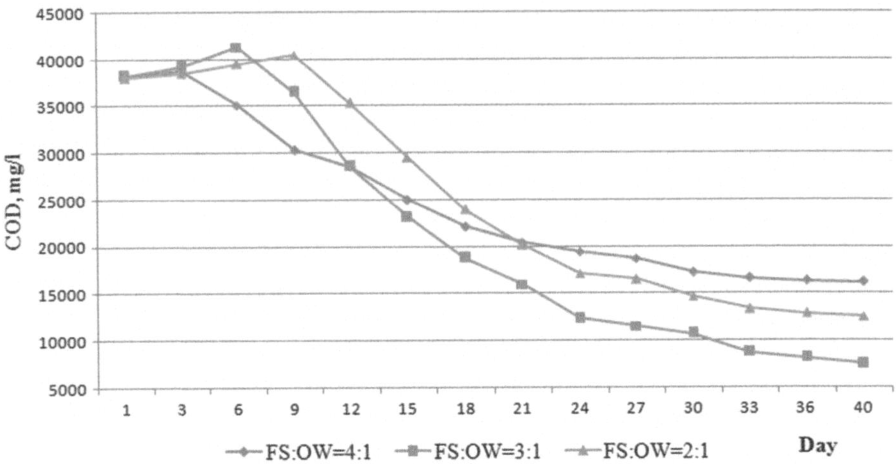

Fig. 4 The change of COD in leachate of the reactors

researches in the first stage of anaerobic digestion, the processes of hydrolysis and acid genesis occur and produce volatile fatty acids (VFAs) such as acetate, propionate, butyrate, and lactate that reduce pH; pH levels in the stages are usually 5.5–6.5, after that, the methanogenic phase occurs that used these acids to form CH_4 and made an increase of pH from 6.5 to 8.2 (Al Seadi et al. 2008). It can be seen that with FS to OW ratio at 2:1 with the largest of organic waste in three ratios, it produced the higher quantity of VFAs, and pH decreased in 6 which was not good for anaerobic digestion. For ratio at 4:1, it had the lowest of organic load so pH was not influent and maintained in the suitable range for anaerobic digestion. Besides, the buffering capacity of FS is quite good; it helps to stabilize pH in the anaerobic digestion process in the different ratios of FS to OW in this experiment, especially in ratio at 2:1 when pH dropped 6; it could slow growth to 6.5 that is relevant for methanogenic phase. It is probably suitable with some researches which illustrated that the high alkalinity of FS was imparted by the formation of ammonia bicarbonate (NH4HCO3) during the hydrolysis of urea (H_2NCONH_2). Therefore FS has high buffering capacity (Montangero and Strauss 2004).

3.2.3 The Change of COD in Leachate of the Reactors

The changes of COD in leachate are shown in Fig. 4.

Figure 4 shows that for FS to OW ratio at 4:1; 3:1, and 2:1, COD in leachate increased in the first stage and decreased after that. It can be seen that in the first stage of anaerobic digestion produced isolate organics such as VFAs (Cavinato 2011; Al Seadi et al. 2008), these substances increased COD in leachate; after that methanogenic microorganisms are feeding volatile fatty acids and release CH_4 and

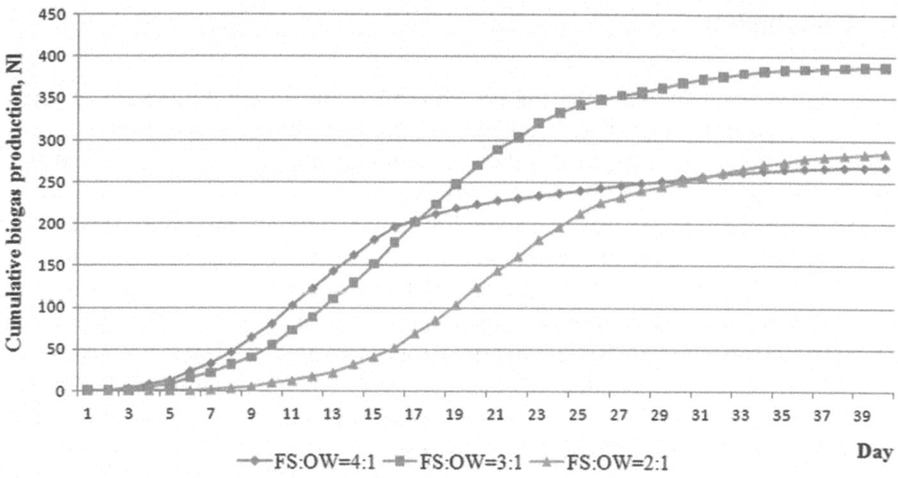

Fig. 5 Cumulative biogas production with time at different mixing ratios

CO_2 (Cavinato 2011; Al Seadi et al. 2008) so COD in leachate decreases. For FS to OW ratio at 3:1, COD rose on the first 6 days and degraded then. And the reduction total of COD at the ratio was highest at 80.5%. At ratio of 2:1, the increase of COD was observed during the first 9 days, and it only began to decrease then. After 40 days, the total COD reduction at this ratio is 67%; COD could continue to drop in the next days; however, the experiment finished on day 40. For FS to OW ratio at 4:1, COD only increased on the first 3 days and then decreased. But after 40 days, COD only declined at 59% the slowest in three ratios. The reason of this could be that the proportion of VFA produced was low so COD in leachate did not decrease as much as other ratios.

3.2.4 Biogas Production in the Reactors

Cumulative gas production at different mixing ratios with time is shown in Fig. 5.

From Fig. 5, it can be seen that biogas production for FS:OW at the ratio 3:1 was 387.98NL after 40 days equivalent to 514.33 Nl/kgVS of feed and was the highest among three mixing ratios used in this study. For FS to OW ratio at 4:1, biogas production was the lowest with 268.27 Nl after 40 days equal to 378.23 Nl/kgVS of feed. At the ratio of 2:1, cumulative biogas production was 285.21 Nl equal to 354.26 Nl/kgVS after 40 days. Therefore, the result of the highest COD decreased in leachate in ratio at 3:1 was the largest cumulative biogas in three mixing ratios of this experiment. And, in ratio at 4:1, the total COD reduction was the lowest so cumulative biogas production was the smallest.

4 Conclusion

This study shown that, when mixing FS with OW in the different ratios, the parameters of the anaerobic digestion changed and cumulative biogas production was very different at the ratios. It found that the maximum quantity of biogas generated at a mixing ratio of 3:1 was 387.98 Nl after 40 days.

FS to MW ratio at 3:1 was considered to be the optimum mix ratio. Because it generated the highest biogas production in addition, the change of parameters of anaerobic digestion process was pretty good, the removal COD in leachate was higher than other ratios. At this ratio, the biogas yield was 514.33 Nl/kgVS of feed, which is equivalent to biogas yield of poultry slurry (0.35–0.60 m3/kgVS) (Al Seadi et al. 2008). Thus, the potential of biogas recovery when mixing fecal sludge and organic waste is high relatively. However, it should be more specific to study about operating mode to achieve more efficiency results.

Acknowledgment This research was supported by Thai Nguyen University of Technology and National University of Civil Engineering, Vietnam.

References

Al Seadi T, Rutz D et al (2008) Biogas handbook, University of Southern Denmark Esbjerg, NielsBohrsVej 9–10, DK-6700 Esbjerg, Denmark

Babel S, Sae-Tang J et al (2009) Anaerobic co-digestion of sewage and brewery sludge for biogas production and land application. Int J Environ Sci Technol 6(1):131–140

Cavinato C (2011) Anaerobic digestion fundamentals 1, Summer school on biogas technology for sustainable second generation biofuel production

Hanoi Urban Environment (2015) The situation and solutions of management, collection and treatment of fecal sludge, International Conference on Management of sludge from the drainage system and sanitation (FS3–2015) – Proceedings sludge management in Vietnam opportunity to improve, pp 25–27

Ho Chi Minh City Urban Drainage Co., Ltd (2015) Sewer and feacal sludge management in Ho Chi Minh City. In: Proceedings of the 3rd international feacal sludge management conference, Focus on Vietnam

Khai NQ (2002) Biogas technology, Labour and Social Publishing

Klingel F, Montangero A et al (2002) Fecal sludge management in developing countries, Eawag/Sandec

Lay JJ, Li YY, Noike T (1997) Influences of pH and moisture content on the methane production in high-solids sludge digestion. Wat Res 31(6):1518–1524

Montangero A, Strauss M (2004) Fecal sludge treatment, Eawag/Sandec

Parkin GF, Owen WF (1986) Fundamentals of anaerobic digestion of wastewater sludges. J Environ Eng 112:867–920

Pohland FG (1968) Thermal energy interchange during anaerobic methane fermentation of waste organic substrates. Am Soc Microbiol 16(10):1518–1523

Sandec Training Tool 1.0 – Module 5 (2008) Faecal sludge management, Eawag/Sandec

Strande L, Ronteltap M et al (2014) Faecal sludge management systems approach for implementation and operation. IWA Publishing, London

Strauss M, Montangero A (2002) Fecal sludge management – review of practices, problems and initiatives, Eawag/Sandec
Strauss M, Drescher S et al (2003) Co-composting of faecal sludge and municipal organic waste – a literature and state – of – knowledge review, Eawag/Sandec
Stronach SM, Rudd T, Lester JN (1986) Anaerobic digestion processes in wastewater treatment. Springer, Berlin
Thai NTK, Nhue TH et al (2013) Fecal sludge management from the sanitation. Science and Technology Publishing, Hanoi
The Prime Minister's Office (2009) The strategy of intergated management of solid waste in Vietnam up to 2025 and the vision to 2050, Hanoi
Vögeli Y, Lohri CR, Gallardo A, Diener S, Zurbrügg C (2014) Anaerobic digestion of biowaste in developing countries, Eawag/Sandec
World Bank (2012) Assessment of the Urban Wastewater Management in Vietnam, Hanoi

Part III
Biofuels and Bioenergy

Bioethanol Production from Waste Breads Using *Saccharomyces cerevisiae*

P. Datta, S. Tiwari, and L.M. Pandey

Abstract In this study, waste breads have been utilized as the only nutrient source for the production of glucose using *Aspergillus niger* and subsequently ethanol production from glucose by *Saccharomyces cerevisiae*. Solid-state fermentation of waste bread by *Aspergillus niger* resulted in the production of a multienzyme solution containing amylolytic and proteolytic enzymes. The glucoamylase and protease enzymes are then extracted, and enzyme activities were quantified. This crude enzyme extract was used for the hydrolysis of waste bread at 55 °C at 300 rpm. After hydrolysis, the amount of glucose was determined using anthrone colorimetric method, and the free amino nitrogen (FAN) was determined by ninhydrin colorimetric method. The resulting solution contains approximately 145 g/l of glucose. The bread hydrolysate was then further used to produce ethanol having concentration of 54 ± 2 g/l. The result depicts yield of ethanol from glucose, $Y_{P/S}$, as 0.37, i.e., a conversion efficiency of 72%. This process is important in the sustainable chemical industry because it converts the waste food into a value-added product like ethanol.

Keywords Glucose production • Bioethanol • Solid-state fermentation • Enzyme extraction • Waste bread

---

P. Datta
Center for the Environment, Indian Institute of Technology, Guwahati, India

S. Tiwari
The Department of Biosciences & Bioengineering, Indian Institute of Technology, Guwahati, India

L.M. Pandey (✉)
Center for the Environment, Indian Institute of Technology, Guwahati, India

The Department of Biosciences & Bioengineering, Indian Institute of Technology, Guwahati, India
e-mail: lalitpandey@iitg.ernet.in

S.K. Ghosh (ed.), *Utilization and Management of Bioresources*,
DOI 10.1007/978-981-10-5349-8_12

1 Introduction

Bioethanol is the most widely used biofuel in the twenty-first century, and it is produced by fermentation and saccharification of various biomasses in facilities called biorefineries where all the bio-products like biofuels and biochemicals are processed (Melikoglu et al. 2013a, b). Bioethanol is basically a renewable source of energy that reduces the detritus greenhouse effect of fossil fuels, and its production from waste food biomass is one of the convenient techniques used for bioethanol production (Moukamnerd et al. 2013). While food waste is emerging as one of the major problems globally, due to increase in population, it is expected to rise twofold times in the next 25 years to come. Food waste also holds major portions of municipal solid waste in countries like the United States, Hong Kong, and Singapore. Among the many food materials that are wasted, bread is most common in the European and Asian countries. In the United Kingdom, food waste is around 1.2 million tons per annum, while in Japan it is around 45.1 million tons annual (Leung et al. 2012; Melikoglu et al. 2013a, b; Moukamnerd et al. 2013; Kiran et al. 2014).

Traditionally, bioethanol is made by different processes of hydrolysis and fermentation which includes numerous steps and processes which makes it a costly affair (Moukamnerd et al. 2013). Therefore, to bring down the cost, a new method involving the enzymatic hydrolysis is utilized, where a portion of waste breads is utilized for solid-state fermentation. Later, the leftover breads are further used for the production of hydrolysate which is then fermented for producing ethanol. This method was developed by the University of Manchester in the United Kingdom (Melikoglu et al. 2013a, b).

For high ethanol production, the glucose produced should also be large in amount which can be achieved by increasing the enzyme concentration, temperature, and agitation speed and also by increasing its hydrolysis time. In comparison to batch culture, using fed-batch culture has theoretically increased the conversion yield of glucose by almost 92%, and alternatively solid-state fermentation can be used to reduce the risk of catabolite repression (Kiran et al. 2014). Although commercially available enzymes can be used for enzymatic hydrolysis, it increases the cost inevitably. Bread can be considered optimum for solid-state fermentation due to its porous structure and nutrient composition. The solid surface of bread which is high in nutrients is susceptible for growth of microorganisms on its surface (Melikoglu et al. 2013a, b).

The enzyme glucoamylase, which is being extensively used for the hydrolysis of starch in biorefineries, is produced by a fungi *Aspergillus*. Solid-state fermentations are very similar to the natural habitat of *Aspergillus* where it also secretes protease enzyme in amounts high enough to assimilate and consume proteinaceous substrates (Leung et al. 2012; Melikoglu et al. 2013a, b). These enzymes are used to hydrolyze gluten-free flour to carbon- and nitrogen-rich streams. The hydrolysate can be easily converted to ethanol by the process of fermentation using brewer's yeast *Saccharomyces cerevisiae* which produces high amount of ethanol (Leung et al. 2012).

2 Materials and Methods

2.1 Microorganism

The fungal strain *Aspergillus niger* was used for the production of glucoamylase- and protease-rich waste breads which were further used in this experiment. The information regarding the storage, sporulation, and inoculum preparation of this strain have been studied in a previous publication. *Saccharomyces cerevisiae* was utilized here for bioethanol production.

2.2 Waste Bread as Raw Material

Waste bread was collected from the Canteen of Indian Institute of Technology, Guwahati (IITG). The composition of the waste bread (moisture content, starch, total organic nitrogen, protein, and total phosphorus) was obtained from this publication (Leung et al. 2012).

2.3 Solid Waste Fermentation

Waste bread slices were cut into approximately 1 cm^3 in size. Fifteen grams of bread cubes was kept into a 10 cm petri dish and sterilized by autoclaving at 120 °C for 30 min prior to solid-state fermentation. The fungal spores were then diluted and spread on the bread surfaces evenly and kept for incubation at 30 °C for 3–4 days.

2.4 Enzymatic Hydrolysis

One liter reactor with stirrer impeller and automatic temperature controller was used for this enzyme hydrolysis purpose. After the experimental setup is fully stabilized with the desired temperature, different amounts of waste bread pieces were blended in 500 ml of water for 20 min. Then the fungal spores from the solid-state fermentation were added into the reaction vessel. The stirring was set at 300 rpm, and the temperature was kept at 55 °C. Sample was taken from this reaction mixture at every hour till 24 h for glucose estimation and FAN calculation. Then the samples were centrifuged at 10,000 rpm at 15 °C, and the supernatant was filtered using Whatman filter paper.

2.5 *Microbial Fermentation*

Saccharomyces cerevisiae was utilized in the fermentation for bioethanol production. The fermentation was carried out with 5% as well as 10% inoculums size to check better ethanol production. The bread hydrolysate which was obtained after the enzymatic hydrolysis was treated as the production media in this stage. After the inoculum has been transferred into the hydrolysate, the flasks were kept in a shaker incubator for fermentation at 28 °C. The rpm was 180 up to 12 h, which was set to 120 afterward. Samples were taken at every 6 h for 24 h to check the optimum time for ethanol production. The pH of every sample was checked simultaneously to observe if there was any acid production. The optical density of every sample was also checked.

2.6 *Enzyme Extraction*

When the solid-state fermentation is completed, the waste breads were suspended in distilled water to prepare a mixture of concentration 50 g/l. The mixture was blended for 30 min and then centrifuged for 10 min at 2,500 g. After this the supernatant was filtered by Whatman filter paper to get the filtrate which contains the enzymes (Melikoglu et al. 2013a, b).

2.7 *Analytical Techniques*

2.7.1 Glucose Estimation by Anthrone Colorimetric Method

Glucose standard was prepared in the range of 1–10 mg/10 ml. Anthrone reagent was prepared by dissolving 200 mg anthrone in 100 ml of concentrated H_2SO_4. Four milliliters of anthrone reagent was added in each of the standard solutions of glucose and mixed well. After the mixture, the color of solution changed to bluish green. Then the samples were kept in boiling water bath for 10 min and cooled to room temperature. Optical density was measured at 620 nm, and standard curve was plotted (Sumbhate et al. 2012).

2.7.2 Free Amino Nitrogen (FAN) Analysis

The FAN concentration was measured by following the ninhydrin colorimetric method. Glycine solution in water was used as the standard solution (MB013-500G). According to the European Brewery Convention Protocol, 0.1085 g of pure glycine ($C_2H_5NO_2$; MW, 75.07; purity, 99%) was dissolved into 100 ml of deionized water in a volumetric flask. This solution has FAN concentration of 200 ppm. From this stock solution, 100 ppm, 50 ppm, and 25 ppm solutions were diluted, and standard curve was prepared. Then 1 ml of ninhydrin solution was added to 5 ml of sample followed by gentle stirring for 4–7 min at 80–100 °C. The reaction of ninhydrin with amino acids of the sample resulted in the formation of purple colored Ruhemann's complex. After cooling to room temperature in a cold water bath, the absorbance was measured at 570 nm, and concentrations of the samples were calculated from the standard curve (Lie 1973).

2.7.3 Estimation of Ethanol by Colorimetric Method

Ethanol standard solutions were prepared in the range of 1.6–12.8 mg/ml. Five milliliters of sodium dichromate solution (4 g/100 ml) was added to each of the 15 ml of standard solutions. Then 25 ml of H_2SO_4 was added into every solution. The pH of the solution was maintained by 0.2 M of acetate buffer having pH value of 4.3. Then mixture was shaken for 1 min and kept in incubation for 2 h. After that, green color product was observed, the O.D. of the solutions was taken at 578 nm, and from the readings, the standard curve was calibrated. The experimental samples were treated in the same way followed by O.D. measurement, and subsequently the concentration of ethanol was determined from the standard curve.

2.7.4 Quantification of Enzyme Activity and Acetone Precipitation

The bread hydrolysate contains mainly glucoamylase and protease enzyme. One unit of glucoamylase activity can be defined by measuring the amount of enzyme required to release 1 g of glucose in 1 min under the assay condition. One unit of proteolytic enzyme activity can be measured by the amount of enzyme required to release 1 mg free amino nitrogen within 1 min under assay condition.

After the solid-state fermentation, the crude extract was suspended in acetone in 1:0.5–1:4 ratios to check at which proportion the precipitation is occurring well. After 1 h when the extract is completely precipitated, the mixture is centrifuged at 10,000 g for 30 min at 4 °C. After harvesting, the precipitated enzymes were collected and suspended in Tris-HCl buffer (0.05 M, pH = 6.5) (Negi and Banerjee 2010).

3 Result and Discussion

3.1 Waste Bread Composition

Waste bread is such a very nutritious food waste which is rich in organic carbon and nitrogen. According to Delgado et al. (2009), 100 g of white bread contains almost 50 g of carbohydrate (47 g is in the form of starch), 37 g water, and around 8 g of protein, which make 95% of the bread total weight.

3.2 Production of Glucose as Generic Fermentation Medium

Figure 1 shows the glucose generation profile with respect to time with 30% (w/v) bread concentration. Glucose production has almost reached to its saturation around 20 h. So it is sufficient to continue this reaction at 55 °C for 24 h for complete breakdown of the nutrients mainly dextrin which is present in bread. The glucose concentration at the end of hydrolysis was found to be 145 ± 10 g/l, i.e., 0.59 g of glucose per gram of bread.

3.3 Data Analysis of Free Amino Nitrogen to Measure Total Assimilable Nitrogen

Figure 2 shows the free amino nitrogen (FAN) concentration profile with time. The ninhydrin colorimetric method measures the total assimilable nitrogen. Very low and also very high concentration of FAN delays the yeast growth and subsequently

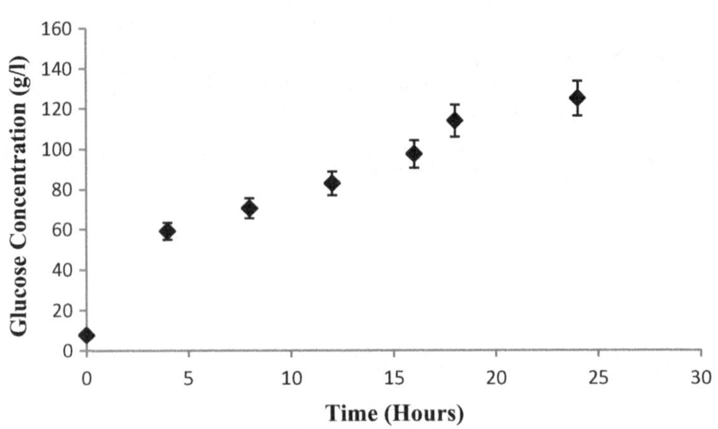

Fig. 1 Formation of glucose with time during enzymatic hydrolysis of waste bread

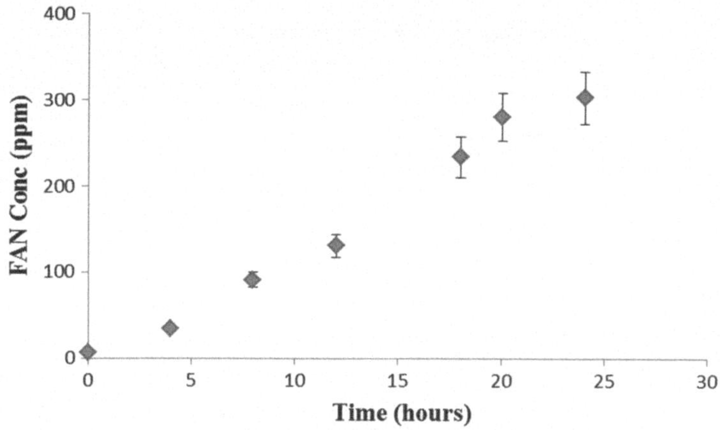

Fig. 2 Formation of FAN with time during enzymatic hydrolysis of waste bread

the ethanol production (Wang et al. 2007). So FAN is a very important parameter when the bioethanol production is the main concern. Here the enzymatic hydrolysis of waste bread results in the production of 270 ± 25 ppm of FAN at 55 °C with 30% bread concentration.

3.4 Bioethanol Production Using Waste Bread Hydrolysate as Substrate

In this fermentation process, yeast and various enzymes convert the fermentable carbohydrate (dextrin, maltotriose, maltose, glucose) of waste bread into ethanol. Other by-products like lactic acid, acetic acid, and carbonic acid are also produced, which may increase the pH sometimes (Wang et al. 2007) (Fig. 3).

3.5 Bioethanol Generation and Subsequent Glucose Consumption Profile

From the data, it is observed that the ethanol production gradually increased with time and the concentration of glucose decreased with time. The final ethanol concentration was found to be 54 ± 2 g/l. This indicates yield of ethanol from glucose, $Y_{P/S}$, as 0.37 as compared to theoretical yield of 0.511, i.e., a conversion efficiency of 72%. In case of ethanol production if the pH profile is considered, there is almost a gradual decrease in pH (see Fig. 4) due to the production of very trace amount of acetic acid, butyric acid, and acetone (Yang et al. n.d.).

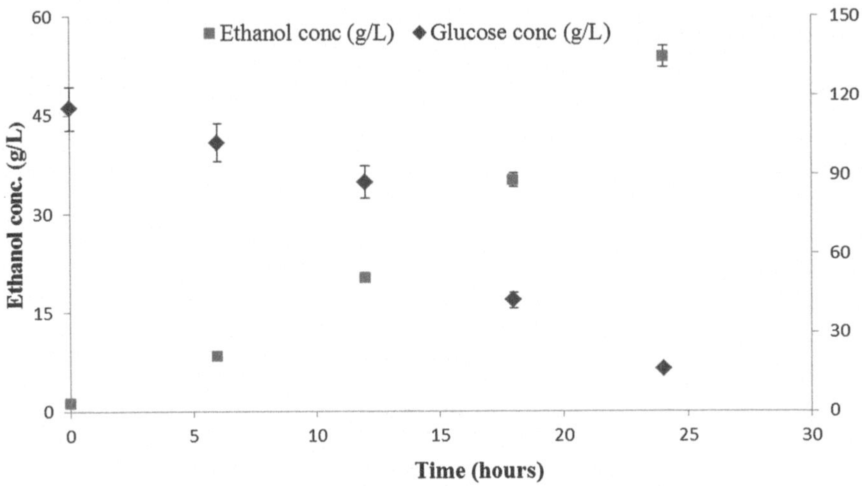

Fig. 3 Formation of ethanol and depletion of glucose with time during fermentation

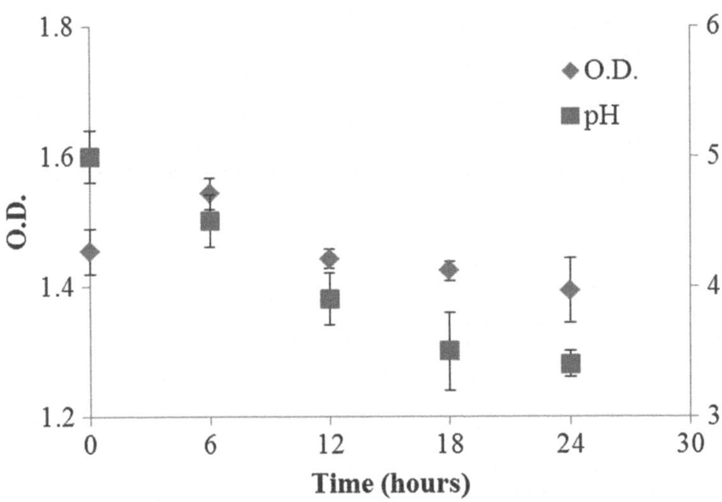

Fig. 4 Change in O.D. and pH during fermentation of ethanol

3.6 *Optical Density and pH Profile with Time*

Figure 4 depicts an initial increase in biomass concentration up to 6 h corresponding ethanol concentration of 8.4 ± 1 g/l, and after 6 h biomass concentration is gradually decreased. The pH value during fermentation is found to decrease with increase in time, confirming the production of acids during anaerobic fermentation of ethanol.

3.7 Enzyme Extraction and Quantification

After the solid-state fermentation of waste breads and the acetone precipitation of crude extracts, the crude enzyme extract was quantified, and it was approximately 5 ± 1 g. The activity of crude enzyme was found to be 0.01 Unit.

4 Conclusion

The paper demonstrated that food waste such as bread can also be used as a sole fermentation feedstock to produce ethanol. Solid-state fermentation of waste bread by *Aspergillus niger* produced crude enzymes, which were used for the hydrolysis of waste bread to produce glucose and FAN. After hydrolysis, 145 ± 10 g/l and 270 ± 20 g/l of glucose and FAN were produced, respectively. The bread hydrolysate was then further used to produce ethanol by yeast fermentation and resulted concentration of 54 ± 2 g/l, i.e., yield of ethanol from glucose, $Y_{P/S}$, as 0.37. The yield can be further enhanced by optimizing the operating parameter. The present study demonstrates that not only ethanol but other value-added products can also be produced from waste bread.

Acknowledgment The authors would like to express their sincere gratitude to the Center for the Environment and Biosciences and Bioengineering Department of Indian Institute of Technology for providing all the required facility to carry on this work.

References

Delgado R, Castro A, Vázquez M (2009) A kinetic assessment of the enzymatic hydrolysis of potato (Solanum tuberosum). LWT-Food Sci Technol 42(4):797–804

Kiran EU, Trzcinski AP, Ng WJ, Liu Y (2014) Bioconversion of food waste to energy: a review. Fuel 134:389–399

Leung CCJ, Cheung ASY, Zhang AY-Z, Lam KF, Lin CSK (2012) Utilisation of waste bread for fermentative succinic acid production. Biochem Eng J 65:10–15

Lie S (1973) The EBC-NINHYDRIN method for determination of free Alpha amino nitrogen. J Inst Brew 79(1):37–41

Melikoglu M, Lin CSK, Webb C (2013a) Kinetic studies on the multi-enzyme solution produced via solid state fermentation of waste bread by Aspergillus awamori. Biochem Eng J 80:76–82

Melikoglu M, Lin CSK, Webb C (2013b) Stepwise optimisation of enzyme production in solid state fermentation of waste bread pieces. Food Bioprod Process 91(4):638–646

Moukamnerd C, Kawahara H, Katakura Y (2013) Feasibility study of ethanol production from food wastes by consolidated continuous solid-state fermentation. J Sustain Bioenergy Syst 3(2):143–148

Negi S, Banerjee R (2010) Optimization of culture parameters to enhance production of amylase and protease from Aspergillus awamori in a single fermentation. Afr J Biochem Res 4(3):73–80

Sumbhate SV, Nayak S, Goupale D, Tiwari A, Jadon RS (2012) Colorimetric method for the estimation of ethanol in alcoholic-drinks. J Anal Tech 1(1):1–6

Wang R, Ji Y, Melikoglu M, Koutinas A, Webb C (2007) Optimization of innovative ethanol production from wheat by response surface methodology. Process Saf Environ Prot 85(5):404–412

Yang J, Lee PJ, Di Gioia AJ (n.d.) Fast HPLC analysis for fermentation ethanol processes

Pretreatment and Production of Bioethanol from *Citrus reticulata* Fruit Waste with Baker's Yeast by Solid-State and Submerged Fermentation

Anup D. Chahande, Vidyadhar V. Gedam, Pranay A. Raut, and Yogesh P. Moharkar

Abstract Excessive use of fossil fuels leads the world to investigate the other alternative sources of energy. Biofuels are considered as one of the emerging and more reliable energy sources. The present study investigates the potential of *Citrus reticulata* (orange) fruit waste for efficient production of bioethanol.

During the study orange peel and pulp waste samples in the ratio of 1:1 were pretreated to acidic and heat treated followed by solid-state and submerged fermentation. Fermented fractions were then subjected to distillation. Distilled fermented fractions were then analyzed by gas chromatography-mass spectrometry (GC-MS). The chromatographic analysis revealed that from *Citrus reticulata* fruit wastes, biodiesel can be produced, and the highest bioethanol yield (6.0029%) was observed via solid-state fermentation compared to submerged fermentation. Positive and promising results of GC-MS show that *Citrus reticulata* fruit wastes can be a feasible alternative for efficient bioethanol production.

Keywords *Citrus reticulata* • Orange pulp and peel • Gas chromatography

1 Introduction

Consumption and dependency on fossil fuels to meet energy demand continued to increase in 2014–2015 (Statistical Review of World Energy 2015). Excessive use of fossil fuels focused the attention of researchers to investigate the other alternative source of energy. One of such alternative is biofuels. It is also necessary to search and explore more promising ways for the production of biofuels from different

A.D. Chahande (✉) • P.A. Raut • Y.P. Moharkar
Department of Biotechnology, PIET, Nagpur, India
e-mail: anup.chahande@gmail.com

V.V. Gedam
WTM Division, CSIR-NEERI, Nagpur, India
e-mail: vgedam@yahoo.com

S.K. Ghosh (ed.), *Utilization and Management of Bioresources*,
DOI 10.1007/978-981-10-5349-8_13

materials rather than the conventional lignocellulosic sources. And fermentation is one of the key important ways for the large production of biofuels (Shanmugam 2009; Mark et al. 2007; Deepa et al. 2015; Singh et al. 2015; Zainab and Fakhra 2014). Fermentation requires raw material which can be used as energy source by microorganism, which then can be converted into more valuable end product (Ylitervo 2008; Periyasamy 2009).

Agro-industrial fruits plays a vital role in worldwide economy. Thus agro-industrial and fruit processing industry waste can be considered as a promising source for the fermentation (Ylitervo 2008; Dhillon et al. 2004; Periyasamy 2009; Deepa et al. 2015; Singh et al. 2015; Zhou et al. 2007). Among all the fruits, citrus fruit production has been increased exponentially in recent decades. It was estimated that nearly 100 million metric tons (MMT) of citrus fruits were consumed each year (Citrus: World Markets and Trade 2016). India had produced nearly 11,147 (Eleven Thousend One Hundred Fourty Seven) metric tons (MT) of citrus fruit in the year 2013–2014 (Indian Horticulture Database 2014; Horticultural Statistics at a Glance 2015). Citrus fruits are consumed fresh, or they are mostly preserved by making citrus juice which may be ready to drink or in concentrated form.

After Citrus fruit juice is extracted, the remaining waste of the fruit also serves as a rich source of lignocellulosic material and also can be used as raw material for the bioethanol fermentation (Ylitervo 2008; Dhillon et al. 2004; Periyasamy 2009; Farid 2008; Mishra et al. 2012). *Citrus reticulata* is composed of carbohydrate polymers (cellulose, hemicellulose) as well as aromatic polymer (lignin). Both the cellulose and hemicellulose fractions are polymers of sugars and can be used as a raw material during fermentation for efficient bioethanol production (Ylitervo 2008; Farid 2008; Wilkins et al. 2007; Jung and Kim 2015; Weiyang Zhou et al. 2007; Shi et al. 2009; Sørensen et al. 2012). But the major hurdle during this process is to get reducible sugar for the fermentation process. Leading microorganism (*Saccharomyces cerevisiae*) is a highest bidder for the commercial production of ethanol (Tesfaw et al. 2014; Dake et al. 2010; Von Schenck 2013) However, it cannot convert lignocellulosic biomass into bioethanol ; it requires biomass pretreatment to liberate the required sugar for further fermentation. There are numerous techniques available in literature for the efficient pretreatment of biomass which includes enzymatic degradation, acidic treatment, and alkaline treatment followed by fermentation which can efficiently produce bioethanol (Asgher et al. 2014; Li et al. 2014; Jung et al. 2013; Ibrahim 2012; Eisenhuber et al. 2014; Jung and Kim 2015; Passoth et al. 2013; Cuevas et al. 2010; Soudham et al. 2014; Steffien et al. 2014). Earlier research suggests that the solid-state fermentation used for the setup has potential to give a high production of bioethanol (Farid 2008; Shanmugam 2009; Deepa et al. 2015; Prasad 2014; McCue and Shetty 2005). The present study focused to investigate the potential and possibility of mandarin orange (*Citrus reticulata*) peel and pulp waste in the ratio of 1:1 for efficient bioethanol production via solid-state and submerged fermentation with baker's yeast.

2 Material and Methods

2.1 *Collection, Sample Preparation, and Total Sugar Determination of* Citrus reticulata *Fruit Waste*

Orange (*Citrus reticulata*) samples were collected from local market of Nagpur (Maharashtra, India). The waste peel and pulp of *Citrus reticulata* were separated out after taking out juice by the fruit processor. The sample was kept for sun drying (6 h for 2 days). The dried samples were grinded to particle size of about two to ten mesh in diameter and were stored in sterile flask till further treatment. Ten grams of substrate (orange peel + pulp in the ratio of 1:1) was subjected to the Benedict's test, so as to estimate the concentration of reducing and nonreducing sugar in the sample.

2.2 *Preparation of Inoculums for Fermentation*

During the study, the commercially available baker's yeast was used for the fermentation. Thirty milliliters of distilled water was taken in sterilized Erlenmeyer flask (250 ml) and was boiled for 20 min. The flask was moved into laminar hood, and 5 g of dry weight of baker's yeast was added to the hot water and stirred properly. Temperature of hot water was not more than 45–50 °C. This was done to rehydrate the baker's yeast. After 10 min, 10 ml of this baker's yeast is used as inoculum for fermentation which was added in aseptic condition. Fresh inoculums were prepared before each batch of fermentation.

2.3 *Pretreatment on the Sample*

2.3.1 Acidic Pretreatment

1N concentrated sulfuric acid and 1N nitric acid were prepared for acid pretreatment and were stored in the reagent bottle. After preparing the reagent, 50 g of *Citrus reticulata* waste sample containing orange peel and pulp in the ratio of 1:1 was prearranged. Then 100 ml of sulfuric acid and 100 ml nitric acid were added in each flask containing *Citrus reticulata* waste. The ratio for acid pretreatment was maintained as 1:2:2. These flasks containing substrate + pretreatment reagents were kept on rotary shaker for 6 h at 90 rpm. After the acid pretreatment, the sample was washed six to seven times by distilled water to separate out any acid residues from the substrate with the help of muslin cloth. For further complete neutralization, sodium hydroxide was also added in the substrate to maintain the required fermentation pH, i.e., 6.5–6.7. The pretreated substrate was then filtered, and the filtered substrate (about 20–25 g) was then transferred into another flask for further steam pretreatment.

2.3.2 Steam Pretreatment

After acidic pretreatment the substrate was subjected to the heat pretreatment. During steam pretreatment the substrate sample was autoclaved for about 60 min at 121 °C. Steam pretreatment was mainly performed to ensure that there is no contamination left before subjecting to the fermentation process.

2.4 Fermentation

Fermentation of these above pretreated substrates was done via two methods, i.e., solid-state fermentation and submerged fermentation. All the experiments were done in triplicates.

2.4.1 Solid-State Fermentation

Ten milliliters of earlier aseptically prepared inoculum was added to each flask containing pretreated 20–25 g of substrate. After adding baker's yeast solution, flasks were covered with balloon, so that the carbon dioxide gas can be stored, which is the indicator of fermentation process. The flasks were then kept for fermentation process for around 74 h at room temperature on rotary shaker at 90 rpm. Once the fermentation period is over, the mixed sample was then subjected to distillation process to separate out the liquid fraction at 75–76 °C. Collected distilled fractions were then analyzed for bioethanol percentage (v/v) by gas chromatography-mass spectrometry (GC-MS).

2.4.2 Submerged Fermentation

Ten milliliters of earlier aseptically prepared inoculums was added to each flask containing pretreated 20–25 g of substrate, and the volume is made up to 250 ml. After adding baker's yeast solution, flasks were covered with balloon. The flasks were then kept for fermentation around 74 h at room temperature on rotary shaker at 90 rpm. After fermentation, distillation process was carried out to separate out the liquid fraction at 75–76 °C. Collected distilled fractions were then analyzed for ethanol percentage (v/v) by GC-MS.

2.5 Determination of Biodiesel by GC-MS

The distilled fractions for the presence of bioethanol were conducted using gas chromatography-mass spectrometry (GC-MS) (Thermo Fisher model 8610C), equipped with a 60 m column (Restec MXT-1, Id 0.53 mm, 5 μM), with column injector and FID conditions: 220 °C; H_2, 25 PSI, equivalent to 12 ml/min; air, 2 PSI, 3,613 ml/min; gain set to medium. The GC was also equipped with an internal air compressor and hydrogen generator and was also coupled with mass spectra. N_2 was used as carrier gas with pressure control (24 PSI constant; equivalent to 27 ml/min). The oven temperature (column and injector temperature) was initially set at 50 °C and then elevated at the rate of 7 °C/min to 100 °C, thus giving a total run time of 7 min. Furthermore, 2 μL sample was injected manually at time 0, using a 5 μl Hamilton syringe, and temperature cycle was started. Syringe was thoroughly washed with ethyl acetate between injections to avoid cross contamination. Each injection was repeated three times; ethanol routinely came out at retention time equivalent to 65 °C. The mass spectra of pure ethanol were also run on GC-MS as a reference.

3 Results and Discussion

The Benedict's test gives enough estimation of sugars that can be subjected to saccharification and fermentation. The presence of reducing sugar in the given substrate after Benedict's test (orange peel + pulp) was found to be 1.5% by comparing Benedict's reagent before acid pretreatment. The acid and steam pretreatment were performed to improve the further fermentation process. Most of the cellulosic components can be converted to reducing sugar after acid pretreatment followed by autoclaving which thereby improve hydrolysis efficiency (Hsu et al. 2011).

The GC-MS analysis of 1N pretreated distilled sample shows that the amount of bioethanol present during submerged fermented sample after comparing with standard gas chromatograms and mass spectra of pure ethanol was 1.94%, whereas for solid-state fermentation, it was observed to be 6.0029%. In *Citrus reticulata* fruit waste samples, the highest bioethanol yield (6.0029%) was obtained via solid-state fermentation. It was observed that *Citrus reticulata* fruit waste sample had the potential for biodiesel production via solid-state fermentation as well as submerged fermentation process.

4 Conclusions

The present study revealed certain interesting facts about bioethanol production from *Citrus reticulata* fruit wastes. The waste sample after acidic and autoclave pretreatment which is a well-established method shows promising results. Though the bioethanol production from the orange peel is a well-exploited area, less work is done on pulp. The study highlights that 1:1 w/w ratio of orange peel and pulp with acidic and autoclave pretreatment followed by submerged and solid-state fermentation can be a promising method for biodiesel productions.

It can be concluded that the waste from the *Citrus reticulata* fruit wastes (peel + pulp) can be an alternative and promising resource for the efficient bioethanol production. Still the feasibility and practicability of this process for efficient biodiesel production in an optimized way need to be explored intensively.

References

Asgher M, Bashir F, Iqbal HMN (2014) A comprehensive ligninolytic pre-treatment approach from lignocellulose green biotechnology to produce bio-ethanol. Chem Eng Res Des 92:1571–1157

Chuan-Liang Hsu, Ku-Shang Chang, Ming-Zhang Lai, Tsan-Chang Chang, Yi-Huang Chang, Hung-Der Jang (2011) Pretreatment and hydrolysis of cellulosic agricultural wastes with a cellulase-producing Streptomyces for bioethanol production. Biomass Bioenergy 35(5):1878–1884

Citrus: World Markets and Trade (2016) World production, markets, and trade reports web: http://www.fas.usda.gov/data/citrus-world-markets-and-trade

Cuevas M, Sánchez S, Bravo V, García JF, Baeza J, Parra C, Freer J (2010) Determination of optimal pre-treatment conditions for ethanol production from olive-pruning debris by simultaneous saccharification and fermentation. Process Biochem 89:2891–2896

Dake M, Jadhv J, Patil N (2010) Variations of two pools of glycogen and carbohydrate in Saccharomyces cerevisiae grown with various ethanol concentrations. J Ind Microbiol Biotechnol 37:701–706

Deepa A, Da Costa M, Hriiyia M (2015) Production of bioethanol from *Citrus limetta* and *citrus maxima* fruit waste by batch fermentation. Int J Pharma Biosci 6:567–578

Dhillon SS, Gill RK, Gill SS, Singh M (2004) Studies on the utilization of citrus peel for pectinase production using fungus *Aspergillus niger*. Int J Environ Stud 61:199–210

Eisenhuber K, Krennhuber K, Steinmüller V, Jäger A (2013) Comparison of different pre-treatment methods for separating hemicellulose from straw during lignocellulose bioethanol production. Energy Procedia 40:172–181

Farid T (2008) Ethanol production from cellulosic biomass by encapsulated *Saccharomyces cerevisiae*. Thesis of University College of Boras, School of Engineering, pp 210–221

Hassan Al-Haj Ibrahim (2012) Pretreatment of straw for bioethanol production. Energy Procedia 14:542–551

Horticultural Statistics at a Glance (2015) Horticulture statistics division department of agriculture, Cooperation & Farmers Welfare Ministry of Agriculture & Farmers Welfare Government of India. Web: agricoop.nic.in/imagedefault/hortstat_glance.pdf

Indian Horticulture Database (2014) web: http://nhb.gov.in/area-pro/NHB_Database_2015.pdf

Mark W, Wilbur W, Karel G (2007) Simultaneous saccharification and fermentation of citrus peel waste by *Saccharomyces cerevisiae* to produce ethanol. Process Biochem 42:1614–1619

McCue P, Shetty K (2005) A model for the involvement of lignin degradation enzymes in phenolic antioxidant mobilization from whole soybean during solid-state bioprocessing by *Lentinus edodes*. Process Biochem 40:1143–1150

Ming Li, Shengli Si, Bo Hao, Yi Zha, Can Wan, Shufen Hong, Yongbo Kang, Jun Jia, Jing Zhang, Meng Li, Chunqiao Zhao, Yuanyuan Tu, Shiguang Zhou, Liangcai Peng (2014) Mild alkali-pretreatment effectively extracts guaiacyl-rich lignin for high lignocellulose digestibility coupled with largely diminishing yeast fermentation inhibitors in Miscanthus. Bioresour Technol 169:447–454

Mishra J, Kumar D, Samanta S, Vishwakarma MK (2012) A comparative study of ethanol production from various agro residues by using *Saccharomyces cerevisiae and Candida albicans*. J Yeast Fungal Res 3:12–17

Passoth V, Tabassum MR, Nair HAS, Olstorpe M, Tiukova I, Ståhlberg J (2013) Enhanced ethanol production from wheat straw by integrated storage and pre-treatment (ISP). Enzym Microb Technol 52:105–110

Periyasamy S, Venkatachalam S, Ramasamy S, Srinivasan V (2009) Production of bio-ethanol from sugar molasses using Saccharomyces Cerevisiae. Mod Appl Sci 3(8)

Shanmugam P (2009) Production of bio-ethanol from sugar molasses using *Saccharomyces cerevisiae*. Can Cent Sci Educ 3:32–37

Shi J, Sharma-Shivappa RR, Chinn M, Howell N (2009) Effect of microbial pretreatment on enzymatic hydrolysis and fermentation of cotton stalks for ethanol production. Biomass Bioenergy 33:88–96

Singh ARP (2014) A comparative study of bioethanol production ability of Bacillus subtilis and Saccharomyces cerevisiae using banana and Orange peels. Int J Sci Eng Res 5:1–5

Singh J, Suhag M, Dhaka A (2015) Augmented digestion of lignocellulose by steam explosion, acid and alkaline pretreatment methods: a review, carbohydrate polymers. Process Biochem 117:624–631

Sørensen A, Teller P, Hilstrøm T, Ahring B (2012) Hydrolysis of miscanthus for bio-ethanol production using dilute acid presoaking combined with wet explosion pre-treatment, enzymatic treatment. Bioresour Technol 99:6602–6607

Soudham VP, Brandberg T, Mikkola J-P, Larsson C (2014) Detoxification of acid pretreated spruce hydrolysates with ferrous sulfate and hydrogen peroxide improves enzymatic hydrolysis and fermentation. Bioresour Technol 166:559–565

Statistical Review of World Energy (2015) Web: http://www.bp.com/en/global/corporate/about-bp/energy-economics/statistical-review-of-world-energy.html

Steffien D, Aubel I, Bertau M (2014) Enzymatic hydrolysis of pre-treated lignocellulose with *Penicillium verruculosum* cellulases. J Mol Catal B Enzym 103:29–35

Tesfaw A, Assefa F (2014). Current trends in bioethanol production by Saccharomyces cerevisiae: substrate, inhibitor reduction, growth variables, coculture, and immobilization. Int Sch Res Not 2014, Article ID 532852

Von Schenck A, Berglin N, Uusitalo J (2013) Ethanol from Nordic wood raw material by simplified alkaline soda cooking pre-treatment. Appl Energy 102:229–240. 38

Weiyang Zhou, Wilbure Widmer, Karel Grohmann (2007) Economic analysis of ethanol production from citrus peel waste. Fla State Hortic Soc 120:310–315

Wilkins M, Widmer W, Grohmann K, Cameron R (2007) Hydrolysis of grapefruit peel waste with cellulase and pectinase enzymes. Bioresour Technol 98(2007):1596–1601

Ylitervo P (2008) Production of ethanol and biomass from orange peel waste by *Mucor indicus*. Thesis of University College of Boras (School of Engineering), pp 181–300

Young Hoon Jung, Kyoung Heon Kim (2015) Acidic pretreatment, pretreatment of biomass. Chapter 3 Book Korea University Graduate School, Seoul, Republic of Korea, pp 27–50

Young Hoon Jung, In Jung Kim, Hyun Kyung Kim, Kyoung Heon Kim (2013) Dilute acid pretreatment of lignocellulose for whole slurry ethanol fermentation. Bioresour Technol 132:109–114

Zainab B, Fakhra A (2014) Production of ethanol by fermentation process by using yeast *Saccharomyces cerevisiae*. Dep Environ Sci 13:24–32

Production of Ethanol from Waste Potato Using Locally Available Biocatalyst

Amit Ganguly, Esha Bala, Priyabrata Banerjee, Richard Blanchard, and Pradip Kumar Chatterjee

Abstract Bioethanol is a renewable energy source produced from the resources which can be easily replenished. The bioethanol production through fermentation may provide an economically competitive source of energy by its incorporation into gasoline. Production of bioethanol from waste food crops like potatoes could be the better substrate, and the waste produced is also biodegradable. Lack of storage facilities and postharvest losses make potato a promising crop which can be used for production of ethanol. Moreover, the conversion of potato starch into glucose by *bakhar* is more cost-effective, and fermentation with baker's yeast *Saccharomyces cerevisiae* yields maximum amount of ethanol. This process of production of ethanol from waste potato would be promising and economically effective for the production of biofuel, called bioethanol.

Keywords Waste potato • Pretreatment • Saccharification • Fermentation • Bioethanol

A. Ganguly (✉) • P. Banerjee • P.K. Chatterjee
CSIR Central Mechanical Engineering Research Institute, Durgapur 713209, India
e-mail: amitganguly022@gail.com

E. Bala
Maulana Azad National Institute of Technology, Bhopal 462003, India

R. Blanchard
Loughborough University, Leicestershire LE11 3TU, UK

S.K. Ghosh (ed.), *Utilization and Management of Bioresources*,
DOI 10.1007/978-981-10-5349-8_14

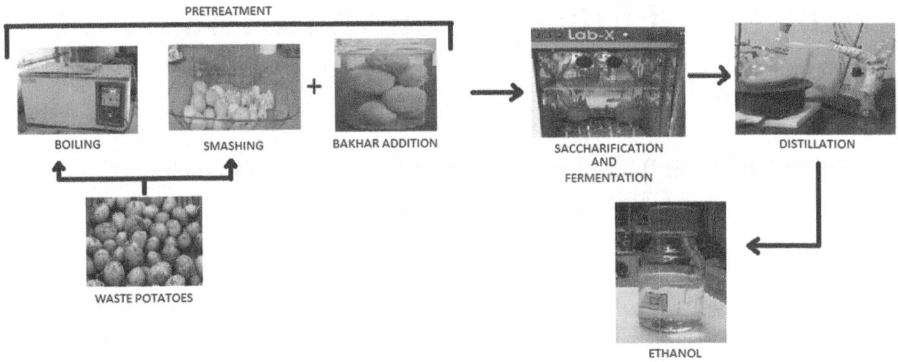

1 Introduction

The fossil fuel depletion and consequent increase in price encourage the study of alternative process and substrates to meet the global energy demand. Besides, fossil fuel also has the direct impact on the atmosphere. The increase in the use of fossil energy causes greenhouse gas emissions that have an adverse effect on the environment. The burning of petroleum-based fuels causes the increase of CO_2 level in the environment which is directly responsible for global warming. Methane, hydrogen and ethanol can be used as potential substitutes to fossil fuel as alternative sources of energy. Among these three, bioethanol is the most favourable alternative liquid fuel. Bioethanol can be derived from agricultural biomass resources or waste which has been considered as the cleanest liquid fuel alternative to the fossil fuel like gasoline. Agricultural material like sugarcane juice, rice, potato starch, molasses, cassava and bagasse can be used for the production of bioethanol. Apart from the known usage of ethanol as a fuel, about 45% of the produced ethanol is being used as portable alcohol and 40% in the industrial sectors, and only the remaining is available for blending with petrol. In the industrial sectors, ethanol is used by chemical, pharmaceutical industries, etc. It is also used to produce ethyl tertiary-butyl ether (ETBE) (Chandel et al. 2007). Bioethanol is made by fermenting from a variety of biomass from their sources. It is acknowledged that bioethanol is a unique transportation fuel with powerful economic, environmental and strategic attributes.

The ethanol from biomass-based waste materials from renewable sources for fuel or fuel additives is fermented to produce bioethanol (Ghosal et al. 2013). It is flammable and volatile, so it evaporates easily when left in an open container. The chemical formula of ethanol is C_2H_5OH. Bioethanol is a high-quality, stable liquid. It is an alternative renewable energy source produced through the anaerobic digestion process. Bioethanol is processed using either free or immobilised cell. Immobilised cells are more advantageous over free cells as it has the ability to separate cell mass and reduce the risk of contamination and better operational stability

Table 1 Potato composition (Rani et al.; Yekta and Ulgar 1994)

Composition	Content (%)
Starch	20
Moisture	80.28
Total protein	2.19
Ashes	0.65
Crude fibre	0.85
Total sugars	0.41
Total lipids	0.12

and cell viability for several cycles of operations enhancing yields (Ghosh et al. 2014; Gohel and Duan 2012). Currently, bioethanol production is based on the fermentation of starch- and sugar-based feedstocks, leading to the production cost fluctuation with the price of these food sources. Based on the utilisation of waste potatoes, ethanol is produced from potatoes. In potato cultivation, waste potatoes are produced from 5 to 20% of crops as by-products (Liimatainen et al.; Ghosal et al. 2013). However, depending on the biological waste for the production of bioethanol, potato and ethanol have some chemical composition which is required to know for the production of ethanol.

Potato is grown almost everywhere in the world. Potato is a tuberous crop from the perennial nightshade *Solanum tuberosum* L. Potato is a cheap substrate for alcohol fermentation which is rich in starch (Table 1) and production with higher starch yield per unit land cultivated. After China and Russian, India is the third largest potato producer. The amount of waste and by-products of potato is estimated to be around 12–20% of their total production. Good-quality alcohol can be produced from potato which can be used as a good fuel. Therefore, it is a need to explore also an attempt to utilise the possibility of ethanol production with maximum efficiency with suitable processing and low-cost materials for conversion of waste potatoes into ethanol.

2 Materials and Methods

2.1 Biomass Feedstocks and Materials

The substrates for ethanol production are recommended to be cheap. Ethanol has been produced from varieties of the substrate. The current industrial ethanol output was derived from the input of sugar- and starch-based feedstock such as sugar cane, sugar beet, sorghum, corn, wheat, cassava, rice and others (Sanchez and Cardona 2008; Mussatto et al. 2010; Hill et al. 2006). Ethanol production from wastes has two major advantages. On the one hand, it reduces or eliminates the cost of waste disposal. On the other hand, wastes are cheap and it reduces the cost of ethanol production. The amount of substrate depends on the waste potato and broken rice

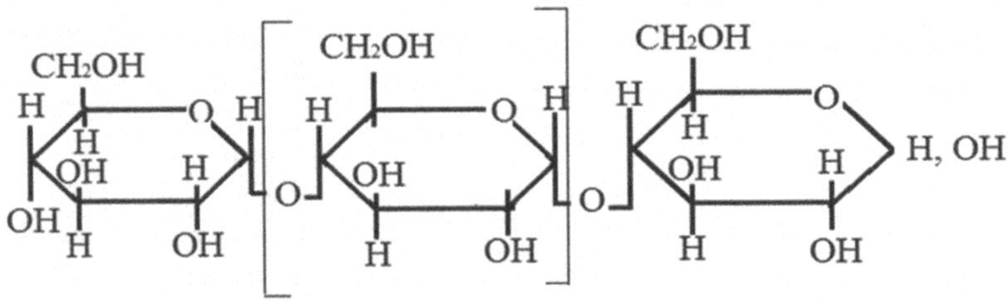

Fig. 1 Structure of starch

which is wasting per year. Waste potatoes and broken rice as biomass are alternative raw materials that have attracted the considerable attention of both the ethanol industry and academic research owing to its large availability, low cost and lack of competition with food production.

In bioethanol production, starch is the main component in potato. Starch (Fig. 1) is a polymeric carbohydrate composed of glucose units. The properties of starch vary with the plant source. Starch is made up of two components, amylose and amylopectin, which are polymers of glucose. Amylose is a linear polymer of glucose units joined by alpha-1, four bonds which consist of up to 20–25% of starch weight. The linear polymer contains up to 6000 glucose units. Amylopectin is a branched polymer in which the glucose units are joined by alpha, 1–4 bonds in the linear molecule sections and by alpha, 1–6 bonds at the branching point which typically occur at every 20–30 glucose units and contain up to 75–80% of most starch types. Amylopectin molecule is one of the largest molecules in nature and has about two million glucose units (Anonymous 1981; Batum 1993; Hoek et al. 1998).

Bakhar plays an important role in fermentation and is an amylolytic enzyme which directly converts the starchy raw material to fermentable sugars and finally alcohol. The principal ingredient of bakhar which is mixed with a traditional starter is low-grade boiled rice (*Oryza sativa* L.). Bakhar is a mixture of old ferments (containing microbial inoculums), parts of different plants (fresh root sand leaves) and rice dust (Dhal et al. 2010; Lin et al. 2012). In the eastern states of India like West Bengal, tribal people prefer rice-based beverages called haria. The traditional starter used for the haria preparation is bakhar or ranu tablet.

Baker's yeast (*Saccharomyces cerevisiae*) can produce a high concentration of ethanol and can grow under the optimum condition without causing harm in its growth (Nigam 2000). *S. cerevisiae* is the microorganism which is used for ethanol production due to its well-known fermentative capabilities (Cot et al. 2007; Rani et al. 2010). Yeast requires nutrients for their growth like minerals, peptides and vitamins and magnesium for glucose conversion. Nitrogen deficiency retards yeast growth and metabolism resulting in arrested sluggish fermentation (Gohel and Duan 2012; Rani et al. 2010). An adequate amount of yeast is required for fermentation of raw materials (rice or potato) for maximum yield of ethanol. Moreover, scientific knowledge about *S. cerevisiae* allows a great potential of metabolic engineering modifications with the aim of boosting its fermentation efficiency.

2.1.1 Temperature

The yeasts at high temperature are ideal for bioethanol productions which are active and tolerant. The enzymatic activity and membrane turbidity of yeast cells are greatly affected by the temperature. The optimum temperature for the fermentation of the *Saccharomyces cerevisiae* is 30–40 °C with a pH range of 4–6. The exponential phase of yeast cells shortened, and ethanol production reduced considerably at a higher temperature (more than 50 °C). Due to change in transport system, accumulation of toxin increases including ethanol in the cell.

2.1.2 pH

For the culture of *Saccharomyces cerevisiae*, 4–6 is the range of optimum pH. The incubation period was continuing for a long time though the ethanol concentration was not reduced significantly due to a decrease in pH below 4, and when the pH was above 6, the concentration of ethanol diminished substantially. So, before fermentation, the pH of the feedstock should be checked or should be optimised in the range of 4–6 for maximum yield of bioethanol.

2.2 *Pretreatment*

Pretreatment is a very important step before saccharification. Preparation of waste potatoes and boiling of the wastes are required before mixing of bakhar. The starch can easily be broken down with the biocatalyst, bakhar into glucose.

Waste potatoes were collected; 500 g of waste potatoes was boiled for 1.5 h in a water bath. The boiled potato was peeled, mashed and mixed very well in 200 ml of water and allowed to cool down. Further, 25 g of bakhar was weighed and mixed in the starch suspension of potato and incubated.

2.3 *Saccharification*

The process of breaking down of a complex carbohydrate such as starch into monosaccharide components such as glucose is termed as saccharification. A saccharifying enzyme such as bakhar provides maximum starch conversion used on a liquefied starch-containing substrate to produce sugars for fermentation. The liquefied

$$C_6H_{12}O_6 \longrightarrow 2(CH_3CH_2OH) + 2(CO_2)$$

Sugar → Ethanol Carbon dioxide

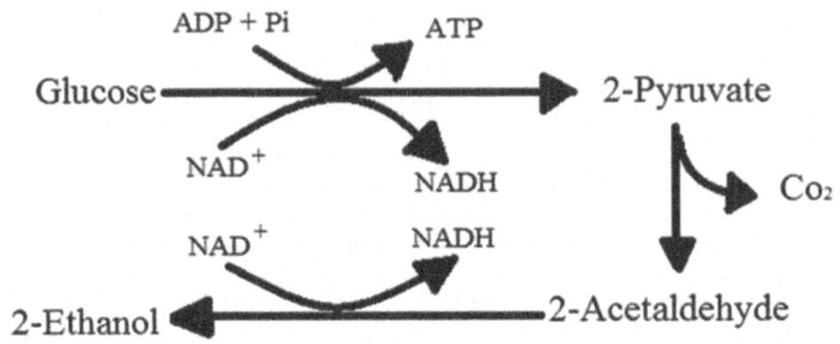

Fig. 2 Biosynthesis process for fermentation of glucose into ethanol

starch-containing substrate was incubated for 48 h at 35 °C in an incubator for the conversion of glucose.

2.4 *Fermentation*

Fermentation is a process of bioconversion of sugars into ethanol using yeast *Saccharomyces cerevisiae.* In this process, 50 g of baker's yeast was added in the saccharified starch suspension and allowed to ferment for 96 h maintaining a pH range of 5–6. Although there are over 150 amylolytic yeast species, their industrial use is limited because of their low ethanol tolerance. The fermentation of glucose into ethanol includes the following biosynthesis process in the form of flow chart given in Fig. 2.

2.5 *Distillation*

Distillation is a process of separating chemical components from a liquid mixture in different boiling points by selective evaporation and condensation. In this process, the ethanol is separated from the fermented starch suspension. The fermented potato starch suspension was distilled for 4–5 h for the yield of ethanol. For further purification, the liquid was further distilled for maximum concentration of ethanol.

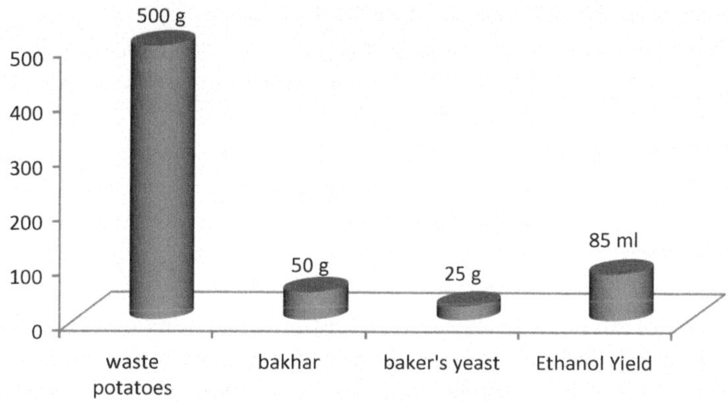

Graph 1 Graphical representation of the ethanol with fermentative materials and feedstocks

3 Results and Discussions

As alternatives to fossil fuels, the use of renewable biofuels is developing in many countries to decrease greenhouse gas emissions and to reduce the dependence on petroleum (Hill et al. 2006; Lin and Tanaka 2006; Zabed et al.). Fermentation of waste potato by baker's yeast (*Saccharomyces cerevisiae*) and sugar production by saccharifying the liquefying starch suspension produce the maximum amount of ethanol. The results showed that starch utilisation, amylolytic activity and ethanol yields are high in *S. cerevisiae*. Starch utilisation of up to 94.8% is observed in 48 h to convert into glucose. By simultaneous saccharification and fermentation at different concentration of enzymes, yeasts and parameters, maximum yield of ethanol was observed. During fermentation, at 96 h maximum yield is observed rather than 24 h and 48 h. Detailed studies and experimentation revealed that bakhar has the potential to convert the starch into glucose within 48 h, but a further extension of saccharification degrades the yield of ethanol. Five hundred grammes of waste potatoes and 25 g of bakhar and 50 g of baker's yeast produce 85 ml of ethanol after saccharification and fermentation. The graphical representation of the ethanol yield with fermentative materials and feedstocks is shown in Graph 1. The yield of ethanol from waste potato could be a turning point to be used as fuel and can also reduce the cost of the ethanol. Fermentation and yield conditions for ethanol production of waste potatoes are described (Table 2).

Table 2 Fermentation and yield conditions for ethanol production

Name of the organism	Temperature (°C)	pH	Bakhar (g)	Time of saccharification (h)	Baker's yeasts (g)	Time of fermentation (h)
Baker's yeast (*Saccharomyces cerevisiae*)	35	5–6	50	48	25	96

4 Conclusion

Currently, a lot of attention has been given for the production of biofuels in order to contribute the rising demands of fuels. Renewable biomass such as sugar, starch and lignocellulosic materials is expected to be one of the dominating renewable biofuels in the transport sectors within coming 20 years. The use of waste potato in biofuel production seems to be potential, as the biofuel production from waste potato is a well-known process. Due to high starch content in potato, it has been proposed as a feedstock for bioethanol production. The analysis reveals that production of ethanol from potato could meet the need of global demand. According to the study and analysis with experiments, it can be concluded that with the help of bakhar, large amount of ethanol can be produced. The experimental result shows 0.17 ml/g of ethanol is produced from waste potato. Bakhar reduced the cost of the ethanol, and the material is also biodegradable. With today's increasing demand and cost of the fuel, this process of production of ethanol could be a breakthrough to meet the need of social, economic and environmental problems.

Acknowledgement This work is supported by the Council of Scientific and Industrial Research-Central Mechanical Engineering Research Institute, Durgapur, West Bengal, India, and the authors express their sincere gratitude to the Director of CSIR for extending their kind support to execute this research work.

References

Anonymous (1981) Novo enzyme in the production of ethanol from starch containing crops. Novo Industry AS, Bagsvaerd-Denmark, p 47

Batum M (1993) Industrial application of enzymes. Enzymes as industrial tools. Ege University Biotechnology Centre, izmir, Turkey, pp 229–271

Chandel AK, Chan ES, Rudravaram R, Narasu ML, Rao LV, Pogaku R (2007) Economics and environmental impact of bio-ethanol production technologies: an appraisal. Bioethanol Mol Biol Rev 2:14–32

Cot M, Loret OM, François J, Benbadis L (2007) Physiological behaviour of Saccharomyces cerevisiae in aerated fed-batch fermentation for high level production of bioethanol. Fed Eur Microbiol Soc 7:22–32

Dhal NK, Pattanayaik C, Reddy S (2010) Bakharstarch fermentation—a common tribal practice in Orissa. Indian J Tradit Knowl 9:279–281

Ghosal A, Banerjee S, Chatterjee S (2013) Biofuel precursor from potato waste. IJRET: International Journal of Research in Engineering and Technology, Delhi, pp 213–219

Ghosh K, Maity C, Adak A, Halder SK, Jana A, Das A, Parua (Mondal) S, Mohapatra Das PK, Pati BR, Mond KC (2014) Ethnic preparation of Haria, a rice-based fermented beverage, in the province of lateritic West Bengal, India. Ethnobotany research and applications. J Plants People Appl Res 12:39–49. pp. 41-50

Gohel V, Duan G (2012) Conventional process for ethanol production from Indian broken rice and pearl millet. Genencor Bio-Products Co. Ltd., Bioprocess Biosystems Engineering, China, pp 1297–1308

Hill J, Nelson E, Tilman D, Polasky S, Tiffany D (2006) Environmental, economic, and energetic costs and benefits of biodiesel and ethanol biofuels. The National Academy of Sciences, Departments of Ecology, Evolution, and Behavior and Applied Economics, University of Minnesota, USA, pp 11206–11210

Hoek PV, Dijken JPV, Pronk JT (1998) Effect of specific growth rate on fermentative capacity of Baker's yeast. Department of Microbiology and Enzymology, Kluyver Institute of Biotechnology, Netherlands, pp 4226–4233

Lin Y, Tanaka S (2006) Ethanol fermentation from biomass resources: current state and prospects. Appl Microbiol Biotechnol 69:627–642

Lin Y, Zang W, Li C, Sakakibara K, Tanaka S, Kong H (2012) Factors affecting ethanol fermentation using Saccharomyces cerevisiae BY4742. Biomass Bioenergy 47:395–401

Mussatto SI, Dragone G, Guimaraes PMR, Silva JPA, Carneiro LM, Roberto IC, Vicente A, Domingues L, Teixeira JA (2010) Technological trends, global market, and challenges of bio-ethanol production. Biotechnol Adv 28(6), ISSN 0734-9750|1873-1899

Nigam JN (2000) Continuous ethanol production from pineapple cannery waste using immobilized yeast cells. J Biotechnol 80:189–193

Rani P, Sharma S, Garg FC, Raj Kushal, Leela Wati (2010) Ethanol production from potato flour by Saccharomyces cerevisiae. CCS Haryana Agricultural University: Department of Microbiology, pp 733–736

Sanchez Oscar J, Cardona Carlos A (2008) Trends in biotechnological production of fuel ethanol from different feedstocks. Colombia, University of Caldas: Department of Engineering, Bioresource. Technology 99, pp 5270–5295

Yekta G, Ulgar G (1994) Enzymatic hydrolysis and production of ethanol from potato starch. GIDA, Age University, Department of Food Engineering, Izmir, pp 89–92

Zabed H, Faruq G, Sahu JN, Azirun MS, Hashim R, Boyce AN. Bioethanol production from fermentable sugar juice. Malaysia: Institute of Biological Sciences, University of Malaya. Sci World J 1–11

Design of a Multi-tank Processor to Produce Biodiesel Using Waste Vegetable Oil in Nigeria

Adeshola Oluremi Openibo and Nurudeen Adekunle Raji

Abstract Biofuels are a promising long-term renewable energy source having potentials which could address both environmental impacts and security concerns posed by current dependence on petroleum-based fuels. Biodiesel fuel is a renewable energy fuel produced from biological materials or biomass. Development of biodiesel processors is still a novel technology in Nigeria where till date; much works has not been recorded on the development of biodiesel processors. A one-stage nine-flask continuous-batch biodiesel reactor is developed to withhold the heat generated during the reaction of hydrocarbons (vegetable oil), alcohols (methanol or ethanol), and the catalyst (sodium or potassium hydroxide) for biodiesel production. The stages include methoxide tank to mix the methanol; the reactor where the transesterification of the mixtures of vegetable oil, alcohol, and catalyst occur; the settling tower where freshly reacted batch of biodiesel is transferred to free up the reactor; first wash tank for wet washing; second wash tank for dry washing; dry tank to free up the wash tanks for the next batch and also dry the fuel much faster with much better results through heating; glycerin tank where the glycerin from the wash tank is drained; biodiesel tank in which the dry biodiesel is stored; and water tank used in washing the biodiesel.

Keywords Biodiesel • Transesterification • Vegetable oil • Wash tank • Hydrocarbons • Alcohols • Catalyst

A.O. Openibo (✉)
Department of Vehicle Inspection Service, Lagos State Ministry of Transportation, Logos, Nigeria
e-mail: adeshola.openibo@yahoo.com

N.A. Raji
Department of Mechanical Engineering, Faculty of Engineering, Lagos State University, Logos, Nigeria

S.K. Ghosh (ed.), *Utilization and Management of Bioresources*,
DOI 10.1007/978-981-10-5349-8_15

1 Introduction

Gasoline contributes to increase hazardous emissions (Abdulkareem et al. 2012), and according to Alamu (2007), diesel with higher carbon numbers contribute to emissions of high particulate matters such as high sulfur dioxide and high polyaromatic hydrocarbons.

Also, the growing energy needs and irregular supply, unstable cost and increasing environmental concern, and protecting the environment and planning for long-term energy security become necessary (ECN 2008). This prompted investigation into alternative fuels with comparative properties to nonrenewable and environmentally unfriendly fossil fuel, and biodiesel-based fuels became a considerable choice because of their renewable, nontoxic, and safe to store characteristics. Biodiesel is one of such renewable alternative fuels that could be domestically produced from new or used vegetable oil and animal fat through a chemical reaction with methanol or ethanol. This reaction requires heat and a strong catalyst (alkalis, acids, or enzymes) to achieve complete conversion of the vegetable oil into the separated esters and glycerin. During the transesterification reaction, glycerin is obtained as a by-product which could be used in the pharmaceutical and cosmetic industries. Not only can biodiesel be used alone in neat form, but it can also be mixed with petroleum diesel fuel in any unmodified diesel engine (Nakpong and Wootthikanokkhan 2010).

It is a known fact that biodiesel is synthesized from a variety of feedstocks, such as soybean, rapeseed, palm, and others which are food grade raw materials. The competition in using these raw materials as both food and energy crops has become a great obstacle in biodiesel production on a commercial scale.

Also, waste cooking oil (WCO) is considered an economic and increasingly available resource for biodiesel production (Jain and Sharma 2010; Kee et al. 2010; Han et al. 2009).

The use of biodiesel as an alternative fuel has been globally accepted, and several research works have been done to first-generation biodiesel production where homogenous catalysts such as NAOH, KOH, etc. are used.

2 Aim and Objective

The broad aim of this research work is to develop a multi-tank biodiesel processor in Nigeria and carry out performance evaluation tests using waste vegetable oil (WVO). Many researchers had worked on one-, two-, and three-flask processor in the past in Nigeria, but this research work is basically on a one-stage nine-flask processor as listed below:

1. Methoxide tank to mix the methanol and lye together before it is injected into the reactor

2. The reactor where the transesterification of the mixtures (vegetable oil + alcohol + catalyst) occurs
3. The settling tower where freshly reacted batch of biodiesel is transferred to free up the reactor for yet another batch
4. First (1st) wash tank where the product is washed with water to decant fuel from soap and glycerin (i.e., wet washing)
5. Second (2nd) wash tank to finally free the fuel from all impurities by passing it through wood shaving (i.e., dry washing)
6. Dry tank to free up the wash tanks for the next batch and also dry the fuel much faster with much better results through heating
7. Glycerin tank where the glycerin from the wash tank is drained
8. Biodiesel tank where the dry biodiesel is stored
9. Water tank used in washing the biodiesel

3 Previous Research Works on Biodiesel Processors

Many researchers have delved into the production of biodiesel and its processor worldwide. In developed nations, many manufacturing companies and home brewers have come up with various designs for biodiesel processor, and it is currently on sale in developed nations. Various types of biodiesel processors ranging from small-scale sizes to the large-scale plants have been built by different researchers in both developed and developing nations, but Nigeria as a nation does not have encouraging records in terms of development of processors and biodiesel production.

- Mayvan et al. (2011) – designed and fabricated batch-type stirred tank reactor (STR) with a capacity of 70 l in which two methods of mechanical and hydraulic mixing were introduced and electrostatic coalescing method in separation of glycerin and dry wash by absorbing column and Magnesol powder as filter aid.
- A Florida-based biodiesel processor manufacturing company partnered with a Nigerian company "Avandith Energy" in 2014 to get a B-60 biodiesel processor for their pilot transesterification facility. The B-60 plant will produce four batches of biodiesel in 24 h and will be used as hands-on educational tool to show students and government agencies how to make renewable energy (Fig. 1).
  - Also, Daniyan et al. (2013) designed a small-scale biodiesel processor capable of transesterifying 20 l of biodiesel per batch as represented in Fig. 2
  - Ramesh et al. (2013) presented a biodiesel plant of 250 l/day capacity at the Tamil Nadu Agricultural University, India (Fig. 3)
  - Tint and Mya (2009) also produced a 120 l biodiesel pilot plant from Jatropha oil (Jatropha curcas); warm water was used for washing and sand was used as filter

This has really shown that everyone knows the advantages of the eco-friendly properties of biodiesel and the need for a processor.

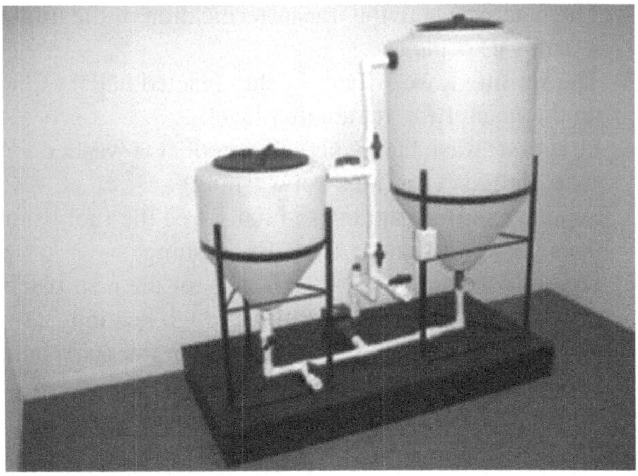

Fig. 1 B-60 biodiesel kit

The research work on focus will see to the design and development of a processor with locally sourced materials to make available a 100-l Nigerian-made biodiesel processor that would be locally produced in Nigeria. It would save cost a great deal when compared with the foreign exchange between Nigeria and other nations on imported ones.

4 Materials and Methods

The design of this biodiesel processor was given utmost considerations on two subheads:

(i) The process
(ii) The processor

5 The Process

This comprises the reaction of methoxide, waste vegetable oil (WVO), and alkali such as sodium or potassium hydroxide under a specific temperature, but in this study, sodium hydroxide is considered, and the reaction would be aided by continuous stirring of the aforementioned in a mechanical reactor. At a given temperature, glycerol tap would be opened to give way for glycerin which goes into its own separate tank. Time is allowed for the reaction to complete, and then the biodiesel is transferred to a settling tank to give way for a new batch. Thereafter, from the settling tank, the product flows into a wet wash tank where it is washed with the introduction of hot water from the water tank. The mixture is stirred, biodiesel floats, and water goes down and drained out through the drain tap. This water is allowed to

Fig. 2 20 l of biodiesel processor by Daniyan et al.

Fig. 3 Biodiesel pilot plant (College & Research Institute Tamil Nadu Agricultural University, India)

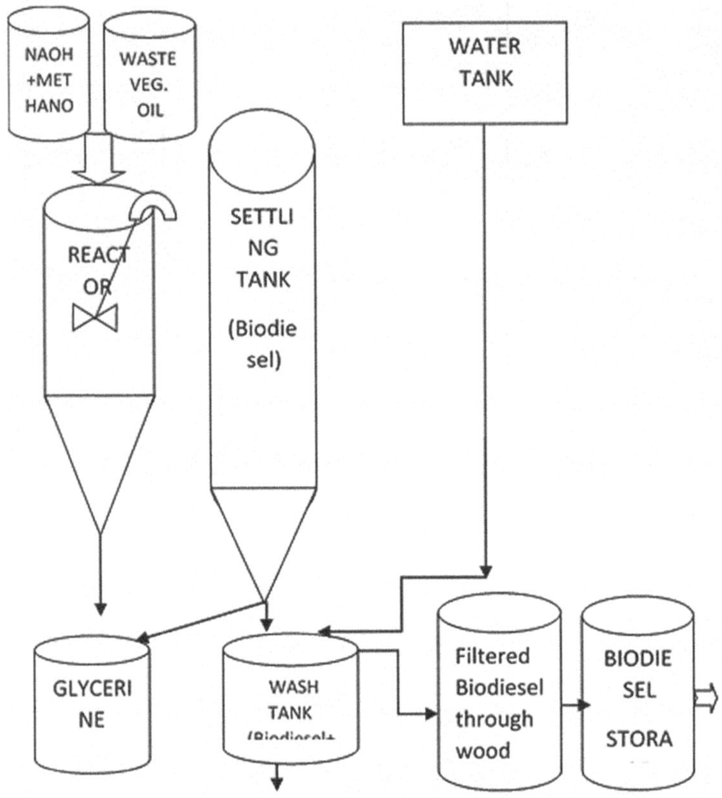

Fig. 4 Schematic drawing

settle and treated with acid to neutralize its alkalinity before it could be disposed in order not to endanger plant and animal life. The biodiesel on the other hand is passed into a dry wash tower through wood shaving to further remove any left impurities. Thereafter, it goes into the final storage tank, ready for dispensing to the final consumer.

As illustrated in Fig. 4, the process comprises the use of dangerous chemicals such as NAOH, methanol, and acids, and as such metallic substances such as copper, zinc, lead, etc. that may catalyze or react with these chemicals were avoided in the design. Timers and thermostat on heating elements shall be installed in order to prevent overheating of oils. Generally, there are two available methods for washing biodiesel:

(i) Wet washing
(ii) Dry washing

Both methods shall be adopted to get a very clean biodiesel. Wet washing employs the use of water to separate glycerol from the final product which shall then be passed through wood shaving in the dry wash tank to remove every other impurity left in the product. The process is as illustrated in Figs. 4, 5, and 6.

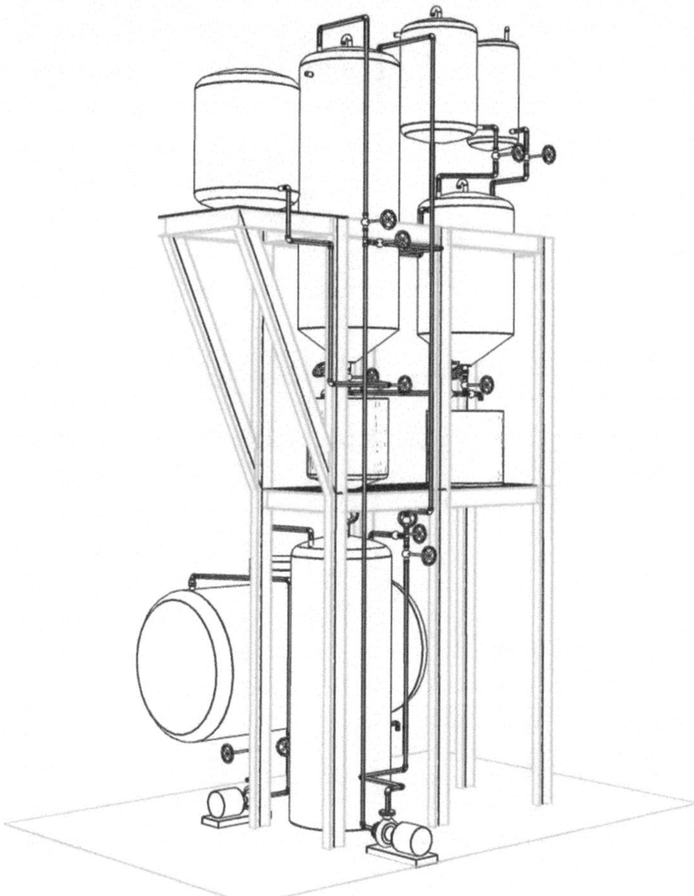

Fig. 5 Framed multiprocessor

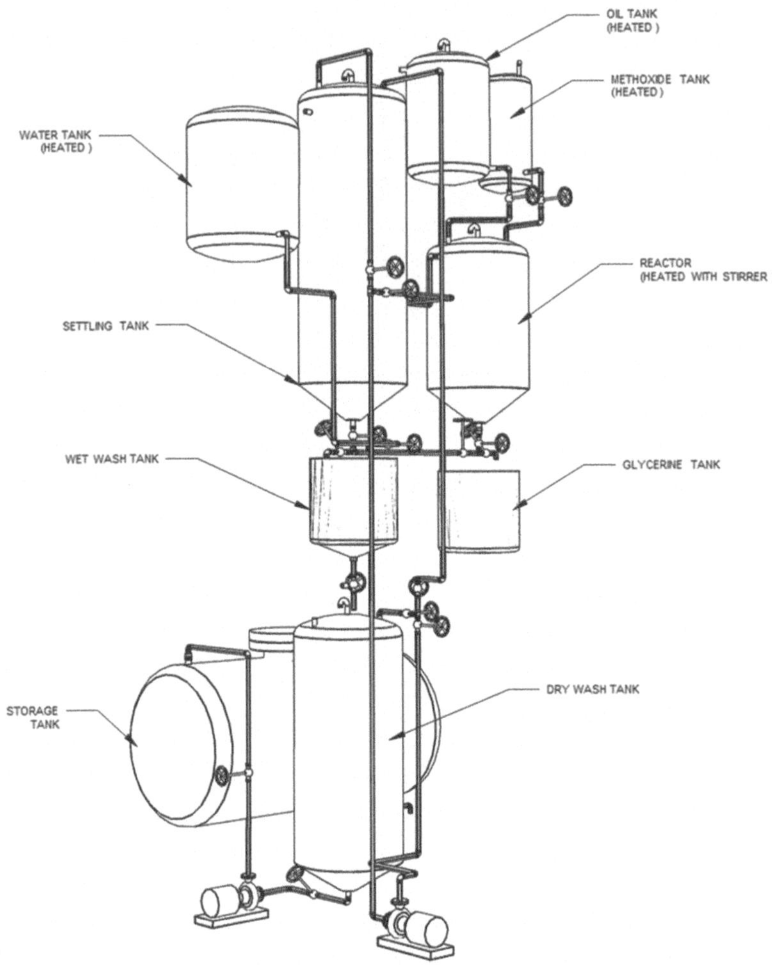

Fig. 6 Detailed drawing

6 Conclusions and Recommendations

Developing a biodiesel processor will contribute to reduction of greenhouse gases in the environment. So far, no research work on 100 l capacity in a nine-drum/flask plant using a combined wet and dry wash process has been recorded in Nigeria. Moreover, it is going to be a bridge between the first- and second-generation biodiesel productions, i.e., using nonedible feedstock/waste oil with homogeneous catalyst to produce a standard biodiesel fuel. Future consideration of designing a processor having two or more methoxide and settling tanks in order to prepare for two to three batches so as to save time and increase quantity produced (Fig. 7).

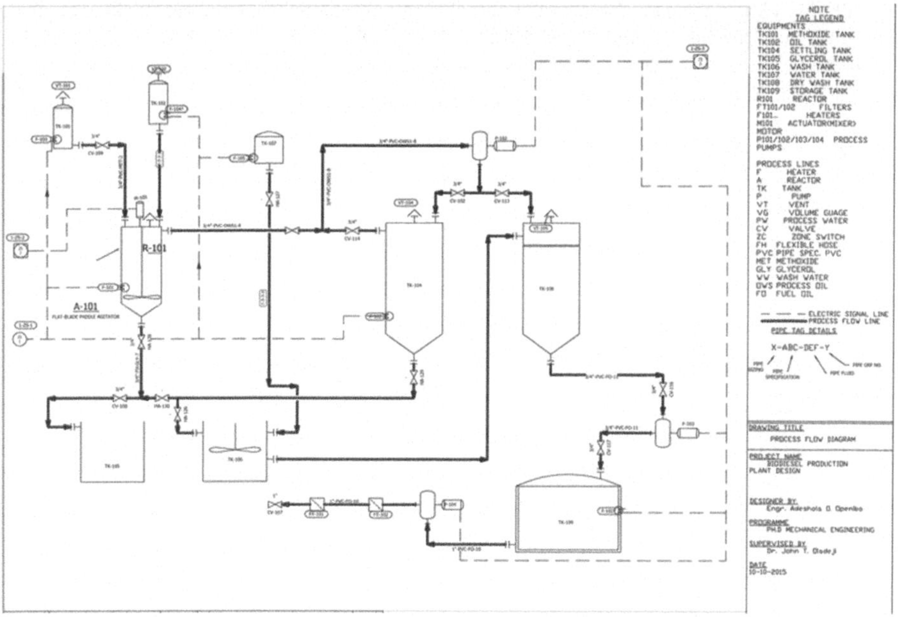

Fig. 7 Flow diagram of the process

References

Abdulkareem AS, Jimoh A, Afolabi AS, Odigure JO, Odili UC (2012) Production and characterization of biofuel from refined groundnut oil. http://dx.doi.org/10.5772/52443

Alamu OJ (2007) Effect of ethanol-palm kernel oil ratio on Alkali-catalyzed biodiesel yield. Pac J Sci Technol 8(2):212–219

Daniyan IA, Adeodu AO, Dada OM, Aribidara AA (2013) Design of a small scale biodiesel processor. J Emerg Trends Eng Appl Sci (JETEAS) 4(4):576–580. © Scholarlink Research Institute Journals, 2013 (ISSN: 2141-7016) jeteas.scholarlinkresearch.org

Energy Commission of Nigeria (2008) National Energy Policy. http://www.energy.gov.ng

Han M, Yi W, Wu Q, Liu Y, Hong Y, Wang D (2009) Preparation of biodiesel from waste. Bioresour Technol 100:2308–2310

Jain S, Sharma MP (2010) Biodiesel production from Jatropha curcas oil. Renew Sustain Energy 14:3140–3157

Kee LM, Teong LK, Rahman MA (2010) Homogeneous, heterogeneous and enzymatic catalysis for transesterification of high free fatty acid oil (waste cooking oil) to biodiesel: a review. Biotechnol Adv 28:500–518

Mayvan AA, Ghobadian B, Najafi G, Omidkhah MR (2011) Design, fabrication and evaluation of a novel biodiesel processor system. SET2011. In: 10th international conference on Sustainable Energy Technologies, İstanbul, 4–7 September 2011

Nakpong P, Wootthikanokkhan S (2010) Roselle (hibiscus sabdriffa L.) oil as an alternative feedstock for biodiesel production in Thailand. Fuel 89:1806–1811

Ramesh D, Samapathrajan A, Venkatachalam P (2013) Production of biodiesel from Jatropha curcas oil by using Pilot biodiesel plant

Tint TK, Mya MO (2009) Production of biodiesel from Jatropha oil (Jatropha curcas) in pilot plant. World Acad Sci Eng Technol 50:477–483

Part IV
Applied Biotechnology and Bioenergy Systems

Microwave-Assisted Transesterification of Waste Cooking Oil for Biodiesel Production

S. Babel, S. Arayawate, E. Faedsura, and H. Sudrajat

Abstract Waste cooking oil (WCO) is considered the most promising biodiesel feedstock despite its drawback of high free fatty acid (FFA). In this study, microwave-assisted transesterification of WCO is carried out in the presence of potassium hydroxide supported on carbonized coconut shell (KOH/CS) catalyst. The effect of reaction temperature on the yield of fatty acid methyl *esters* (*FAME*) is studied. Conventional transesterification is also performed for comparison. The results show that reaction temperature of 80 °C is optimum for FAME production. The properties of the produced biodiesel satisfy the criteria according to ASTM D6751. Acid-catalysed esterification of WCO before transesterification leads to higher production of FAME due to reduction of FFA. At 80 °C, reaction time of 40 min, alcohol to oil ratio of 12:1 and 5 wt.% catalyst, a FAME production of 91.3% is achieved. At 65 °C, reaction time of 90 min, alcohol to oil ratio of 12:1 and 5 wt.% catalyst, conventional transesterification results in slightly higher FAME production (92.1%). However, conventional transesterification using WCO without esterification as feedstock yields lower FAME. Pretreatment of feedstock with esterification is found to be more beneficial and applicable in the case of biodiesel production through conventional transesterification, allowing production of FAME with higher yields. This furthermore indicates that although microwave heating decreases the reaction time, it does not necessarily lead to increased FAME. The production of FAME depends on not only the type of transesterification process but also the type of feedstock used. Overall, the proposed methodology allows the use of high FFA content feedstock. However, this will need careful selection of feedstock as well as additional treatment prior to transesterification process; otherwise the yield may be reduced.

Keywords Biodiesel • Transesterification • Fame • FFA • Waste cooking oil

---

S. Babel (✉) • S. Arayawate • E. Faedsura • H. Sudrajat
School of Biochemical Engineering and Technology, Sirindhorn International Institute of Technology, Thammasat University, Krung Thep Maha Nakhon, Pathum Thani, Thailand
e-mail: sandhya@siit.ac.th

S.K. Ghosh (ed.), *Utilization and Management of Bioresources*,
DOI 10.1007/978-981-10-5349-8_16

1 Introduction

For over decades, many different ways have been developed to gain as much energy without using the limited resources. Every year millions of dollars have spent on green energy research to find a new resource of energy. Biodiesel is one of the resources which will eventually take place over petrol diesel. Biodiesel is an animal- or vegetable oil-based diesel fuel (Oliveira and Da Silva 2013). The animal or vegetable oil is converted into biodiesel when a mole of triglyceride reacts with three moles of alcohol to produce a mole of glycerol and three moles of mono-alkyl esters. Since biodiesel is produced from animal and vegetable oils which are renewable biomass sources, it has lower emission of pollutant. The major drawback of biodiesel production is the cost of production. Since new vegetable oils and expensive catalysts are typically used, the cost of production increases tremendously. In this context, the use of waste cooking oil (WCO) instead of virgin oil to produce biodiesel is an effective way to reduce the raw material cost.

However, it has been reported that WCO contains large amount of free fatty acids (FFAs) along with moisture which are required to be removed to prevent soap formation as FFA of WCO are sensitive to alkali catalyst (Canakci and Van Gerpen 2001). The high FFA content (>1% w/w) causes soap formation and low yield of biodiesel (Ghadge and Raheman 2005). When dealing with feedstocks containing high FFAs like WCO, alkali catalysts cannot directly catalyse the transesterification of oils containing high FFA levels (Atadashi et al. 2012). To deal with this drawback, a two-step transesterification process appears to be a promising route for transesterification of high FFA oils. In this technique, the first step is an acid-catalysed process which involves esterification of the FFAs to FAMEs, followed by a second step, alkali-catalysed transesterification. For the second step, there are several heating systems employed. An alternative heating system, that is, microwave irradiation, was reported to be superior to conventional system (Hsiao et al. 2011). By using microwave irradiation, the reaction time can be shorter.

The purpose of this research is to obtain biodiesel from WCO through transesterification with microwave heating. Conventional transesterification was used for comparison. A two-step process involving esterification of WCO with H_2SO_4 as catalyst in the first step was also applied. The effect of reaction temperature during microwave heating on the FAME production was investigated. KOH supported on carbonized coconut shell carbon (KOH/CS) was employed as heterogeneous alkali catalyst. Use of heterogeneous alkali catalyst in this study is intended to alleviate the problems associated with homogeneous catalysis. KOH/CS could be recycled and reused several times with better separation of the final product, minimizing material and processing cost. Apart from these advantages, the activity of a heterogeneous alkali catalyst was also found to resemble a homogeneous counterpart at the same operating condition (Kim et al. 2004).

2 Materials and Methods

2.1 *Materials*

New oil and waste cooking oil used in the study were collected from Useful Food Co., Ltd., Bangkok, Thailand. Density, viscosity, FFA and fatty acid methyl *esters (FAME)* of NCO and WCO were determined according to standard methods (Helrich 1990; Azcan and Yilmaz 2013).

2.2 *Catalyst Preparation*

The heterogeneous base catalysts selected in the study are potassium hydroxide (KOH) on the carbonized coconut shell. For carbonization of coconut shell, the coconut shell was first cleaned with water and dried at the room temperature for 24 h. It was then crushed with blender, sieved to 600–1000 μm and carbonized in the furnace at 700 °C for 2 h. The obtained black charcoal, having BET surface area of 0.8 m^2/g and mean pore size of 15.7 nm, was furthermore subjected to impregnation with KOH.

For impregnation with KOH, 10 g of the obtained CS was cleaned with deionized water, filtered and dried in an oven at 105 °C for 12 h. Subsequently, 37.5 g of KOH was dissolved in 150 ml of deionized water and the CS was impregnated in this KOH solution at room temperature under agitation in shaker at 180 rpm for 24 h. The obtained product was filtered, dried in an oven at 150 °C for 24 h and kept in a close vessel to avoid the reaction with humidity. The amount of KOH impregnated in CS was determined by back titration method (Babić et al. 1999) and found to be 31.2 g in 10 g of CS.

2.3 *Esterification*

The homogeneous acid catalyst chosen in this study was sulphuric acid (H_2SO_4). H_2SO_4 is the most common acid catalyst used due to its good catalytic activity and simplicity in H_2SO_4/methanol preparation as concentrated H_2SO_4 can be added directly to methanol. It was used in the feed pretreatment process (esterification) to reduce the amount of FFA content of WCO and to reduce the chance of foam formation during subsequent transesterification process. For acid-catalysed esterification, 120 g of oil was first poured into 250-ml three-neck round-bottom flask and heated to reach the desired temperature of 65 °C. Then, 27.2 g of methanol and 0.6 g of H_2SO_4 were poured and mixed in the flask under continuous stirring. The reaction was kept at 65 °C for 2 h and allowed to cool down to room temperature. The mixture was poured into a separating funnel. The ester layer located in the upper layer was separated by gravity. The glycerol, extra methanol and undesired products were in the lower layer and were decanted. The ester layer was washed several times with a small amount of hot water until the washings were neutral. The ester layer was then dried over sodium sulphate, filtered and kept in the refrigerator for transesterification.

2.4 Microwave-Assisted Transesterification

The microwave-assisted transesterification was performed in a microwave (MARS 6, 240/50). First, 10 g of oil, 0.5 g of catalysts and 4.35 g of methanol were poured into the vessel and put in a microwave for 40 min with a ramping time of 10 min. The operating temperature was varied (60, 70, 80 and 90 °C). After 40 min, the mixture was filtered and poured into a separating funnel. The ester layer was separated by gravity and located in the upper layer. The glycerol, extra methanol and undesired products were in the lower layer and were decanted. The ester layer was washed several times with a small amount of hot water each until the washings were neutral. The ester layer was then dried over sodium sulphate, filtered and kept in the refrigerator to reduce the chance of oil decomposition for further analysis of ester amount by gas chromatography (Perkin-Elmer 850) equipped with flame ionization detector and HP-INNOWax column. Standard method for the analysis was based on EN14103 with methyl heptadecanoate (C17) as internal standard and heptane (C_7H_{16}) as solvent. The %FAME was calculated from:

$$\%\text{FAME} = \frac{\sum A - A_{\text{EI}}}{A_{\text{EI}}} \times \frac{C_{\text{EI}} \times V_{\text{EI}}}{m} \times 100 \quad (1)$$

Where,

$\sum A$ = total peak area of FAME (C14–C24)
A_{EI} = peak area of standard (C17)
C_{EI} = concentration of standard (C17, mg/mL)
V_{EI} = volume of standard solution (mL)
m = mass of product (mg)

2.5 Conventional Transesterification

The conventional transesterification process is an open system performed in the glass reactor with the condenser to reduce the loss of methanol. The transesterification was carried out in a 250-ml three-neck round-bottom flask. The reactor was placed in a heated oil bath. 1.5 g of catalyst was put into the flask containing 30 g of oil. The flask was heated to 65 °C under continuous stirring, and 13.1 g of methanol was then poured into the flask. The reaction was kept at a desired temperature for 2 h and allowed to cool down to room temperature. The mixture was filtered and subjected to separation process which was the same with that for microwave-assisted transesterification.

3 Results and Discussion

3.1 Oil Properties

Table 1 summarizes the properties of NCO and WCO. As can be seen, the density of NCO and WCO is not that different. However, WCO has slightly higher density due to the small particles from food processing and some part of food contaminants. Figure 1 shows the appearance of NCO and WCO. The colour of WCO is darker than that of NCO. Comparing the viscosity of NCO and WCO, WCO has a higher value. The higher viscosity of WCO is due to the same reason for higher density, that is, the presence of food contaminants. This higher viscosity of WCO may also

Table 1 Properties of NCO and WCO

Property	NCO	WCO
Density (g/cm^3)	0.90172	0.90574
Viscosity (mm^2/s)	75.5575	78.6338
Acid number (mg KOH/g oil)	0.5589	1.3602
FFA (%)	0.2794	0.6801
FAME (%)	1.013474	1.305325

Fig. 1 Appearance of NCO (**a**) and WCO (**b**)

Table 2 Properties of TWCO

Property	TWCO
Density (g/cm^3)	0.90338
Viscosity (mm^2/s)	70.2314
Acid number (mg KOH/g oil)	0.8058
FFA (%)	0.40293
FAME (%)	4.675523

be due to the reaction occurring during food processing such as oxidation and polymerization which can increase viscosity of the oil. Although WCO has a higher FAME, its FFA is also higher than that of NCO. Thus, the amount of FFA needs to be reduced through esterification with homogeneous acid catalyst, which was done in this study.

3.2 *Esterification of WCO*

Table 2 shows the properties of WCO after being esterified with H_2SO_4. As shown, after esterification, density, viscosity and FFA decrease while FAME increases. These changes are beneficial for subsequent transesterification process. In particular, the FFA content in treated waste cooking oil (TWCO) is smaller than those in NCO and WCO. This indicates that FFA can be changed to methyl ester through the esterification with H_2SO_4. Based on this finding, acid-catalysed esterification is proven to be suitable for feedstocks with high FFAs like WCO, which are low-grade oil.

3.3 *Transesterification*

In this research, two transesterification processes, microwave-assisted transesterification and conventional transesterification, were performed and compared. NCO, WCO and TWCO were used as feedstocks for transesterification. For transesterification with microwave heating, the reaction temperature was varied. In fact, transesterification can occur at different temperatures, depending on the properties of oils. It could be at ambient temperature or at a temperature close to the boiling temperature of methanol. However, high reaction temperature typically speeds up the reaction and shortens the reaction time.

Figure 2a shows the effect of temperature on the FAME production from NCO at the methanol to oil ratio of 12:1 in the presence of KOH/CS with microwave-assisted process. When increasing the temperatures from 60 °C to 80 °C, the FAME

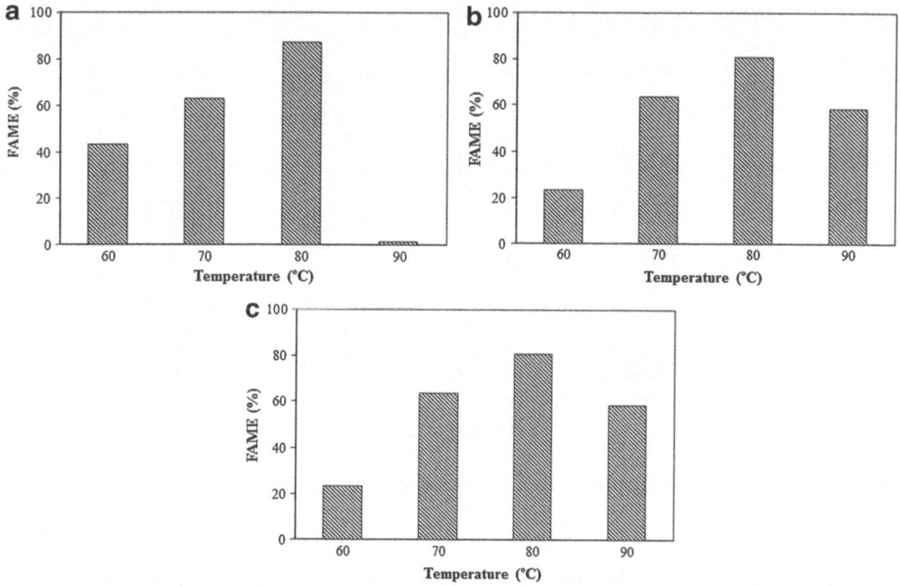

Fig. 2 Effect of reaction temperature on the FAME production with microwave-assisted transesterification of NCO (**a**) and WCO (**b**)

production increases more than 44%. However, when the temperature increases to 90 °C, there is a reduction of the FAME production. This may be due to three factors. First, high temperature enhances not only transesterification but also saponification which is detrimental to FAME formation. Second, at the beginning of reaction, the FAME production increases because reactant contacts fresh catalyst. Then, it decreases due to slightly deactivated but stabilized catalyst. Third, the temperature of 90 °C is quite far from the boiling point of methanol (65 °C). Hence, chances of methanol loss are high.

Moreover, the trend of FAME production using WCO is also the same with that using NCO (Fig. 2b). At 60 °C, the least conversion (23.21%) is observed. At 80 °C, the reaction conversion is highest (80.89%). When increasing the temperature to 90 °C, a decrease of FAME production is observed. In all the cases for three different oils, the optimum reaction temperature is found to be 80 °C.

Comparing the oil used in the microwave-assisted transesterification with KOH/CS, use of TWCO enables higher yields of FAME in all the temperatures applied (Fig. 3). This indicates that the two-step process involving acid-catalysed esterification in the first step can be a better option for producing biodiesel from feedstocks with high FFAs like WCO. The FFAs in the feedstock can be reduced first through esterification before the feedstock is subjected to alkali-catalysed transesterification.

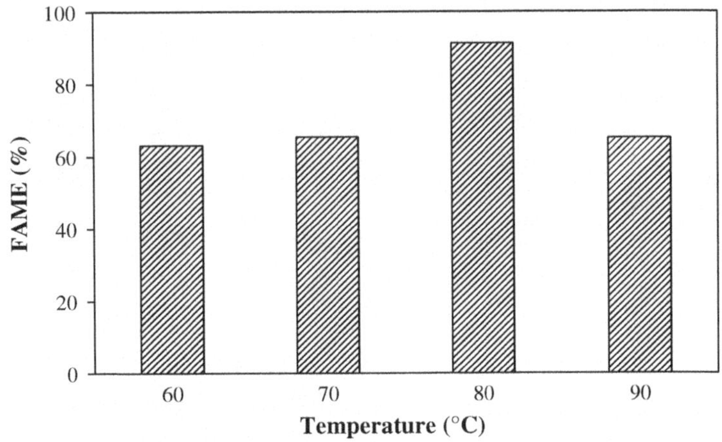

Fig. 3 Effect of reaction temperature on the FAME production with microwave-assisted transesterification of TWCO

For comparison, conventional transesterification was also performed. The results show that NCO is a better feedstock compare to WCO since it leads to higher yields of FAME (Fig. 4). This is obviously because NCO contains a lower FFA (Table 1). Moreover, TWCO is found to be the best feedstock due to the same reason. It contains the lowest FFA, allowing effective production of FAME. These results are similar to those of microwave-assisted transesterification. Reduction of FFA through acid-catalysed esterification is proven to result in higher yield of FAME. Furthermore, using TWCO as feedstock, conventional transesterification yields slightly higher FAME but with longer reaction time. Long reaction time is actually a main drawback in conventional process as compared to microwave-assisted one.

3.4 Biodiesel Properties

The properties of biodiesel obtained from the two-step process with acid-catalysed esterification in the first step and alkali-catalysed conventional transesterification in the second step are summarized in Table 3. It can be seen that most of its properties meet the standard criteria of biodiesel according to ASTM D6751. The density of the biodiesel obtained is less than those of NCO, WCO and TWCO because of the change in chemical structure from triglyceride to methyl ester. The molecules pack tightly and require less volume. The density of biodiesel is also within the standard. Furthermore, comparing the viscosity of produced biodiesel with those of the raw material, the viscosity of the biodiesel is extremely lower. This low viscosity is a good property for fuel. The viscosity is very close to the standard. Regarding the

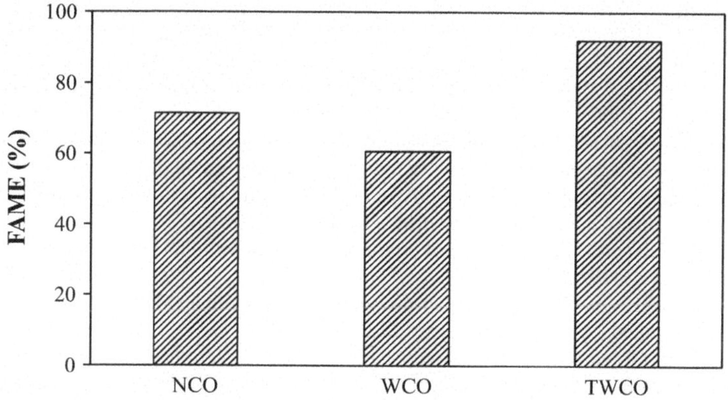

Fig. 4 FAME production with conventional transesterification

Table 3 Comparison of produced biodiesel with ASTM D6751 standard

Property	Biodiesel from this work	ASTM D6751
Density, @ 15.5 °C (g/cm^3)	0.85956	0.87473
Kinematic Viscosity, 40 °C (mm^2/s)	6.6009	1.9–6.0
Acid Number (mg KOH/g oil)	0.2171	0.50 max

acid number, it is in the range of standard. Actually, if it is required, the acid number can still be decreased by neutralization reaction or washing the biodiesel.

4 Conclusion

Transesterification reaction of WCO in the presence of KOH/CS catalyst is successfully carried out through microwave heating. Reaction temperature plays a critical role in the FAME production. The highest FAME production (91.3%) is obtained using 5 wt.% catalyst, alcohol to oil ratio of 12:1 at 80 °C and 40-min reaction time. On the basis of this result, microwave heating is shown to be a promising technique for biodiesel production in a short reaction time with appreciable FAME yields. Moreover, the performance of conventional transesterification is also comparable to that of microwave-assisted one. However, pretreatment with acid-catalysed esterification must be done before transesterification process to reduce the FFA content in the feedstock. Such pretreatment is appropriate for feedstocks with high content of FFAs. These results are encouraging for the future analysis of transesterification assisted by microwaves using low-quality oils like WCO.

References

Atadashi I, Aroua M, Aziz AA, Sulaiman N (2012) Production of biodiesel using high free fatty acid feedstocks. Renew Sust Energ Rev 16:3275–3285

Azcan N, Yilmaz O (2013) Microwave assisted transesterification of waste frying oil and concentrate methyl ester content of biodiesel by molecular distillation. Fuel 104:614–619

Babić B, Milonjić S, Polovina M, Kaludierović B (1999) Point of zero charge and intrinsic equilibrium constants of activated carbon cloth. Carbon 37:477–481

Canakci M, Van Gerpen J (2001) Biodiesel production from oils and fats with high free fatty acids. Trans ASAE 44:1429

Ghadge SV, Raheman H (2005) Biodiesel production from mahua (Madhuca indica) oil having high free fatty acids. Biomass Bioenergy 28:601–605

Helrich K (1990) Official methods of Analysis of the AOAC. Volume 2, Association of Official Analytical Chemists Inc

Hsiao M-C, Lin C-C, Chang Y-H (2011) Microwave irradiation-assisted transesterification of soybean oil to biodiesel catalyzed by nanopowder calcium oxide. Fuel 90:1963–1967

Kim H-J, Kang B-S, Kim M-J, Park YM, Kim D-K, Lee J-S, Lee K-Y (2004) Transesterification of vegetable oil to biodiesel using heterogeneous base catalyst. Catal Today 93:315–320

Oliveira L, Da Silva M (2013) Comparative study of calorific value of rapeseed, soybean, jatropha curcas and crambe biodiesel, in: International Conference on Renewable Energies and Power Quality (ICREPQ'13) Bilbao (Spain), 20–22

Isolation, Characterisation of Novel *Pseudomonas* and *Enterobacter* sp. from Contaminated Soil of Chandigarh for Naphthalene Degradation

Bulbul Gupta, Kunal, Anita Rajor, and Jaspreet Kaur

Abstract Naphthalene an organic pollutant arises from industrial operations and natural events such as forest fires. According to US EPA list of priority pollutants, naphthalene is considered as possible carcinogen due to its low solubility and bioavailability. Degradation of this recalcitrant can be carried out using physical and chemical methods but it leads to various products, most of them are toxic to the environment. Therefore, bioremediation using selected microorganisms remains the most suitable solution to treat such pollutants.

In the present study, bacterial strains utilising naphthalene as sole carbon and energy source were isolated from crude oil-contaminated soil. Out of the eight isolates initially screened, two bacterial isolates S3 and F3 were selected on the basis of their best growth in M9 minimal medium containing naphthalene, as the sole source of carbon. Isolate S3 and F3 were characterised through biochemical, physiological and phylogenetic analysis (16S rRNA) that revealed their significant similarity to *Pseudomonas aeruginosa* sp. and *Enterobacter cloacae* sp. The sequence was submitted to NCBI database and the assigned accession numbers were JX254648.1 and JX480546. Isolates S3 and F3 were then utilised for the degradation of naphthalene in oil-contaminated soil, and decrease in 77.77 % and 61.11 % of concentration, respectively, after 7 days of incubation, was observed as determined by HPLC.

Keywords Aromatic hydrocarbons • Biodegradation • *Enterobacter* • Naphthalene • *Pseudomonas*

B. Gupta • J. Kaur (✉)
Biotechnology Branch, University Institute of Engineering and Technology,
Panjab University, Chandigarh, India
e-mail: jaspreet_virdi@yahoo.com

Kunal • A. Rajor
School of Energy and Environment, Thapar University, Patiala, Punjab, India

S.K. Ghosh (ed.), *Utilization and Management of Bioresources*,
DOI 10.1007/978-981-10-5349-8_17

1 Introduction

Petroleum hydrocarbons are common contaminants in soil, air, freshwater and various other habitats, often occurring as a result of oil spillage, industrial and domestic wastewater discharge, incomplete combustion of fossil fuels and organic matter, urban runoff, vehicular exhaust etc. (Tecon et al. 2010; Naidu et al. 2012). Continuous release of large amount of mono-aromatic hydrocarbons such as benzene, toluene, ethyl benzene and xylene (BTEX) as well as polycyclic aromatic hydrocarbons (PAH) such as naphthalene in the ecosystems has resulted in considerable increase in concentration of contaminants (Meckenstock et al. 2012; Andreoni and Gianfreda 2007). Owing to slow dissolution and limited attenuation potentials, these contaminants are expected to remain in the systems for hundreds and thousands of years. A number of physiochemical processes including adsorption, incineration and absorption can be used to treat PAHs, but the costs for chemicals and fuels as well as further treatment or disposal of secondary wastes are increasingly inhibiting adoption of these solutions. The currently accepted disposal methods of incineration or burial in secure landfills (USEPA 1993; International Tanker Owners Pollution Federation Limited (ITOPF) 2006) can become prohibitively expensive when the amounts of contaminants are large. This often results in cleanup delays while the contaminated soil continues to pollute groundwater resources (Pye and Patrick 1983), necessitating speedy removal of the contaminants. Hence, there is a need to develop a new in situ biodegradation approach for cleaning up of these contaminants instead of using the conventional remediation strategies which require investments and that face technical difficulties (Bombach et al. 2010; Steliga et al. 2012; Mazzeo et al. 2012). Biodegradation can achieve complete and cost-effective elimination of aromatic pollutants through harnessing diverse microbial metabolic processes (You et al. 2013; Li et al. 2012). Nonavailability of rapid and reliable on-site tools for measuring the bioavailable hydrocarbon fractions further enhances the role of biodegradation in environmental cleanup. A very few studies have been reported for biodegradation of aromatic hydrocarbons (Jin et al. 2013; Djokic et al. 2011; Ntougias et al. 2015). Thus, biodegradation of monocyclic, polycyclic and chlorinated aromatic hydrocarbons should be carried out for complete removal of the contaminants from the environment (Zhou et al. 2016).

Naphthalene, a bicyclic polyaromatic hydrocarbon, is in the list of simplest PAHs defined by the International Union of Pure and Applied Chemistry (IUPAC) (G.P Moss, IUPAC nomenclature for fused ring systems). The US Environmental Protection Agency (EPA) has designated 32 PAH compounds as priority pollutants which also include naphthalene. It serves as a model for understanding the properties of higher class of environmentally prevalent petroleum hydrocarbons. Further, naphthalene being the most soluble polyaromatic hydrocarbon and the genes responsible for degradation are plasmid-encoded, making it the most suitable compound for studying microbial metabolisms. All bioremediation projects require the use of specific microorganisms under optimal conditions that facilitate the biodegradation of the contaminants. An evaluation of indigenous microorganisms is also important to correlate the bacterial community with their biodegradation ability towards a targeted pollutant.

In the present study, microbial population isolated from fuel filling stations of Chandigarh was acclimatised in the laboratory and checked for their growth in the presence of naphthalene as a sole carbon source. The isolates showing best growth at higher concentration of naphthalene were selected and characterised through morphology and biochemical testing and at genomic level. The biodegradation potential of the isolates in oil-contaminated soil sample was also analysed under optimum conditions.

2 Materials and Methods

2.1 Collection of Soil Samples

For the isolation of naphthalene-degrading bacteria, soil samples were collected from crude oil-polluted soil near fuel filling stations in Chandigarh.

2.2 Screening of Naphthalene-Degrading Bacteria

One gram of contaminated soil samples was dispensed in 50 ml of nutrient broth (NB) and incubated for 2 days at 37 °C on rotary shaker at 150 rpm. After growth, serial dilutions were made and 50 μl of the aliquot was surface-spread on nutrient agar plates and incubated at 37 °C for 72 h. On the basis of morphological differences, the bacterial colonies were picked and streaked on M9 medium agar plates containing naphthalene (100 ppm). Out of these, the best grown isolates were selected and were maintained in M9 medium containing K_2HPO_4 2.5 g l^{-1}, KH_2PO_4 2.5 g l^{-1}, $(NH_4)_2HPO_4$ 1.0 g l^{-1}, $MgSO_4.7H_2O$ 0.20 g l^{-1}, $FeSO_4.7H_2O$ 0.01 g l^{-1}, $MnSO_4.7H_2O$ 0.007 g l^{-1} and naphthalene (100 mg l^{-1}), as the sole source of carbon at pH 6.95.

2.3 Identification of Selected Bacterial Isolates

2.3.1 Morphological and Biochemical Characterisation

Selected isolates were characterised by their morphology on nutrient agar. Gram staining and morphological characterisation were done according to Cappuccino and Sherman (Cappuccino and Sherman 2010). Additional biochemical tests were performed for taxonomic characterisation which included gelatin liquefaction, lipase screening, casein hydrolysis, H_2S production, lipase test, xylanase test, pectinase screening, methyl red Voges-Proskauer (MR-VP), citrate utilisation, carbohydrate utilisation and amino acid utilisation (Aneja 2008).

2.3.2 Molecular Characterisation Based on 16S rRNA Gene

For 16S rRNA gene sequencing, DNA was isolated from bacterial strain by using QIAamp DNA purification kit (Qiagen). The 16S rRNA gene was amplified by using PCR method with universal bacterial primers 8F (5′-AGAGTTTGA TCMTGGCTCAG) and 1492R (5′-GGYTACCTTGTTACGACTT-3′). Amplified PCR product was purified using Qiagen MinElute Gel extraction kit according to the manufacturer's protocol. Sequencing of the purified 16S rRNA gene was performed using BigDye Terminator v3.1 Cycle Sequencing Kit (Applied Biosystems, USA) as recommended by the manufacturer. The purified sequencing reaction mixture was electrophoresed automatically using ABI 3730*1 Genetic Analyser (Applied Biosystems, USA). The 16S rRNA gene sequence of the selected strains was further analysed using BLAST tool and compared to the corresponding neighbour sequences from GenBank-NCBI (National Centre for Biotechnology Information) database.

The consensus sequence (~1402 bp) was imported into the Multalin program and multiple sequence alignment was performed with related species (GenBank-NCBI database). The results obtained were further imported into the MEGA software for the construction of a phylogenetic tree using bootstrap analysis with 1,000 replicates, the substitution method used was Kimura 2-Parameter model and the statistical method used was neighbour joining.

2.3.3 Physiological Characterisation

The growth of the strains under different pH (6–10), temperature (10, 20, 28, 37 and 44 °C) and NaCl concentration (4, 8, 12, 16 and 20 %) was investigated. Cultures were incubated at different temperatures (10–44 °C) and the growth was determined after 48 h of incubation. For pH optimisation, the experiment was performed at different pH (6–10) at their optimised temperature. The effect of salt concentrations (4–20 %) on the growth of the strains was determined at their optimum pH and temperature by determination of OD values.

2.4 Biodegradation Studies of Naphthalene

Biodegradation of naphthalene using selected isolates was evaluated in clean uncontaminated and oil-contaminated soil samples which were dried, sieved and sterilised by autoclaving. Naphthalene (100 ppm) dissolved in methanol was added to the clean soil sample and mixed thoroughly, acting as control. Five grams of clean soil and contaminated soil was added in 150 ml conical flasks containing 50 ml of M9 medium. The medium was inoculated with cultures of $O.D_{600\ nm}$ 1. All vials were incubated at 37 °C in a rotary shaker at 150 rpm. The whole sample contained in the individual vials was extracted at 0 times and after 1 week for the purpose of measuring residual concentration (Bishnoi et al. 2009):

$$\text{Biodegradation efficiency}\,(\%) = \left[(\text{Co} - \text{Ce}) / (\text{Co})\right] \times 100$$

where

Co: initial concentration of naphthalene (μg/g)
Ce: final concentration of naphthalene (μg/g)

2.5 Analytical Methodology

Aliquots (5 ml) were centrifuged at 10,000 g for 10 min and the supernatant was filtered by 0.2 μm filter and kept for analysis. Analysis was performed by high-performance liquid chromatography (HPLC); the analytical column C18, 150* 4.6 mm, 5 μm was used. A gradient elution was done. Solvent A (deionised water) and solvent B (acetonitrile) were used as mobile phases. Gradient elution (0 min, 50 % B; 0–20 min, linear change from 50 to 100 % B; 20–25 min, 100 % B; 25–26 min, 50 % B; 26–30 min, 50 % B; 30.10 min, stop; run time, 30 min) was performed with a 0.7 ml/min at constant flow rate. The column oven temperature was 40 °C, and the injection volume was 20 μl, for all standards and samples (Smoker et al. 2010). The concentration of naphthalene was determined at λ254 nm by comparison to standard curve. Concentration of PAHs was calculated by comparing peak areas of sample chromatogram with the peak area of standard chromatogram:

$$\text{Concentration of PAHs in sample}\,(\mu\text{g} / \text{g}) = \frac{\text{Peak area of chromatogram of sample}}{\text{Peak area of chromatogram of standard PAHs compounds}} \times \text{Concentration of standard PAH compound}$$

3 Results

3.1 Selection of Naphthalene-Degrading Bacteria

After serial dilutions, a total of eight bacterial strains were isolated from oil-contaminated sites and screened for their growth on naphthalene as sole carbon source. Two bacterial isolates S3 and F3 were further selected on the basis of their best growth in M9 minimal medium containing naphthalene (100 ppm). So these two strains were used for biodegrading studies of naphthalene.

Table 1 Biochemical characterisation of selected bacterial isolates

Strains>>biochemical tests ↓	**F3**	**S3**	**Strains>>amino acid ↓**	**F3**	**S3**
Pectinase screening	+	+	Tyrosine	+	+
Lipase screening	+	–	Valine	+	+
Xylanase screening	–	–	Glycine	+	+
Starch hydrolysis	–	–	Arginine	+	+
Cellulase activity	–	–	Alanine	+	+
Gelatin liquification	+	+	Leucine	+	+
Hydrogen sulphide production	–	–	Threonine	+	+
Citrate utilisation	+	–	**Strains>>carbohydrate ↓**	F3	S3
Citrate catalyses	+	–	Sucrose	+	+
Methyl red test	–	–	Glucose	+	+
Voges-Proskauer	–	+	Mannitol	+	+
			Lactose	+	+

3.2 Identification of Selected Bacterial Isolates

3.2.1 Morphological and Biochemical Characterisation

Morphological Characterisation

The two bacterial isolates F3 and S3 were analysed taxonomically. The colony of strain F3 was off white in colour with irregular margins and rough surface, whereas strain S3 was greenish in colour with irregular margins and smooth surface. Both the isolates were gram negative and rod shaped. Spore staining revealed the presence of spores in both the strains.

Biochemical Characterisation

Biochemical tests revealed that both the isolates showed positive results for gelatin liquefaction and amino acid and carbohydrate utilisation indicating hydrolysis of lactose (ferment either of the monomers released, usually only the glucose) and sucrose. Both the isolates showed negative results for starch hydrolysis and cellulase activity.

In the methyl red test, yellow colour was recorded as an end point for both strains, indicating little or no acid remains in the medium, but for Voges-Proskauer, the test result was positive for S3, while a negative reaction was showed by F3 strain, indicating the formation of non-acidic end product by S3, while for F3, the presence of acid in the medium was shown. Other biochemical characteristics of both strains are also shown in Table 1.

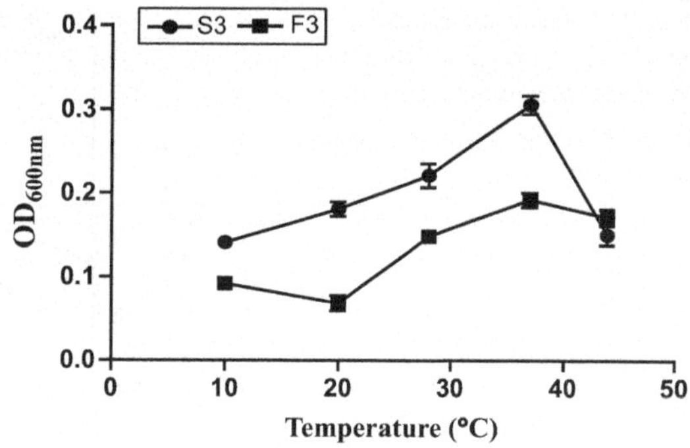

Fig. 1 Growth of strains S3 and F3 in M9 medium with 100 ppm naphthalene at different temperatures

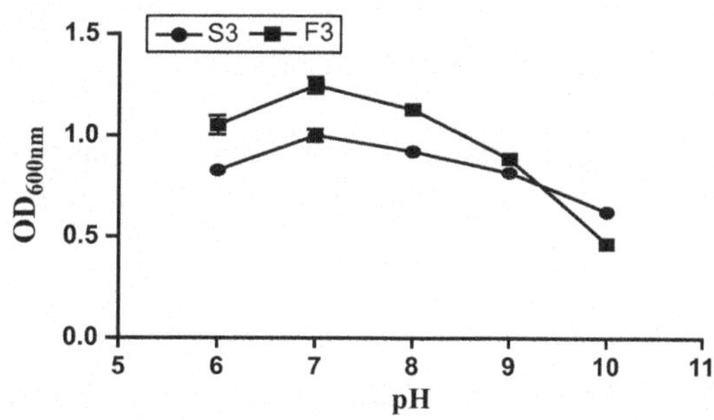

Fig. 2 Growth of strain S3 and F3 in M9 medium with 100 ppm naphthalene at different pH

3.3 Physiological Characterisation

The optimum temperature, pH and salt concentration in M9 medium for the growth of both strains was determined.

3.3.1 Effect of Temperature

The isolates showed the maximum growth at 37 °C with $O.D_{600}$ of 0.298 (S3) and 0.186 (F3) after 48 h of incubation, whereas decrease in growth was observed both at 28 and 44 °C (Fig. 1), indicating that isolates are mesophilic in nature.

3.3.2 Effect of pH

The maximum growth for both isolates was observed at pH 7 with an $O.D_{600}$ of 0.981(S3) and 1.220 (F3), after 48 h of incubation. Good growth was also observed at pH 8 but it decreases gradually up to pH 10, as shown in Fig. 2.

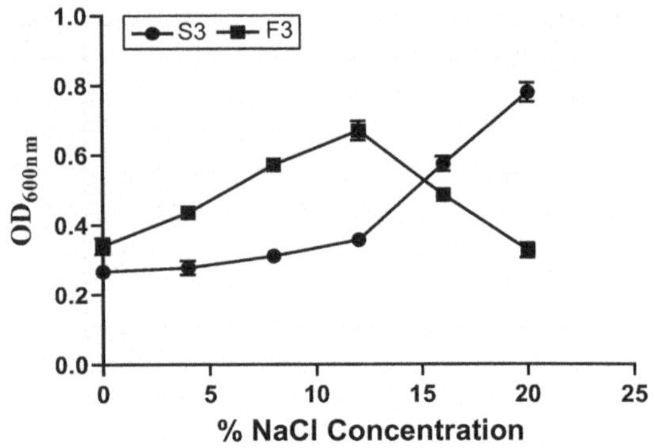

Fig. 3 Growth of strain F3 and S3 in M9 medium with 100 ppm naphthalene at different salt (NaCl) concentrations

3.3.3 Effect of Salt Concentration

For F3 and S3 cultures, maximum bacterial growth occurred at 12 % and 20 % of the salt (NaCl) concentration with $O.D_{600}$ (0.67) and (0.78), indicating the bacteria to be moderately and extremely halophilic in nature (Fig. 3).

3.4 Molecular Characterisation Based on 16S rRNA Gene

The S3 and F3 bacterial isolates were further identified on the basis of their sequence analysis. The obtained 16S rRNA sequence was submitted to BLAST in order to find a homology with other 16S rRNA sequences. Comparing the 16S rRNA sequence of the isolates with the sequences in the GenBank revealed that isolate S3 showed 98 % homology to *Pseudomonas aeruginosa* strain DSM 50071, and phylogenetic analyses based on 16S rRNA sequence of S3 also showed that it is closely related to *Pseudomonas aeruginosa* strain DSM 50071 (Fig. 4). Similarly for F3, the results showed 98 % homology of the isolate to *Enterobacter cloacae* strain 279-56, and phylogenetic analyses based on 16S rRNA sequences indicate its close relatedness to *Enterobacter cloacae* strain 279-56 and *Enterobacter cloacae* strain LMG 2683 (Fig. 5). The GenBank accession numbers assigned to the submitted nucleotide sequences of S3 and F3, by NCBI, are JX254648.1 and JX480546.

3.5 Biodegradation Studies of Naphthalene

Biodegradation study was conducted with naphthalene-adapted bacterial strains F3 and S3 at optimised conditions. Biodegradation was done in spiked soil sample (100 ppm) and contaminated soil sample. The samples were extracted at 0 day and after 7 days of incubation. The peak area of the different chromatograms was used for calculating the biodegradation efficiency of both strains (Figs. 6 and 7). The

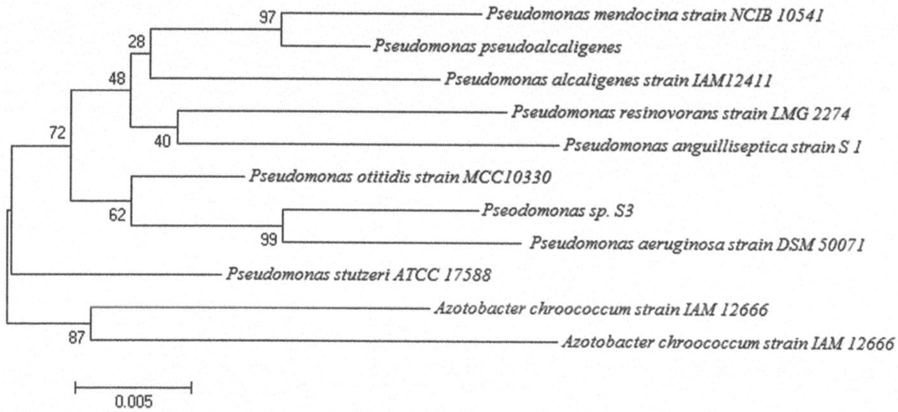

Fig. 4 Neighbour-joining tree showing strain S3 and related *Pseudomonas* strains. Numbers below tree nodes represent the percentage bootstrap support for 1,000 replicates, respectively. Bar, approximately 0.005 % nucleotide divergence

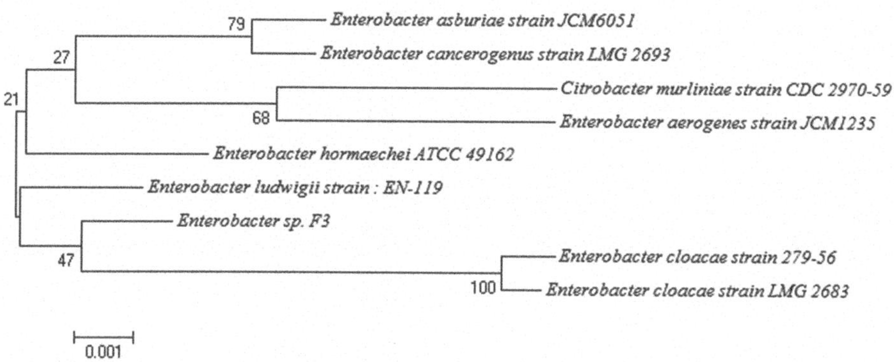

Fig. 5 Neighbour-joining tree showing strain F3 and related *Enterobacter* strains. Numbers below tree nodes represent the percentage bootstrap support for 1,000 replicates, respectively. Bar, approximately 0.001 % nucleotide divergence

biodegradation efficiency of both strains after 7 days of incubation was found to be 77.77 % (S3) and 61.11 % (F3), respectively.

4 Discussion

Thus, from this study, it is summarised that at optimum conditions (pH (7) and 37 °C), strain S3 (*Pseudomonas* sp. S3) showed better result for degradation of naphthalene after 7 days of incubation. With strain S3 (*Pseudomonas* sp. S3), the detectable concentration of naphthalene (163.63 μg g^{-1}) was reduced to 36.36 μg g^{-1} as compared to strain F3 (*Enterobacter* sp. F3) where the concentration was

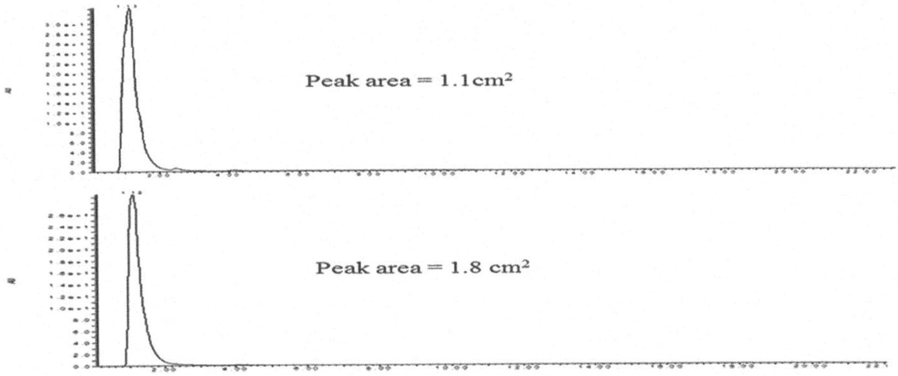

Fig. 6 Chromatogram of spiked soil (100 μg g^{-1}) and contaminated soil (163.63 μg g^{-1}) at 0 day of incubation

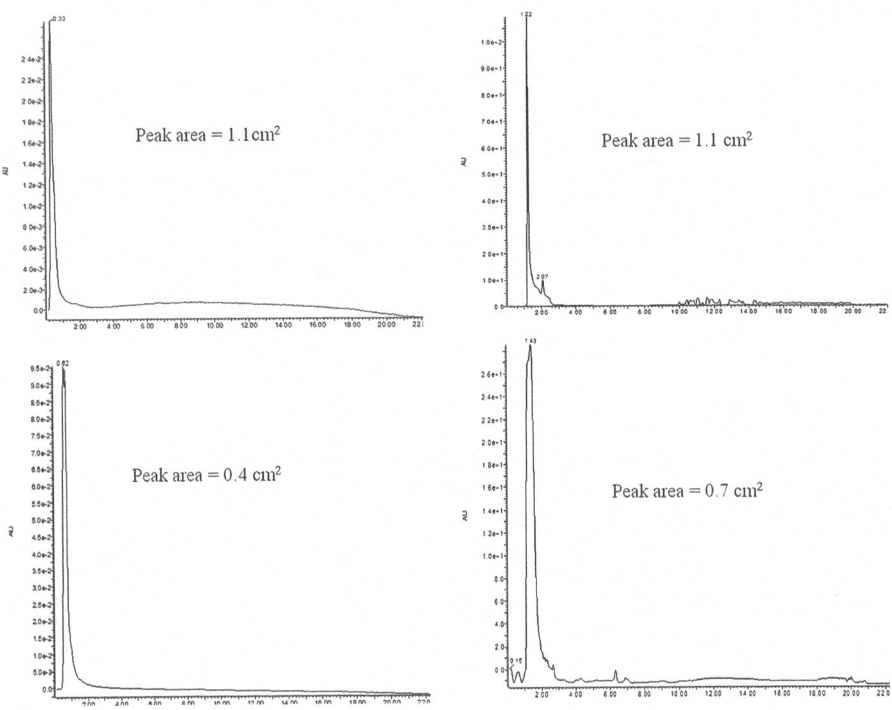

Fig. 7 Chromatogram of spiked soil and contaminated soil after 7 days of incubation with strain S3 and F3

reduced to 63.63 μg g^{-1}. Both strains being moderately and extremely halophilic in nature provided an advantage of degrading the pollutants in PAH-contaminated environment with high salinity. *Pseudomonas stutzeri* P-16 and *P. saccharophila* P-15 isolates from creosote-contaminated soil were also reported to degrade naphthalene, 2-methylnaphthalene and 1-methylnaphthalene (Stringfellow and Aitken 1995). It indicates that *Pseudomonas* sp. can be an important part of the oil-degrading microbial community in the areas exposed to petroleum oil. In addition to microbial potential, the right ecological conditions are also very important for microbial activity for biodegradation. A report showed maximum degradation of naphthalene by *Pseudomonas* sp. HOB1 in the pH and temperature ranges of 7.5–8.5 and 35–37 °C, respectively (Pathak et al. 2009).

5 Conclusion

Under optimum conditions, the isolates (S3 and F3) can metabolise 127.27 $\mu g\ g^{-1}$ and 100 $\mu g\ g^{-1}$ of naphthalene within 7 days of incubation, showing ability to degrade the compound with high efficiency. Salt tolerance by both strains up to a concentration of 12 % (F3) and 20 % (S3) makes the strain suitable for degradation in the environment with high salinity. Thus, the use of these isolates alone or in combination could result in effective degradation of PAHs and related compounds present in the contaminated environment. These findings indicate good potential of the acclimatised indigenous bacteria for oil pollutant degradation.

References

Andreoni V, Gianfreda L (2007) Bioremediation and monitoring of aromatic-polluted habitats. Appl Microbiol Biotechnol 76:286–308

Aneja KR (2008) Experiments in microbiology, plant pathology and biotechnology. New Age International Publisher

Bishnoi K, Sain U, Kumar R, Singh R, Bishnoi NR (2009) Distribution and biodegradation of polycyclic aromatic hydrocarbons in contaminated site of Hisar (India). Indian J Exp Biol 47:210–217

Bombach P, Richnow HH, Kastner M, Fischer A (2010) Current approaches for the assessment of in situ biodegradation. Appl Microbiol Biotechnol 86:839–852

Cappuccino J, Sherman N (2010) Microbiology: a laboratory manual. Pearson Education, Limited, 9

Djokic L, Narancic T, Nikodinovic-Runic J, Savic M, Vasiljevic B (2011) Isolation and characterisation of four novel gram-positive bacteria associated with the rhizosphere of two endemorelict plants capable of degrading a broad range of aromatic substrates. Appl Microbiol Biotechnol 91:1227–1238

International Tanker Owners Pollution Federation Limited (ITOPF) (2006) Disposal of oil and debris. In: Response strategies

Jin HM, Choi EJ, Jeon CO (2013) Isolation of a BTEX-degrading bacterium, Janibacter sp. SB2, from a sea-tidal flat and optimization of biodegradation conditions. Bioresour Technol 145:57–64

Li H, Zhang Q, Wang XL, Ma XY, Lin KF et al (2012) Biodegradation of benzene homologues in contaminated sediment of the East China Sea. Bioresour Technol 124:129–136

Mazzeo DE, Matsumoto ST, Levy CE, de Angelis Dde F, Marin-Morales MA (2012) Application of micronucleus test and comet assay to evaluate BTEX biodegradation. Chemosphere 90:1030–1036

Meckenstock RU, Lueders T, Griebler C, Selesi D (2012) Microbial hydrocarbon degradation at coal gasification plants, handbook of hydrocarbon and lipid microbiology. Springer, Berlin/Heidelberg, pp 2293–2312

Naidu R, Nandy S, Megharaj M, Kumar RP, Chandalavada S et al (2012) Monitored natural attenuation of a long-term petroleum hydrocarbon contaminated sites: a case study. Biodegradation 23:881–895

Ntougias S, Melidis P, Navrozidou E, Tzegkas F (2015) Diversity and efficiency of anthracene degrading bacteria isolated from a denitrifying activated sludge system treating municipal wastewater. Int Biodeterior Biodegrad 97:151–158

Pathak H, Kantharia D, Malpani A, Madamwar D (2009) Naphthalene degradation by Pseudomonas sp. HOB1: in vitro studies and assessment of naphthalene degradation efficiency in simulated microcosms. J Hazard Mater 166:1466–1473

Pye V, Patrick R (1983) Groundwater contamination in the United States. Science 221:713–718

Smoker M, Tran K, Smith RE (2010) Determination of Polycyclic Aromatic Hydrocarbons (PAHs) in shrimp. J Agric Food Chem 10:1021

Steliga T, Jakubowicz P, Kapusta P (2012) Changes in toxicity during in situ bioremediation of weathered drill wastes contaminated with petroleum hydrocarbons. Bioresour Technol 125:1–10

Stringfellow WT, Aitken MD (1995) Competitive metabolism of naphthalene, methylnaphthalenes and fluorene by phenanthrene-degrading pseudomonads. Appl Environ Microbiol 61:357–362

Tecon R, Beggah S, Czechowska K, Sentchilo V, Chronopoulou PM et al (2010) Development of a multistrain bacterial bioreporter platform for the monitoring of hydrocarbon contaminants in marine environments. Environ Sci Technol 44:1049–1055

USEPA (1993) Provisional guidance for quantitative risk assessment of polycyclic aromatic hydrocarbons. USEPA/600/R-93/ 089

You Y, Shim J, Cho CH, Ryu MH, Shea PJ et al (2013) Biodegradation of BTEX mixture by Pseudomonas putida YNS1 isolated from oil-contaminated soil. J Basic Microbiol 53:469–475

Zhou H, Wang H, Huang Y, Fang T (2016) Inter biodeterior & biodegrad characterisation of pyrene degradation by halophilic Thalassospira sp. strain TSL5-1 isolated from the coastal soil of Yellow Sea, China. Int Biodeterior Biodegrad 107:62–69

Biodegradation of Azo Dye Using the Isolated Novel Bacterial Species: *Acinetobacter* sp.

Uttariya Roy, Papita Das, Avijit Bhowal, and Siddhartha Datta

Abstract Dyes present in water affect the balance of aquatic life as well as human being due to its toxicity. Insufficient sunlight impairs the process of photosynthesis of aquatic plants and phytoplankton, and, thus, they die without having sufficient food. After the death of these plants and animals, the numbers of zooplanktons and other higher organisms present in that aquatic system are automatically reduced. Finally, in this way the aquatic ecosystem loses its balance. The mixing of dye-colored wastewaters into the aquatic ecosystem and, thus, its accumulation in wildlife through food chain can cause many negative ecotoxicological effects and public health hazards.

Congo red dye is a type of azo dye which can cause harmful effects on environment. This present study aims to investigate the degradation of azo dye (Congo red) using bacterial species, *Acinetobacter* sp., at different variable parameters (pH, temperature, salinity, agitation, dye concentration, and inoculum volume). Optimum pH, temperature, salinity, agitation, dye concentration, and inoculum volume for this study are pH 7, 37 °C, 3 mg/L, 120 rpm., 50 mg/L, and 50 ml/L, respectively, to degrade the Congo red dye by using *Acinetobacter* sp. Optimum dye decolorization efficiency was found to be 87.89%.

Keywords Biodegradation • Azo dye • Congo red • *Acinetobacter* sp. • Salinity

1 Introduction

Dyes present in surface water at very low concentration are visible, and it is unfit for human consumption. It prevents sunlight and thus affects photosynthetic activity of aquatic life. The mixing of dye-colored wastewaters into the aquatic ecosystem and, thus, its accumulation in wildlife through food chain can cause many negative ecotoxicological effects and public health hazards (Kodam et al. 2005). Dye compounds have the ability to increase the chemical and biological oxygen demand and

U. Roy (✉) • P. Das • A. Bhowal • S. Datta
Department of Chemical Engineering, Jadavpur University, Kolkata, India
e-mail: uttariyar@gmail.com

S.K. Ghosh (ed.), *Utilization and Management of Bioresources*,
DOI 10.1007/978-981-10-5349-8_18

suspended solids. The dyes may also be carcinogenic and impart toxicity to aquatic life and human beings (Ghosh and Saha 2013).

Azo dyes are the most diverse group of synthetic dyes and are widely used in different industries such as textile, cosmetics, food, leather tanning, printing, paint, pigment, pharmaceutical, solvents, acrylic, paper, and pulp. Basically, dyes are stable to heat, oxidizing agents, and light because of their complex molecular structure. So, it is very hard to remove the dye compounds from environment (Senthilkumar et al. 2013). The conventional methods for the elimination of dye compounds are chemical precipitation, ultrafiltration, ion exchange, reverse osmosis, electrodialysis, etc. All these methods have many limitations or disadvantages (Chowdhury et al. 2011). In this context, dye degradation by different microbes is more efficient and advantageous process (Saha et al. 2012; Chowdhury and Saha 2010, 2011).

2 Materials and Methods

Congo red used in this study was of commercial grade. Molar mass and melting point of this dye are 696.66 g/mol and more than 360 °C, respectively. Other chemicals used were of analytical grade and obtained from Merck, India.

A bacterial strain was isolated from soil of textile industry; after serial dilution, the culture was prepared and then streaked onto nutrient agar plate aseptically in a laminar and incubated at 37 °C for 3 days in an incubator. After 3 days of incubation, the strain was inoculated into Luria-Bertani broth medium from the plate in a laminar aseptically, and then the broth medium containing strain was incubated overnight at 37 °C and at 120 rpm in an incubator cum shaker. On the other hand, that strain was selected for further characterization studies. Gram staining had been performed to study the morphological characteristics of the strain. Under the microscope thin, rod-shaped, red-colored bacteria had been visualized.

Cultures of that particular strain incubated overnight were used for the dye degradation experiments. Certain volume of culture was added to the minimal medium (component – K_2HPO_4, KH_2PO_4, $FeSO_4.7H_2O$, $MgSO_4.7H_2O$, $CaCl_2$) containing Congo red dye of different concentrations. The others parameters are pH, salinity, temperature, and agitation. The minimal medium containing Congo red dye with bacterial culture was studied for 15 days in an incubator cum shaker. During this study, aliquots of culture medium were taken away aseptically after specified time intervals and centrifuged at 10,000 rpm for 15 min. The supernatant was collected in each case, and its absorbance was measured using UV/VIS spectrophotometer at 497 nm. The % of dye removal was calculated using the formula:

$$\%\text{Removal} = \frac{(\text{Initial concentration} - \text{final concentration})}{\text{Initial concentration}} \times 100$$

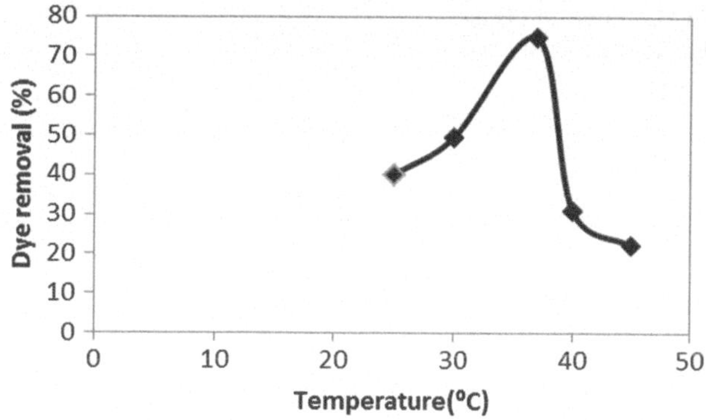

Fig. 1 Effect of temperature on biodegradation of Congo red dye

3 Results and Discussions

Different batch studies had been done to optimize the experimental parameters. These are:

3.1 Effect of Temperature

A batch study varying the temperatures (25, 30, 37, 40, and 45 °C) was done by fixing other parameters (pH, 7; salinity, 3 mg/l; dye concentration, 50 ppm; inoculum volume, 5 ml; and agitation, 120 rpm). The different temperatures were set up one by one for 15 days each in an incubator cum shaker. The dye samples were studied on a daily basis. During this time period, 2 ml of sample was taken out and then centrifuged and finally recorded its absorbance in UV/VIS spectrophotometer to calculate the percentage of dye degradation. The percentage of dye removal was highest at 37 °C in which temperature *Acinetobacter* sp. can grow better. It means that the efficiency of dye degradation of *Acinetobacter* sp. is highest at 37 °C, and so, at 37 °C *Acinetobacter* sp. can better use Congo red dye as their carbon sources, and thus, this bacterial species can degrade better at this temperature (Fig. 1).

3.2 Effect of pH

A batch study varying pHs (1, 3, 5, and 7) was done by fixing other parameters (temperature, 37 °C; salinity, 3 mg/L; dye concentration, 50 mg/L; inoculum volume, 50 ml/L; and agitation, 120 rpm). The different pHs were set up for 15 days in an incubator cum shaker. The dye samples were studied on a daily basis. During this time period, 2 ml of sample was taken out and then centrifuged and finally recorded its absorbance in UV/VIS spectrophotometer to calculate the percentage

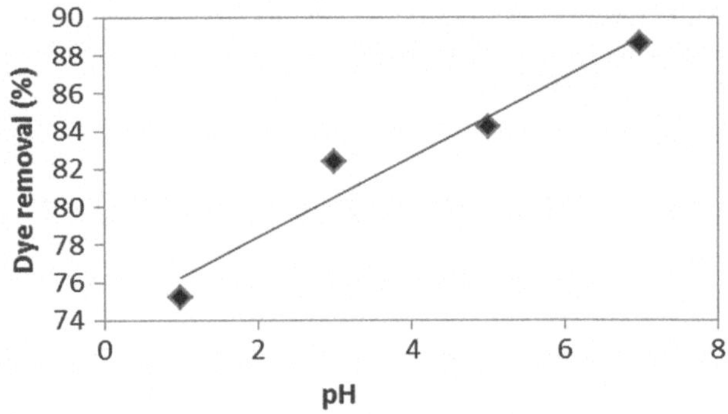

Fig. 2 Effect of pH on biodegradation of Congo red dye

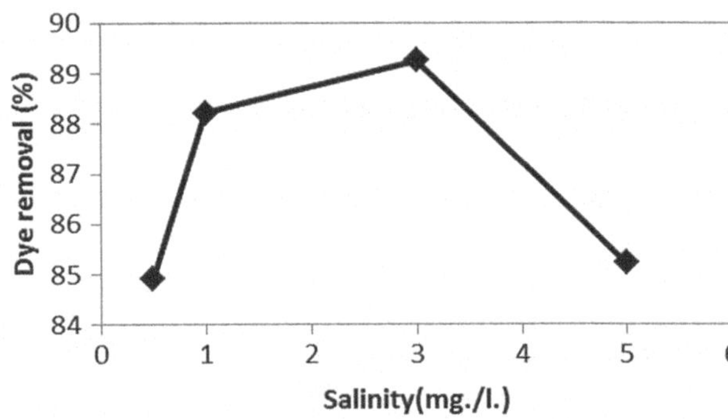

Fig. 3 Effect of salt concentration on biodegradation of Congo red dye

of dye degradation. The percentage of dye degradation was highest at pH 7 at which pH *Acinetobacter* sp. can grow better. It means that the efficiency of dye degradation of *Acinetobacter* sp.is highest at pH 7, and thus this bacterial species can degrade better at pH 7 (Fig. 2).

3.3 *Effect of Salinity*

A batch study varying the salinity (0.5, 1, 3, 5 mg/L) was done by fixing other parameters (temperature, 37 °C; pH, 7; dye concentration, 50 mg/L; inoculum volume, 50 ml/L; and agitation, 120 rpm). The different salt concentrations were set up for 15 days in an incubator cum shaker. The dye samples were studied on a daily basis. During this time period, 2 ml of sample was taken out and then centrifuged and finally recorded its absorbance in UV/VIS spectrophotometer to calculate the percentage of dye degradation. The percentage of dye removal was highest at 3 mg/L at which salt concentration *Acinetobacter* sp. can grow better. It means that the efficiency of dye degradation of *Acinetobacter* sp.is highest at salt concentration of 3 mg/L, and thus this bacterial species can degrade better at the salinity of 3 mg/L (Fig. 3).

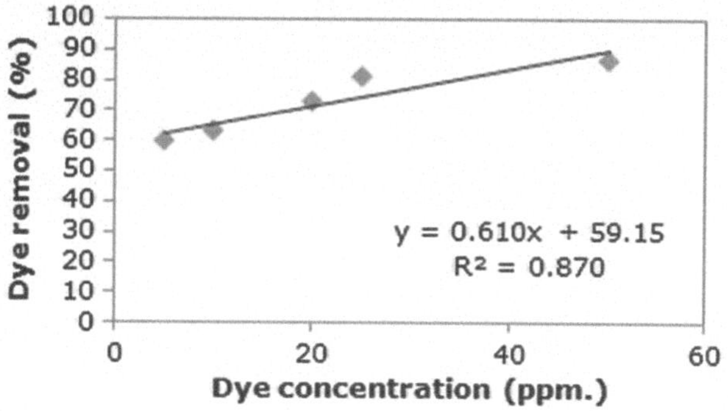

Fig. 4 Effect of dye concentration on biodegradation of Congo red dye

3.4 *Effect of Dye Concentration*

A batch study varying the dye concentrations (5, 10, 20, 25, 50 ppm.) was done by fixing other parameters (temperature, 37 °C; pH, 7; salinity, 3 mg/L; inoculum volume, 50 ml/L; and agitation, 120 rpm). The different dye concentrations were set up for 15 days in an incubator cum shaker. The dye samples were studied on a daily basis. During this time period, 2 ml of sample was taken out and then centrifuged and finally recorded its absorbance in UV/VIS spectrophotometer to calculate the percentage of dye degradation. The percentage of dye degradation was highest at dye concentration of 50 ppm in which dye concentration *Acinetobacter* sp. can grow better. It means that the efficiency of dye degradation of *Acinetobacter* sp.is highest at dye concentration of 50 mg/L (Fig. 4).

3.5 *Effect of Inoculum Volume*

A batch study varying the inoculum volumes (0.5, 1, 2, 5 ml) was done by fixing other parameters (temperature, 37 °C; pH, 7; salinity, 3 mg/L; dye concentration, 50 mg/L; and agitation, 120 rpm). The various inoculum volumes were set up for 15 days in an incubator cum shaker. The dye samples were studied on a daily basis. During this time period, 2 ml of sample was taken out and then centrifuged and finally recorded its absorbance in UV/VIS spectrophotometer to calculate the percentage of dye degradation. The percentage of dye removal was highest at inoculum volume of 5 ml/100 ml (50 ml/L) (Fig. 5).

3.6 *Effect of Agitation*

A batch study varying the agitation speeds (80, 120, 150 rpm.) was done by fixing other parameters (temperature, 37 °C; pH, 7; salinity, 3 mg/L; dye concentration, 50 mg/L; and inoculum volume, 50 ml/L). The different agitation speeds were set up one by one for 15 days each in an incubator cum shaker. During this time period,

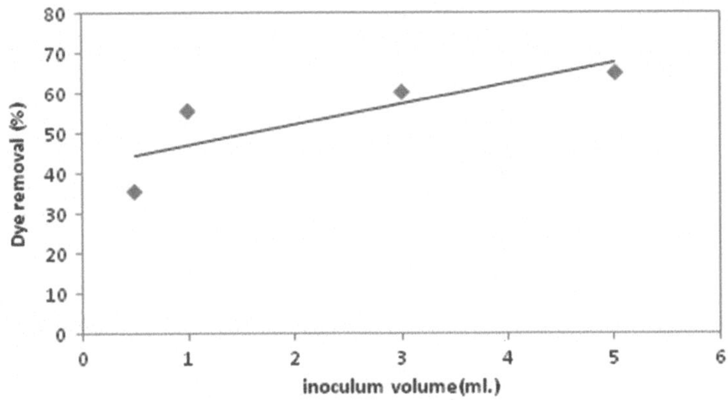

Fig. 5 Effect of inoculum volume on biodegradation of Congo red dye

Fig. 6 Effect of agitation speed on biodegradation of Congo red

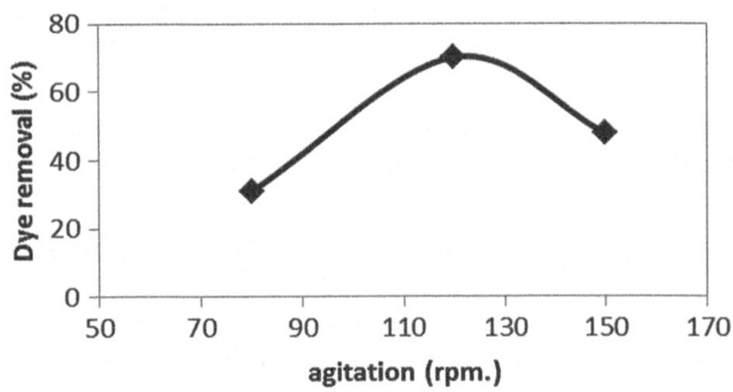

2 ml of sample was taken out and then centrifuged and finally recorded its absorbance in UV/VIS spectrophotometer to calculate the percentage of dye degradation. The percentage of dye removal was highest at the speed of 120 rpm in which speed *Acinetobacter* sp. can grow better. It means that the efficiency of dye degradation by *Acinetobacter* sp.is highest at 120 rpm, and so, at this speed *Acinetobacter* sp. can better uses Congo red as their carbon sources, and, thus, this bacterial species can degrade better at 120 rpm. The dye removal efficacy is better above 120 rpm (at 150 rpm) compared with the agitation speed below 120 rpm (at 80 rpm). So, the biodegradation of Congo red is surely dependent on agitation (Fig. 6).

4 Conclusions

On the basis of these results obtained in this investigation, the following conclusions can be drawn:

Acinetobacter sp. can be used as a dye degradable species to remove Congo red dye from aqueous solutions.

Biodegradation of Congo red azo dye was dependent on temperature, pH, salinity, dye concentration, inoculum volume, and agitation speed.

The levels of the six variables, experimental temperature, 37 °C; solution pH, 7; salinity of the sample, 3 mg/L; initial dye concentration,50 mg/L; initial inoculum volume, 50 ml/L; and agitation speed, 120 rpm, were found optimum for maximum Congo red dye removal.

The corresponding dye decolorization efficiency was found to be 87.89%.

Acknowledgment All authors are thankful to Department of Biotechnology, West Bengal for the financial support (Research Grant-in Aid).

References

Chowdhury S, Saha P (2010) Sea shell powder as a new adsorbent to remove basic green 4 (malachite green) from aqueous solutions: equilibrium, kinetic and thermodynamic studies. Chem Eng J 164:168–177

Chowdhury S, Saha P (2011) Adsorption kinetic modelling of safranin onto rice husk bio matrix using pseudo-first- and pseudo-second-order kinetic models: comparison of linear and non-linear methods. Clean Soil Air Water 39:274–282

Chowdhury S, Mishra R, Saha P, Kushwaha P (2011) Adsorption thermodynamics, kinetics and isosteric heat of adsorption of malachite green onto chemically modified rice husk. Desalination 265:159–168

Ghosh A, Saha PD (2013) Optimization of copper bioremediation by Stenotrophomonas maltophilia PD2. J Environ Chem Eng 1:159–163

Kodam KM, Soojhawon I, Lokhande PD, Gawai KR (2005) Microbial decolourization of reactive azo dyes under aerobic conditions. World J Microbiol Biotechnol 21:367

Saha PD, Chakraborty S, Chowdhury S (2012) Batch and continuous (fixed-bed column) biosorption of crystal violet by Artocarpus Heterophyllus (jackfruit) leaf powder. Colloids Surf B 92:262–270

Senthilkumar S, Prabhu HJ, Perumalsamy M (2013) Response surface optimization for biodegradation of textile azo dyes using isolated bacteria pseudomonas sp. Arab J Sci Eng 38:2279–2291

Study of Some Predominant Arsenic Resistance Bacteria from Soil Samples of Industrial Zones of West Bengal, India

P. Dutta, I. Mallick, A. Ghosh, and M. Basu

Abstract Untreated industrial effluents discharged into the environment pose a serious problem to the soil living organism and human beings also. Almost all the traditional physicochemical methods do not provide effective solutions for the elimination of metals from industrial effluents. Industrial waste soil samples collected from different industrial belts of West Bengal were analyzed for physicochemical and microbiological characteristics. Eighteen arsenic (As)-tolerant bacterial strains were isolated from arsenic-contaminated industrial soil. Among them, three bacterial strains, viz., MsfL2, HsR_21, and HlR_211, exhibited higher As resistance capacity and showed 50% relative growth in LB medium incorporated with 1500 μg/ml of arsenate (V) and 200 μg/ml of arsenite (III). Antibiotic susceptibility of the isolates was also done, and results indicate variation with respect to the tested strain. Additionally it was found that the strains were sensitive to gentamycin, tetracycline, and bacitracin. Biochemical analysis and 16S rRNA sequencing were done to identify and determine the phylogeny of the selected arsenic-tolerant strains. Since arsenic-induced stress has often been known to have correlation with osmotic as well as oxidative stress, the selected strains under study might be metabolically adapted to arsenic-induced oxidative and osmotic stress. Physiological test and PCR results indicate that the isolates probably have developed resistance via arsenic reduction mechanism. From all the above facts, it is evident that the selected isolate(s) is promising candidate(s) for arsenic bioremediation in polluted industrial environment.

Keywords Arsenic • 16S rRNA • Bioremediation

P. Dutta • M. Basu (✉)
Post Graduate Department of Microbiology, Barrackpore Rastraguru Surendranath College, Barrackpore, Kolkata, West Bengal, India
e-mail: dr.malinibasu@gmail.com

I. Mallick • A. Ghosh
Bose Institute, Kolkata, West Bengal, India

S.K. Ghosh (ed.), *Utilization and Management of Bioresources*,
DOI 10.1007/978-981-10-5349-8_19

1 Introduction

Arsenic is included in the category of metalloid that is present in trace amounts in water, soil, rocks, air, and all living matter (Patel et al. 2007). Sources of contamination include erosion of local rocks, human activities through farming, mining, industrial effluents, wood preservatives and arsenical herbicides, various commercial processes, and combustion of fossil fuels. Contamination of groundwater is widespread affecting many areas of the world including Bangladesh, India, Vietnam, Thailand, and Chile (Nordstrom 2002). Arsenic in the environment is a global health problem because of its carcinogenicity. It is responsible for bladder, kidney, liver, lung, and skin cancer and is listed as a class A or class I human carcinogen by the US Environmental Protection Agency (USEPA). The average concentration of arsenic in terrestrial environment is around 1.5–3 mg/kg, but in contaminated area it raises up to 57–83 mg/kg.

Environmental arsenic exists in both organic and inorganic states. Mostly arsenic is present in the forms of arsenite (III) and arsenate (V). Arsenic itself is not soluble in water, but when it combines with different elements, its solubility in water increases (Aitio and Becking 2001). Both arsenite and arsenate are soluble over a wide range of pH and routinely observed, though arsenite is reported to be more toxic (Cervantes et al. 1994). The toxicity of different arsenic species were often reported to vary in an order of arsenite > arsenate > mono-methylarsonate (MMA) > dimethylarsinate (DMA) (Penrose 1974; Stugeron et al. 1989).In some mammals, including human, the As(III) is methylated and converted to some relatively less harmful substance like monomethyl arsenic acid (MMAA) and dimethyl arsenic acid (DMAA) and excreted through urine (Hughes 2002).

Although arsenic is toxic to many bacteria, some bacteria are resistant to arsenic either due to the presence of a strictly phosphate-specific transport system, which prevents the uptake of arsenate which is analogous to phosphate (Willsky and Malamy 1980), or due to an efflux system mediated by the plasmid or chromosomally encoded *ars* operon (Cervantes et al. 1994; Diorio et al. 1995; Cai et al. 1998).The arsenic detoxification mechanisms have been investigated in various microorganisms both anaerobic and aerobic in nature(Niggemyer et al. 2001). Arsenate-reducing bacteria including *Desulfomicrobium*, *Clostridium*, *Bacillus*, *Sulfurospirillum*, *Citrobacter*, *Wolinella*, etc. were identified (Oremland et al. 2004).

Bacterial ability to tolerate osmotic and oxidative stress contributes to their arsenic resistance. Like other heavy metals, arsenic causes oxidative stress by inducing reactive oxygen species (ROS), thereby indicating a distinct correlation between such as oxidative stress, heavy metal stress, etc. (Pichereau et al. 2000; Liu et al. 2001).

In this study, isolated arsenic-tolerant bacteria have been characterized in the following aspects: (1) degree of tolerance to arsenic, (2) biochemical and phylogenetic analysis, and (3) probable mechanism of resistance and thereby their biosorption potential.

2 Material and Method

2.1 Study Site and Primary Screening

Soil samples were collected from different industrial sites of West Bengal. Tribeni, Ichapur, Halisahar, Asansol, Haldia, and Howrah were chosen for the present study. For study of arsenic contamination, six soil samples were collected from six different sampling spots at a depth of 10 cm from top layer with seasonal variation and collected in sterile falcon tubes. After collection, soil samples were stored at 4 °C until further characterization. One gram of soil sample was serially diluted, and 0.1 ml of diluted sample was spread over the Luria agar (LA) plates which were then incubated at 37 °C for 24–48 h. Isolated bacterial colonies were picked and maintained in LA slant.

For primary screening of the isolates for As tolerance, overnight-grown isolates were streaked onto Luria agar (LA) plates incorporated with 100–500 μg/ml of As(V) and 50–200 μg/ml of As(III) and control plate incubated at 37 °C for 24–48 h for visible growth (Abbas and Edward 1989).

2.2 Secondary Screening

Degree of resistance of the selected isolates was also evaluated in the LB medium. Filter sterile metal solutions were added aseptically to LB medium to attain metal concentrations ranging from 500 to 1,500 μg/ml for As(V) and 50 to 200 μg/ml for As(III). Growth of the isolates was determined by measuring the optical density at 540 nm. The relative growth was expressed as a percentage of those obtained in untreated control cultures at the same time which was taken as 100%.

2.3 Heavy Metal Analysis of Soil

The total concentration of heavy metals of those soils was estimated by digestion of 1 g air-dried soil in 10 mL of HNO_3:HCl (1:3), and digested samples were transferred into 50 ml micro-Kjeldahl flask. Soil was subjected to acid digestion using standard method (Basu and Paul 2008), and the concentrations of Fe, Cr, Hg, Cu, Cd, As, Mn, and Zn were determined by atomic absorption spectroscopy (Agilent spectra).

2.4 Biochemical Characterization

Different biochemical properties of the bacterial isolates such as enzyme activity (catalase, urease, and oxidase), denitrification test, arginine dehydrolase, utilization of gelatin, cellulose, casein, starch, and lipid were tested following the standard methods (Cappuccino and Natalie 2012) to characterize the isolates.

2.5 Identification of Isolates

16S rDNA sequence homology was determined for tentative identification of the organism. Genomic DNA was isolated following standard protocol of Sambrook et al. (1989). The 16S rDNA fragments were amplified from the bacterial genomic DNA using 16S rRNA bacterial specific forward and reverse primer sets. The sequence data of 16S rDNA fragments obtained as amplicons were aligned and analyzed using NCBI Blast function (Altschul et al. 1990) and Ribosomal Database Project (Maidack et al. 1997). The phylogenetic lineages of isolate MsfL2, HsR_21, and HlR_211 were built from 16S rDNA sequence using MEGA 6.0 (Tamura et al. 2013).

2.6 Antibiotic Sensitivity

To determine the antibiotic sensitivity of the metal-resistant isolates, antibiotic-impregnated discs (G-VIII-plus; Himedia) [ampicillin (10 mcg), tetracycline (5 mcg), penicillin-G (10 mcg), streptomycin (10 mcg), gentamycin (10 mcg), polymyxin B (300 mcg), chloramphenicol (30 mcg), co-trimoxazole (25 mcg), nitrofurantoin (300 mcg), bacitracin (10 mcg), ciprofloxacin (10 mcg), colistin (10 mcg)] were placed on freshly prepared lawns of each isolate on LA plates. Plates were incubated at 37 °C for 24 h. Depending on inhibition zones, the isolates were categorized as sensitive, intermediate, and resistant as per manufacturer manual. The multiple antibiotic resistance index for each isolate was also calculated (MAR index = a/b where a = number of resistant antibiotics, b = total number of antibiotics exposed) as recommended by Krumpermann (1983).

2.7 Bacterial Growth Under Oxidative Stress and Osmotic Stress

The bacterial isolates were tested for their resistance to oxidative stress by measuring inhibition zones for H_2O_2, As(III), and As(V) on TYEG agar (Huang et al. 2010). Briefly, overnight-grown cultures were plated in TYEG agar medium and were allowed to grow for 4 h. With a cork borer, a hole was drilled into which 25 μl of 3% (*v*/*v*) H_2O_2, As(III) (w/v), and As(V) (w/v) was added. Plates were incubated at 22 °C, and the diameters of the inhibition zones were measured after 24 and 48 h. The experiments were carried out in duplicate.

Three bacterial isolates tolerating 1500 μg/ml of sodium arsenate in liquid culture were used in this study. Degree of NaCl tolerance of selected isolates were evaluated in Luria broth containing 5, 10, and 20% NaCl. Growth of the isolates in LB was determined by measuring the optical density at 540 nm using the uninoculated broth as blank. Relative growth of the isolates was expressed as the percentage of those obtained in untreated control which was taken as 100%.

2.8 Mechanism of as Resistance

The ability of the bacterial isolates to reduce and oxidize arsenic (V) and (III), respectively, was tested by using silver nitrate method (Valenzuela et al. 2009). Isolates were streaked on the Luria agar plate containing either As(V) 1000 μg/ml or As(III) 100 μg/ml and incubated at 37 °C for 72 h. After incubation, small amount of silver nitrate was added to the media. If the media turned brown, it confirms the presence of silver arsenate, and the appearance of bright yellow indicates the presence of silver arsenite.

To determine As(V) reduction property of the isolates, PCR amplification was carried out using genomic DNA as template and aerobic arsenate reductase *arrA (*Kulp et al. 2006*)*. Polymerase chain reaction was performed using a mixture containing 0.5 μM of each primer, 0.25 mM of each dNTPS, 1 U Taq DNA polymerase in 1X PCR buffer, and template DNA (15–20 ng) in a Pro-flex PCR system (Applied biosystem, USA).

2.9 Biosorption of Arsenic

Isolates were inoculated in the Luria broth media incorporated with 1000 μg/ml and 2000 μg/ml arsenate and 100 μg/ml and 200 μg/ml arsenite solution and incubated at 37°C. After 24 and 48 h of incubation, media were centrifuged at 10,000 rpm for 10 min to separate the bacterial biomass from the media. The total arsenic concentration of the media and microbial biomass were estimated by digestion of samples separately in 5 mL of HNO_3:HCl (1:3), and digested samples were transferred into 100 ml micro-Kjeldahl flask. Samples were subjected to acid digestion using standard method (Abbas and Edward 1989), and the concentrations of As were determined by atomic absorption spectroscopy (Surowitz et al. 1984).

3 Results

3.1 Isolation of Bacteria and Primary Screening of Isolates

The highest cfu/ml was recorded in the samples collected from Halisahar and Asansol region while the lowest in the sample from Tribeni (Table 1). One hundred twenty-five different bacterial colonies were isolated from six different sites. Among them 18 isolates showed growth almost equivalent to control in LA spiked with 100–500 μg/ml As(V) and 50–200 μg/ml of As(III).

3.2 Metal Analysis of Soil Samples

Among the soil samples collected for this study, those from Haldia and Ichapur region contained 57.30 and 43.43 μg/mg total As; however, Fe and Mn were the most abundant metals in the soils, while Zn and Hg showed comparatively lower levels (Table 2). It was observed from the heavy metal profile analysis that samples from Haldia and Ichapur are more toxic.

Table 1 Microbiological analysis of waste sediments from various industrial zones of West Bengal

Sampling sites	Number of normal flora (CFU/ml)	% of As-tolerant isolates
Tribeni	18×10^2	26.9
Ichapur	78×10^2	11.1
Halisahar	172×10^1	30
Asansol	420×10^2	13
Haldia	68×10^1	5.7
Howrah	64×10^1	9

Table 2 Metal profile analysis of soil samples

	Metals (μg/mg of soil)							
Sampling sites	Fe	Zn	Cu	Mn	Cd	Cr	Hg	As
Tribeni	60.5	0.88	4.14	800	60	78	15	14.54
Ichapur	180.12	320.12	20.15	200	70	20	55	43.43
Halisahar	154.94	4.14	12.02	900	70	50	10	13.31
Asansol	88.72	442.56	1,178.24	3800	80	544	68	<10.0
Haldia	56.56	3.36	3.58	500	60	58	32	57.30
Howrah	139.16	11.6	15.06	500	60	54	56	10.5

Table 3 Relative growth of arsenic-tolerant isolates

	(A)			(B)		
	As(V), μg/ml			As(III), μg/ml		
Isolates	500	1000	1500	50	100	200
MSfL2	68.75	59.37	57.81	69.9	61	50
$HsR_2 1$	78.08	61.64	53.42	84.7	60.8	65.2
$HlR_2 11$	83	68	56	92.7	67.07	50

Table 4 Biochemical characterization of isolates

Isolates	Morphology and gram character	SH	CD	LH	CH	GH	AD	UR	DN	CA
MSfL2	Gm+ve, rod	–	–	+	–	–	+	–	–	+
$HsR_2 1$	Gm+ve, cocci	+	–	–	–	–	+	+	–	+
$HlR_2 11$	Gm+ve, rod	+	–	+	+	–	+	+	–	+

SH starch hydrolysis, *CD* cellulose degradation, *LH* lipid hydrolysis, *CH* casein hydrolysis, *GH* gelatin hydrolysis, *AD* arginine dehydrolase, *UR* urease, *DN* denitrification, *CA* catalase

3.3 *Secondary Screening*

The selected 18 isolates showing fairly high tolerance to arsenic in metal incorporated solid agar plates were subjected to liquid screening. Relative growth (%) was calculated (Table 3) which revealed that three isolates showed ≥50% relative growth in LB containing 200 μg/ml of As(III) and 1500 μg/ml of As(V).

3.4 *Biochemical Characterization*

All isolates are Gram-positive in nature. They showed positive response for catalase and arginine dehydrolase. All isolates were negative for cellulose degradation, gelatin hydrolysis, and denitrification test. Two isolates $HsR_2 1$ and $HlR_2 11$ showed positive responses to starch hydrolysis and urease (Table 4).

3.5 *Phylogenetic Study*

Determination of phylogeny for the 16S bacterial identification data was done, and tree reliability was tested by Bootstrap method for the Test of Phylogeny option in MEGA 6 software. Best probable similarity of MsfL2 was shown with *Brevibacillus brevis* strain MI2_LS01, $HsR_2 1$ with *Staphylococcus arlettae* strain IAB14, and $HlR_2 11$ with *Bacillus pumilus* strain B6 (Fig. 1).

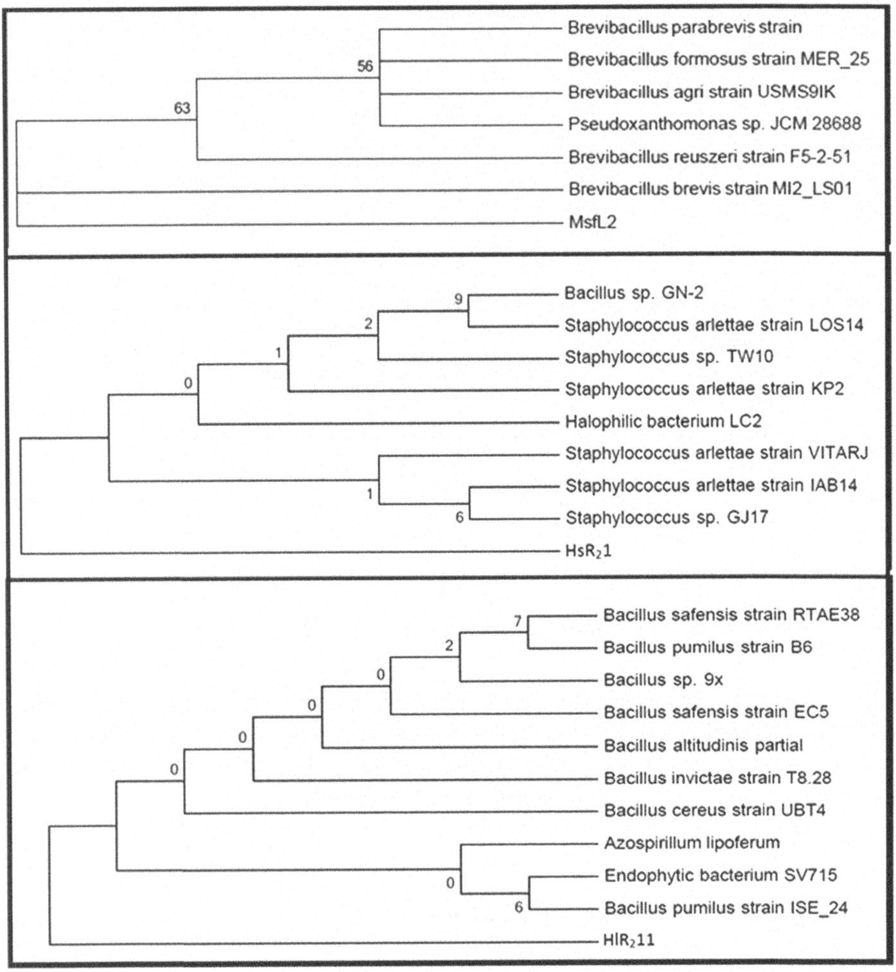

Fig. 1 Phylogenetic analysis of arsenic-tolerant isolates. Based on the maximum identity scored, phylogenetic tree was constructed using MEGA 6

3.6 Antibiotic Sensitivity

As shown in Table 5, all isolates were resistant to penicillin, whereas they were highly sensitive to bacitracin, gentamycin, and tetracycline. MsfL2 and $HlR_2 11$ were resistant to polymyxin B and chloramphenicol. MsfL2 showed intermediate response in case of ciprofloxacin, and $HsR_2 1$ showed intermediate response in case of co-trimoxazole. Multiple antibiotic resistance (MAR) index of the isolates ranged from 0.16 to 0.58 (Table 5). From the MAR index of MsfL2, it may have emerged from an environment contaminated with indiscriminate use of antibiotics.

Table 5 Antibiotic sensitivity profile of arsenic-resistant isolates

	Antibiotics*												
Isolates	Te	B	C	Cot	P	PB	Gen	Nit	CIP	Cl	Amp	S	MAR index
MsfL2	S	S	R	S	R	R	S	R	I	R	R	R	0.58
HsR_21	S	S	S	I	R	S	S	R	S	S	S	S	0.16
HIR_211	S	S	R	S	R	R	S	ND	S	S	ND	S	0.3

"S," "R," "I," and "ND" denote sensitive, resistant, intermediate response and not determined, respectively. *Amp* ampicillin (10 mcg), *Te* tetracycline (5 mcg), *P* penicillin-G (10 mcg), *S* streptomycin (10 mcg), *Gen* gentamycin (10 mcg), *PB* polymyxin B (300 mcg), *C* chloramphenicol (30 mcg), *Cot* co-trimoxazole (25 mcg), *Nit* nitrofurantoin (300 mcg), *B* bacitracin (10 mcg), *CIP* ciprofloxacin (10mcg), *Cl* colistin (10 mcg). *MAR:* multiple antibiotic resistance

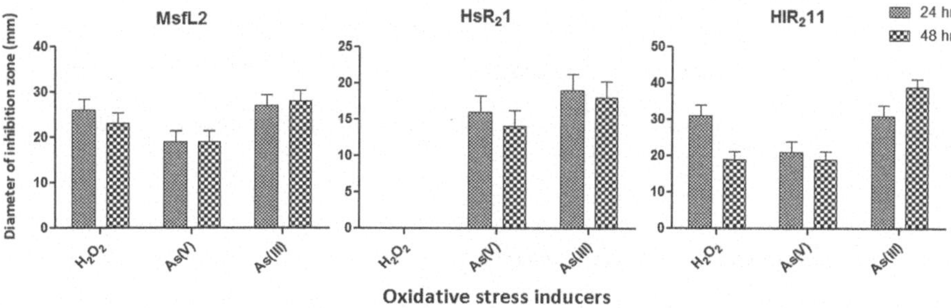

Fig. 2 Diameter of the growth inhibition zone of three arsenic-resistant isolates (Data are means of triplicate and standard errors are calculated)

3.7 *Bacterial Growth Under Oxidative Stress and Osmotic Stress*

Since arsenate is also known to cause oxidative stress in bacteria, the isolates were tested for oxidative stress by measuring the inhibition zones of H_2O_2, As(V), and As(III) (Fig. 2). Under the experimental conditions, no inhibition zone was observed for H_2O_2 with respect to HsR_21 unlike MsfL2 and HIR_211showing 20–30 mm inhibition zone diameter after 24 h which decreased about 50% after 48 h. The isolates showed contrasting response of As(V) and As(III) induced stress indicating As(III) to be more potent oxidative stress inducer than As(V).

The osmotic stress-tolerant pattern of the isolates was determined on the basis of tolerance to different concentration of NaCl. From the relative growth in LB incorporated with sodium chloride, it is evident (Fig. 3) that the growth of the isolates decreased gradually with the increase in concentration of NaCl; ≥50% relative growth was observed in medium supplemented with 5% NaCl. However, HIR_211 significantly showed comparable growth both in 10 and 20% NaCl unlike other isolates whose growth drastically reduced with increase in sodium chloride concentration. On the basis of relative growth in the presence of NaCl, all isolates have been categorized as moderately halophilic.

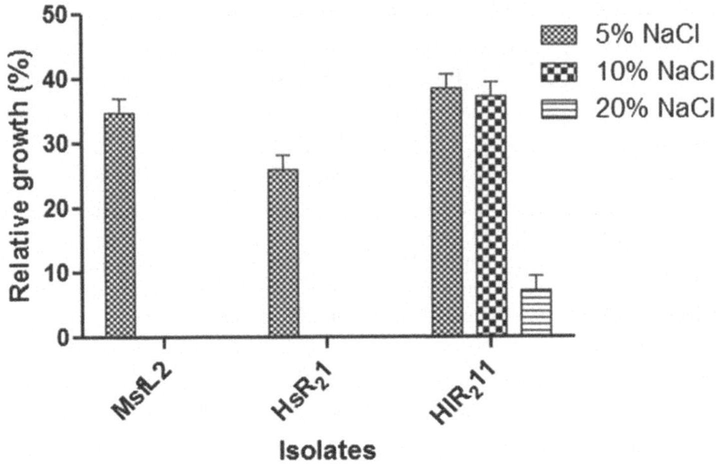

Fig. 3 NaCl tolerance of bacterial isolates. All values represent the average of triplicates

3.8 *Mechanism of As Resistance*

The change of color (brown to yellow) of 72 h old culture grown on LA plate containing As(V) after the addition of silver nitrate indicates the presence of silver arsenite which shows that all the isolates could reduce arsenate (Fig. 4). But no change of coloration was observed when the isolates were grown on As(III)-incorporated LA plates. Hence none of the bacteria has the ability to oxidize arsenite to arsenate. From the observation, it is evident that the isolates probably adopted the enzyme-mediated reduction as resistance mechanism. This was further confirmed by the generation of amplicon (~550 bp) by PCR reaction using As(V) reductase gene primers with respect to $HIR_2 11$ isolate although it is yet to be confirmed for the other two isolates.

3.9 *Biosorption*

It was found that the biosorption capacity of all the isolates in general increased with prolonged incubation for both the concentrations of the metal used [1000 μg/ml and 2000 μg/ml of As(V), 100 μg/ml and 200 μg/ml of As(III)] and was irrespective of the valency of the metal (Fig. 5). However, only isolate $HIR_2 11$ exhibited better biosorption (73.38 μg/mg) at As(III) 100 μg/ml after 24 h incubation. Isolate MsfL2 showed maximum As biosorption potential (247.5 μg/mg As(V) 2000 μg/ml and 35.25 μg/mg As(III) 200 μg/ml after 48 h). Such findings indicate that the selected isolates could serve as effective bioremedial tools of varying capacity.

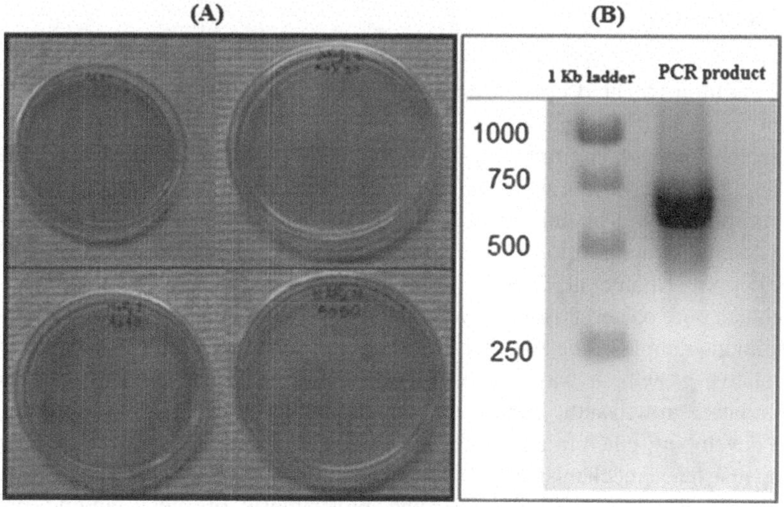

Fig. 4 Arsenic resistance by the bacterial isolates. (**a**) *Top left*, control with 50 μg/ml of As(V); *top right*, MsfL2 with 50 μg/ml of As(V); *bottom left*, HsR_21 with 50 μg/ml of As(V); *bottom right*, HIR_211 with 50 μg/ml of As(V); (**b**) PCR product of arsenate reductase gene

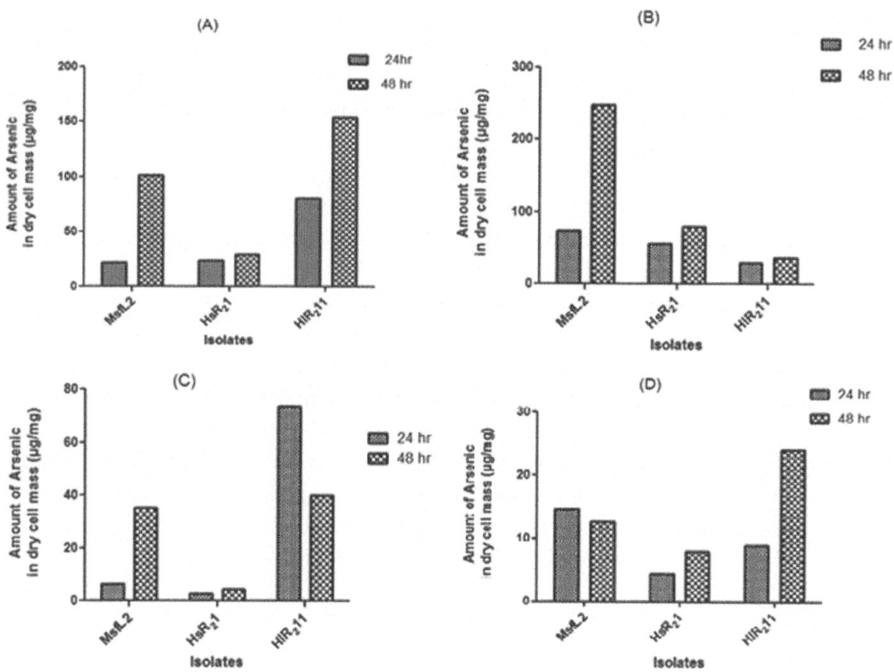

Fig. 5 Bioaccumulation of arsenic by the bacterial isolates. (**a**) As(V) 1,000 μg/ml, (**b**) As(V) 2000 μg/ml, (**c**) As(III) 100 μg/ml, and (**d**) As(III) 200 μg/ml

4 Discussion

Soil containing elevated concentration of metal(s) is a potential source of metal-tolerant bacteria. It is because the environmental condition promotes adaptation of those isolates in such environment. From the metal profile analysis, it was observed that there are wide variations in the level of metal content of the sediment samples, the maximum being arsenic. This reflects the natural condition quite obviously where arsenic concentration of the soil samples was above the permissible limit, i.e., 0.05 ppm (Khan et al. 2000).

Eighteen pure bacterial isolates were obtained from the microbiological analysis of the sediment samples. The As tolerance capacity of these isolates were assessed by their relative growth. It was observed that among the isolates tested three showed $\geq$50% relative growth at the highest concentration of the metal (Table 3). This might be due to developing inherent ability for As tolerance of soil microorganisms. Bacteria develop resistance mechanisms to As by enzymatic transformation of toxic As species, precipitation by oxidation/reduction, and biosynthesis of metal-binding proteins (Valenzuela et al. 2009; Srinath et al. 2002). Among the As-resistant isolates, two, viz., MsfL2 and $HlR_2$11, are Gram-positive rod, and $HsR_2$1 is Gram-positive coccus in nature. The thick cell wall of the Gram-positive cells possibly provide resistance to the toxic arsenic to enter inside the cell which may cause damage to the internal cell organelles and DNA (Berlanga et al. 2009). Although Na^+ ions are not essential for the growth of the isolates, they showed about 30–40% relative growth in LB supplemented with 5% of sodium chloride. Similar results have been reported by Brettar et al. (2002) and Pal et al. (2013) for *Rheinheimera baltica* and *Bacillus flexus*, respectively.

From the comparative study of As absorption potential of the isolates during varying time period, it was found that the removal of As(V) and As(III) increased with time although it was dependent on the valency state of the metal as well (Fig. 5) except $HlR_2$11. Hence, all the isolates were more or less potent in detoxifying arsenic by reduction (Valenzuela et al. 2009). Arsenic-resistant microbes have been reported to reduce the arsenic concentration from the media by developing a number of detoxifying mechanisms including metal reduction, metal efflux, bacterial cell membrane binding, adsorption of heavy metals onto cell surface, and complexation of the metal with exopolysaccharides (Anyanwu and Ugwu 2010).

Bacterial defenses against oxidative burst include antioxidant enzymes such as superoxide dismutase and catalase (Huang et al. 2010; Hassett and Cohen 1989). Inhibition of growth due to oxidative stress varied with the different agents as indicated by the inhibition zone diameter (Fig. 3). Hydrogen peroxide is known to be a strong oxidative stress inducer. However arsenic especially the trivalent form has been found to create higher oxidative stress in soil environment than H_2O_2.

Multiple antibiotic resistance (MAR) index is a tool that reveals the spread of resistance in a given population (Krumpermann 1983). A MAR index greater than 0.2 implies that the strains of such bacteria originate from an environment where several antibiotics are used. Since two of the isolates exhibited more or less fairly high MAR index (Table 5), it is evident that there is a likelihood for exposure of the genera to several antibiotics which might have occurred due to indiscriminate use in the environment (Paul et al. 1997).

5 Conclusion

From this study, it can be concluded that three bacterial isolates MsfL2, HsR_21, and HlR_211 which were identified as *Brevibacillus brevis* strain MI2_LS01, *Staphylococcus arlettae* strain IAB14, and *Bacillus pumilus* strain B6, respectively, can tolerate As and showed ≥50% relative growth at highest concentration of the metal. Isolates are moderately halophilic and can tolerate oxidative stress. All of the isolates have the ability to reduce arsenate to arsenite. According to the results of biosorption, isolates might have evolved mechanisms to tolerate high concentration of arsenic via arsenic-resistant genes. The study thereby reveals that these As-tolerant bacteria can be used for bioremediation of environmental waste sediments contaminated with the metal. However, further elucidation of optimum conditions for bioremediation and its field application is needed.

Acknowledgment The authors would like to express their appreciations to Department of Biotechnology, Government of West Bengal, and Department of Microbiology, Barrackpore Rastraguru Surendranath College, for the financial support and research facilities for this work, respectively. We are also thankful to Sri Prabir Barui, Scientist, Pollution Control Board, Government of West Bengal, for his guidance in sampling as well as AAS facility.

References

Abbas A, Edward C (1989) Effects of metals on a range of *Streptomyces* species. Appl Environ Microbiol 55:2030–2035

Aitio A, Becking G (2001) Arsenic and arsenic compounds. In: Environmental health criteria 224. World Health Organization, Geneva

Altschul SF, Gish W, Miller W, Myers EW, Lipman DJ (1990) Basic local alignment search tool. J Mol Biol 215:403–410

Anyanwu CU, Ugwu CE (2010) Incidence of arsenic resistant bacteria isolated from a sewage treatment plant. Int J Basic Appl Sci 10:64–78

Basu M, Paul AK (2008) Influence of environmental factors on the uptake of chromium by Pseudomonas stutzeri TEM-317 isolated from tannery sludge. J Mycopathol Res 46:289–295

Berlanga GA, Persans MW, Eubanks TM, Lowe KL (2009) Characterization of arsenic-tolerant bacterial cultures from the lower Laguna Madre of south Texas. Tex J Sci 61:1–24

Brettar I, Christen R, Hofle MG (2002) Rheinheimera baltica gen. Nov., sp. nov., a blue-coloured bacterium isolated from the central Baltic Sea. Int J Syst Evol Microbiol 52:1851–1857

Cai J, Salmon K, DuBow MS (1998) A chromosomal *ars* operon homologue of *Pseudomonas aeruginosa* confers increased resistance to arsenic and antimony in *Escherichia coli*. Microbiology 144:2705–2713

Cappuccino JG, Natalie S (2012) Microbiology laboratory manual, 7th edn. Pearson

Cervantes C, Ji G, Ramirez JL, Silver S (1994) Resistance to arsenic compounds in microorganisms. FEMS Microbiol Rev 15:355–367

Diorio C, Cai J, Marmor J, Shinder R, DuBow MS (1995) An *Escherichia coli* chromosomal *ars* operon homolog is functional in arsenic detoxification and is conserved in gram-negative bacteria. J Bacteriol 177(8):2050–2056

Hassett DJ, Cohen MS (1989) Bacterial adaptation to oxidative stress: implications for pathogenesis and interaction with phagocytic cells. FASEB J 3:2574–2582

Huang A, Teplitski M, Rathinasabapathi B, Ma L (2010) Characterization of arsenic resistant bacteria from the rhizosphere of arsenic hyperaccumulator Pteris vittata. Can J Microbiol 56:236–246

Hughes MF (2002) Arsenic toxicity and potential mechanisms of action. Toxicol Lett 133:1–16

Khan AH, Rasul SB, Munir A, Habibuddowla M, Alauddin M, Newaz SS, Hussan A (2000) Appraisal of a simple arsenic removal method for groundwater of Bangladesh. J Environ Sci Health 35:1021–1041

Krumpermann PH (1983) Multiple antibiotic resistance indexing of *Escherichia coli* to identify high risks sources of faecal contamination of foods. Appl Environ Microbiol 46:165–170

Kulp TR, Hoeft SE, Miller LG, Saltikov C, Murphy JN, Han S, Lanoil B, Oremland RS (2006) Dissimilatory arsenate and sulfate reduction in sediments of two hypersaline, arsenic-rich soda lakes: mono and Searles Lakes. California Appl Env Microbiol 72:6514–6526

Liu SX, Athar M, Lippai I, Waldren C, Hei TK (2001) Induction of oxyradicals by arsenic: implication for mechanism of genotoxicity. Proc Natl Acad Sci U S A 98(4):1643–1648

Maidack BL, Olsen GJ, Larson N, Overbeek R, McCaughey MJ, Woese CR (1997) The RDP (ribosomal database project). Nucleic Acids Res 25:109–111

Niggemyer A, Spring S, Stackebrandt E, Rosenzweig RF (2001) Isolation and characterization of a novel As(V)-reducing bacterium: implications for arsenic mobilization and the genus *Desulfitobacterium*. Appl Environ Microbiol 67:5568–5580

Nordstrom DK (2002) Worldwide occurrences of arsenic in ground water. Science 296:2143–2145

Oremland RS, Stolz JF, Hollibaugh JT (2004) The microbial arsenic cycle in Mono Lake, California. FEMS Microbiol Ecol 48:15–27

Pal KC, Mondal NK, Chatterjee SN, Ghosh TS, Datta JK (2013) Characterization of fluoride-tolerant halophilic Bacillus flexus NM25 (HQ875778) isolated from fluoride-affected soil in Birbhum District, West Bengal, India. Environ Monit Assess 186:699–709

Patel PC, GoulhenF BC, Gault AG, Charnock JM, Kalia K, Lloyd JR (2007) Arsenate detoxification in a pseudomonad hypertolerant to arsenic. Arch Microbiol 187:171–183

Paul S, Bezbaruah R, Roy MK, Ghosh AC (1997) Multiple antibiotic resistance (MAR) index and its reversion in Pseudomonas aeruginosa. Lett Appl Microb 24:169–171

Penrose WR (1974) Arsenic in the marine and aquatic environments. Analysis, occurrence and significance. CRC Crit Rev Environ Control 4:465–482

Pichereau V, Hartke A, Auffray Y (2000) Starvation and osmotic stress induced multiresistances. Influence of extracellular compounds. Int J Food Microbiol 55(1–3):19–25

Sambrook J, Fritsch EF, Maniatis T (1989) Molecular cloning, a laboratory manual. Cold Spring Harbor Laboratory Press, Cold Spring Harbor, New York

Srinath T, Verma T, Ramteke PW, Garg SK (2002) Chromium (VI) biosorption and bioaccumulation by chromate resistant bacteria. Chemosphere 48:427–435

Stugeron RE, Siu KWM, Willie SN, Berman SH (1989) Quantification of arsenic species in a river water reference material for trace metals by graphite furnace atomic absorption spectrometry technique. Analyst 114:1393–1396

Surowitz KG, Titus JA, Pfister RM (1984) Effects of cadmium accumulation on growth and respiration of a cadmium-sensitive strain of Bacillus subtilis and a selected cadmium resistant mutant. Arch Microbiol 140:107–112

Tamura K, Stecher G, Peterson D, Filipski A, Kumar S (2013) MEGA6: molecular evolutionary genetics analysis version 6.0. Mol Biol Evol 30(12):2725–2729

Valenzuela C, Campos V, Yañez J, Zaror C, Mondaca M (2009) Isolation of arsenite-oxidizing bacteria from arsenic enriched sediments from Camarones River, Northern Chile. B Environ Contam Tox 82:593–596

Willsky GR, Malamy MH (1980) Effect of arsenate on inorganic phosphate transport in *Escherichia coli*. J Bacteriol 144(1):366–374

Characteristics of Municipal Solid Waste Biochar: Its Potential to be Used in Environmental Remediation

Y. Jayawardhana, P. Kumarathilaka, S. Mayakaduwa, L. Weerasundara, T. Bandara, and M. Vithanage

Abstract Pyrolysis is the foremost thermal conversion process that can be successfully used to transform biomaterials into a value-added product. The estimated biodegradable portion is prominent and denotes as 60% from total waste generation in Asian developing countries. There have been several studies on exploring the pyrolysis of complex organic fraction of municipal solid waste (MSW) streams as a sustainable MSW management technique. The objective of this research was to evaluate physicochemical characteristics of MSW biochar (MSW-BC) produced from organic MSW to observe its potential for landfill contaminant removal with case studies from Sri Lanka. Biochar was pyrolyzed from the MSW in an onsite pyrolyzer. For physicochemical properties of biochar, pH, point of zero charge, electrical conductivity, proximate analysis, ultimate analysis, heavy metal composition, bioavailable heavy metal composition and BET surface area were acquired. In addition, surface functional groups and structural identification were determined by FTIR analysis and scanning electron microscopy (SEM) analysis, respectively. Adsorption capacities for the pollutants (benzene and toluene) were examined by batch sorption experiments. Furthermore, sorption isotherms were fitted using non-linear models for better understanding of the sorption capacities of the materials. Ultimate analysis data suggested high-temperature pyrolysis of MSW. Further, low values for both H/C and polarity index depict the strongly carbonized and highly aromatic structure in BC. Additionally, FTIR suggested a loss of labile, aliphatic compounds and functional groups during pyrolysis and the formation of more recalcitrant, aromatic constituents, whereas BET and SEM data revealed a well-developed porous structure and surface properties, which indicates MSW-BC to be a potential sorbent. Further, the reported total and bioavailable heavy metal content was low in MSW-BC; hence, it can be easily mixed with compost and used as a fertilizer. At the same time, MSW-BC will potentially be used to remediate heavy metals in the landfill leachate. Therefore, MSW-BC shows high potential to be used as a material to remediate

Y. Jayawardhana • P. Kumarathilaka • S. Mayakaduwa • L. Weerasundara • T. Bandara
M. Vithanage (✉)
Faculty of Applied Sciences, University of Sri Jayewardenepura, Nugegoda, Sri Lanka
e-mail: meththika@sjp.ac.lk

S.K. Ghosh (ed.), *Utilization and Management of Bioresources*,
DOI 10.1007/978-981-10-5349-8_20

contaminants simultaneously that minimizes MSW volume. Thus, conversion of MSW to BC and chemical and thermal modification of MSW-BC would allow effective engineering to optimize their properties as a potential material in landfill covers and permeable reactive barriers and integrate in leachate treatment techniques.

Keywords Adsorption • Pyrolysis • MSW-BC • Porous structure • Pollutants

1 Introduction

Most of the developing countries utilize landfills as final disposal technique since landfills are considered as the simplest, cheapest and most cost-effective method of disposing of waste (Barrett and Lawlor 1995). However, the improper solid waste management (SWM) practices generate severe threats on the environment. Therefore, current interventions are aiming to use different technological application to produce sustainable value-added products from waste. Pyrolysis can be considered as an alternative to the conventional SWM methods such as incineration, landfills, bioreactors, aerobic/anaerobic digestion, open-air burning and composting (Serio et al. 2001), which can also produce value-added products from waste. Pyrolysis process has the capability to generate large number of components such as char, tar (oil) and gas during the thermal degradation process (Fagbemi et al. 2001),while reducing the amount of solid waste present (Onyango 2014; Malkow 2004). Literature revealed that the primary objective of the pyrolysis process is the transformation of waste biomass to physically and biologically stable components such as solids, liquids and gases (Roy 1988). The municipal solid waste biochar (MSW-BC) is such value-added product produced from MSW through pyrolysis process.

The temperature gradient, waste composition, catalyst involvement and activation agents are significant factors which influence physical and chemical properties of MSW-BC. Even though various pyrolysis methods are available, the slow pyrolysis is more effective for production of MSW-BC since the properties of MSW-BC that resulted from slow pyrolysis process are remarkable (Chen et al. 2014; Zornoza et al. 2016). The cellulosic portion of MSW tends to gain significant attention as an adsorbent (Bernardo et al. 2012; Sharypov et al. 2002). Thermo-gravimetric studies revealed that the properties of the final product are different based on the degradation pattern of different fibres (lignin, cellulose and hemicellulose), pyrolysis temperatures and nature of the raw material (Yang et al. 2007; Zhou et al. 2013). The properties of BC are closely related to the waste composition and its conditions. If the waste bulk is rich with paper-like materials, the ash content of the BC can be high, whereas plastics and industrial wastes derived BC consist of both toxic compounds and ash (Van Zwieten et al. 2010; Enders et al. 2012). Therefore, the BC produced from industrial wastes, polythene and plastic like compounds will be rich with toxic gases and xenobiotic compounds (i.e., benzene derivatives, polycyclic aromatic compounds). The elemental composition of the lignocelluloses in the BC is the portion that contributes to the efficient sorption process (Venderbosch and Prins 2010).

Therefore, this study discusses the characteristics of BC, transformed from organic material of MSW in the Gohagoda landfill site, Kandy, Sri Lanka, to understand the potential of waste to be reused and recycled and at the same time the potential determination for pollution remediation via a material for capping and leachate treatment.

2 Materials and Methodology

2.1 *Biochar Production and Characterization*

The segregated organic fraction of the MSW obtained from the dumpsite at Gohagoda, Kandy, Sri Lanka, has been used for producing MSW-BC. Pyrolysis was performed in batch reactor built in Gohagoda dumpsite under slow pyrolysis. The furnace temperature was controlled in 450 °C for pyrolysis reaction and the holding time was 30 min. Proximate, ultimate and heavy metal analysis was conducted (muffle furnace Nabertherm N17/HR, Germen, Vario MAX CN, elementar, Germany; Perkin elmer Optima 4300 DV ICP-OES, USA). At the same time, chemical and physical characterization including pH, Electrical conductivity (EC), Cation exchange capacity (CEC), and point of zero charge (pHzpc) and surface characteristics based on BET, FTIR, etc. have been conducted for MSW-BC.

2.2 *Case Studies*

Humic acids (HAs) are dissolved organic compounds (DOCs), and toluene and benzene are volatile organic compounds (VOCs) which cannot easily be removed by conventional landfill leachate treatment methods.

At first, the preselected amount of MSW-BC was added to each amber colour bottle. The solution pH was managed at the range of 2-9 by addition of either 0.1 M HCl or NaOH. Equilibrated concentration of each supernatant was passed through the 0.45 μm membranes. Then, samples were analysed for HA by using UV spectrophotometer (UV-160A Shimadzu) at 254 nm wavelength. For sorption experiments, the dosage selected was 0.5 g/L.

2.2.1 VOC Adsorption (Benzene and Toluene)

The batch sorption technique was employed for interaction identification on MSW-BC and VOCs. The sorbent dosages (1–10 g/L) were investigated and best suited 1 g/L combination was used for further analysis. For benzene and toluene as sorbate, different concentrations (30–300 μg/L) were investigated for isotherm experiments, whereas kinetics were evaluated at initial concentration of 50 μg/L for different time intervals (0.5, 1, 2, 4, 12, 18 and 24 h). Non-linear curve fitting was employed for mechanism determination. Quantitative analysis of benzene and toluene was performed using static headspace equipped gas chromatography-coupled mass spectrometer (Shimadzu GCMS 2010 ultra).

3 Results and Discussion

3.1 General Nature of MSW-BC

Table 1 shows the respective char yield and temperature programme of previous studies with MSW. The yield results can be supported to biomaterial composition where high amount of mass loss under the pyrolysis temperature of 450 °C. During pyrolysis process, the fixed-carbon content of MSW-BC is higher than that of sludge pyrolysis, and MSW shows effective char preparation (Chen et al. 2014). Similar characters of high amount of ash content readily available via pyrolysis in MSW as same as all other literature were observed due to processed biomaterials like paper waste. This ash composition helps to create alkaline nature that facilitate the metal removal ability of MSW-BC (Agrafioti et al. 2014).

3.2 Proximate Results

The thermal degradation resulted disappearances of functional groups which belong to cellulose and hemicelluloses and dehydrogenation of hydroxyl groups due to the higher temperature (Zhang et al. 2015). After the pyrolysis of fibres at high temperature, the results were 46.5% of high-degree fixed-carbon content and 15.6% of ash content (Sørum et al. 2001). The fixed carbon of 63% present indicates moderate pyrolysis temperature of MSW-BC which discloses the fixed aromatic formation during pyrolysis and helps to improve hydrophobicity and exhibit the organic adsorption (Jin et al. 2014; Agarwal et al. 2015). The present study shows 46% fixed-carbon content which depicts high-temperature pyrolysis which can easily be used for organic adsorption. Table 2 shows the proximate results of the MSW-BC from Gohagoda, Sri Lanka.

Table 1 Slow and fast pyrolysis of municipal solid waste and respective derived biochar with their product yield

Type of biomass	Temperature/gradient	Char yield %	References
MSW	300 °C – 5 °C/min	65.6	Zornoza et al. (2016)
	400 °C – 5 °C/min	56.3	
	500 °C – 5 °C/min	47.7	
	700 °C – 5 °C/min	39.6	
MSW	550 °C – 4 °C/min	18 ± 2.0[a]	Velghe et al. (2011)
	900 °C – 10 °C/min	15.86	He et al. (2010)
	900 °C – 10 °C/min	14.92[b]	
	400 °C – 10 °C/min	49.8	Buah et al. (2007)
	700 °C – 10 °C/min	32.3	
Separated (woody material, textile industry residues, cardboard)	350–700 – 10 °C/min	34–19	Phan et al. (2008)
MGW	500 °C – 1 °C/min	48.4	Kabir et al. (2015)
MSW	450 °C – NS	29.8[c]	Yuan et al. (2015)
MSW	500 °C – NS	50	Ateş et al. (2013)
MSW	500 °C	63	Chen et al. (2014)
	600 °C	60	
	700 °C	58	
	800 °C	54	
	900 °C	53	
MSW (organic fraction)	450 °C – 7 °C/min	36.1	(This study)

NS not specified, [a]pyrolysis liquid (water rich/oil/wax), [b]under the catalyst condition, *MSW* municipal solid waste, [c]as a torrefied coal, *MGW* municipal green waste

3.3 Ultimate and Elemental Results

The analytical results for the MSW-BC are presented in Table 2. Ultimate analysis demonstrates low O/C molar ratio of the char and that corroborates the high temperature resulted MSW-BC. Pyrolysis temperature on organic compound adsorption can be distinguished according to Bornemann et al. (2007), and it shows the high-temperature-derived BC favourable for organic compound adsorption. The value of $(N + O)/C$ is closer to 0.2 and compatible on literature and may sorb considerable amount of nutrients like sugars (Zai-ming et al. 2013). Even though, still retain a significant proportion of the polar surfaces in BC. Therefore, non-polar surfaces absorb weakly polar substances, whereas polar groups enrich polar adsorption. However, polar substances in MSW-BC are able to associate with the soil

Table 2 Analytical data for MSW-BC

Proximate analysis

pH	EC(μS/cm)	Moisture (%)	Volatiles (%)	Ash (%)	Carbon (%)
9.7 ± 0.05	310 ± 20	6.3 ± 0.1	31.6 ± 2.2	15.6 ± 3.3	46.5 ± 4.0
–	–	0.4[+]	26.2[+]	50.1[+]	23.3[+]
8.31[*]	–	–	–	–	–
8.0[b]	–	–	26.4[b]	9.2[b]	63.8[b]

Ultimate analysis

C (%)	*H* (%)	*O* (%)	*N* (%)	*S* (%)	Molar *H*/*C*	Molar *O*/*C*	Molar [(*O* + *N*)/*C*]
60.8 ± 0.12	2.79 ± 0.05	14.6 ± 0.02	1.33 ± 0.01	0.16 ± 0.03	0.04 ± 0.01	0.24 ± 0.02	0.26 ± 0.002
80.2[+]	7.4[+]	8.5[+]	2.8[+]	–	0.09[+]	0.24[+]	0.19[+]
64.42[*]	0.22[*]	12.40[*]	2.42[*]	0.29[*]	0.003[*]	0.19[*]	0.23[*]
59.5[b]	9.1[b]	20.8[b]	1.4[b]	0.0[b]	0.15[b]	0.34[b]	0.37[b]
17.46[c]	0.70[c]	10.45[c]	1.54[c]	–	0.48[c]	0.44[c]	0.68[c]

Elemental parameters

Element (mg/kg)	**As**	**Cd**	**Co**	**Cr**	**Cu**	**Fe**	**Mn**	**Ni**	**Pb**	**Sb**	**V**	**Zn**
Total	nd	nd	nd	9.27	10.90	1810.00	305.00	1.81	2.48	nd	nd	82.80
$CaCl_2$ extract	nd	nd	nd	0.08	1.40	11.94	3.50	nd	2.04	nd	nd	14.26
DTPA extract	nd	nd	nd	nd	1.96	18.44	6.61	nd	4.69	nd	nd	9.08
Total metal sludge biochar (a)	–	–	–	92.2	125	–	–	–	67.5	–	–	749
Total metal MSW biochar (b)	12	5	–	64	101	–	–	143	10	–	–	213
Total metal MSW biochar (c)	–	3.37	–	100.3	202.4	31,000	749		51.5	–	–	–

[+] Li et al. (2015), * Liu et al. (2015), [a] Liu et al. (2014b), [b] Jin et al. (2014), [c] Chen et al. (2014), *nd* is not detectable colonization and microbial formation (Liu et al. 2015) and that helps to development of effective cover material amendment with BC (Sadasivam and Reddy 2015; Liu et al. 2015).

The concentration of total and available heavy metal on the MSW-BC and total heavy metal in municipal solid waste and sewage sludge BC is represented in Table 2. Compared to the other MSW-BCs, the amount of heavy metal content is very low in the present MSW-BC (Jin et al. 2014; Chen et al. 2014; Liu et al. 2014a). Higher toxic As and Cd are not detected in MSW-BC, and that condition is favourable for field-level application. Therefore, potential utilization of MSW-BC as sorbent can be implemented without any constraint on the environment.

3.4 pH, Electrical Conductivity and Potentiometric Titration

The electrical conductivity and pH of the MSW-BC are listed in Table 2. Higher pyrolytic temperature further implies the less functional groups specially in acidic groups such as carboxylic and phenolics. At the same time, increment of basic groups can be seen due to separation of alkali salts from organic fraction (Mukherjee et al. 2011). The pHzpc was found to be 6.7 for the MSW-BC. The pHzpc denotes where the MSW-BC pH is neutral. Beyond pH 6.7, the MSW-BC can be used for cationic sorption, while it suits for anion removal at high pH than 7. If the solution pH exceeds the pzc of char, the surface charge of the BC will be positive (Essandoh et al. 2015). Potentiometric titration window is shown in Fig. 1.

3.5 Surface Characterization

The surface properties of the BC are expressed in Table 3. Visualization of deep porous holes in MSW-BC became more prominent. The MSW-BC exhibits substantial porous structure development as well as void establishment (Li et al. 2015). MSW literature revealed that the higher surface is requirement for metal removal from MSW-BC which requires modification and activation. However, the present study exhibited higher surface area for MSW-BC without modification. Similar surface area for the medium temperature-generated BC was observed for azo dye removal from cellulosic MSW study (Agarwal et al. 2015). There was significant improvement of BET surface area (108.47 m^2g^{-1}) for derived MSW-BC. The FTIR spectra (Fig. 2) illustrated the appearance of MSW-BC surface functional groups with the aromatic cycles indicating the ability of a material which can perform surface complexation. Previous studies elaborated the involvement of surface functional groups for cationic metal ion adsorption that is an important mechanism for

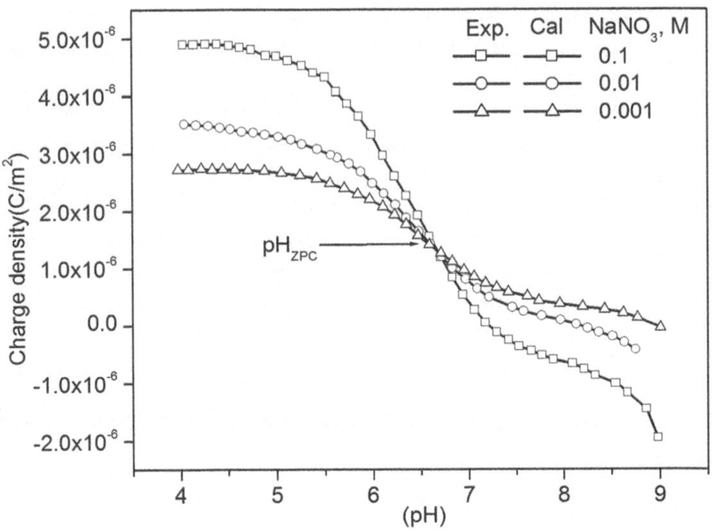

Fig. 1 Potentiometric curves of MSW-BC surface charge against pH of the media (active range of pH 4–9)

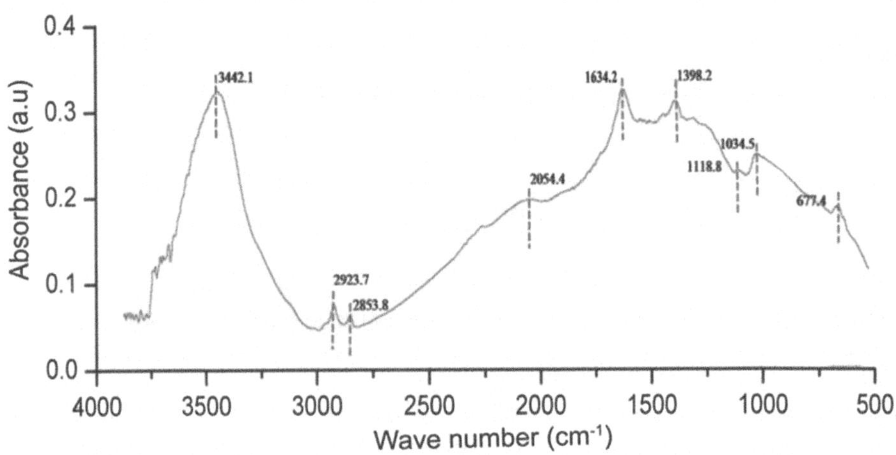

Fig. 2 FTIR spectra of the MSW-BC

MSW-BC (Jin et al. 2014). The peaks at wave numbers 3442 and 1034 show the surface carbon structures; those structures consist of functional groups such as OH^- and aromatic rings. Literature reveal that the MSW-BC have an ability to provide π

Table 3 Surface characteristics for MSW-BC

BET surface area (m^2g^{-1})	Langmuir surface area (m^2g^{-1})	Pore volume (Cm^3g^{-1})	Pore size (nm)
108.47 ± 0.04	212.95 ± 0.41	0.013 ± 0.008	13.057 ± 0.06
2.0[a]	–	0.002[a]	
29.1[b]	–	0.039[b]	–
25.4[c]	–	0.05[c]	3.7[c]

[a] Li et al. (2015), [b] Jin et al. (2014), [c] Chen et al. (2014)

electron and potentially bond heavy metals (Chen et al. 2014). Therefore, literature suggests the potential of MSW-BC in moderate temperature for adsorption of more organics due to carbonization as well as cations by functionality.

3.6 Case Studies

3.6.1 HA Adsorption

Although BC itself contributes to HA, still it has the potential of adsorbing more into its structure. Sorption capacities of the HA into BSW-BC were evaluated at different pHs, and it was observed that the adsorption was higher in acidic region while lower in alkaline region. The maximum sorption (40.36 mg/g) was observed at pH 5–6. Major functional groups in HA as carboxylic and phenolics may release the OH^- at low pH facilitating hydrophobic adsorption with negatively charged MSW-BC surface. At high pH, these anionic species may dissociate and compete with aqueous OH^- to occupy the active BC sites (Omri et al. 2014).

3.6.2 VOC Adsorption

For benzene and toluene, respectively, pH 9.0 and 8.3 were favourable for adsorption, whereas highest adsorption at 24 h reaction time on pH 9.0 and 8.3 was recorded as 85.4 and 87.0% (42.7, 43.5 μg/g), respectively, for benzene and toluene. Freundlich fitting could explain isotherm data with good accuracy for both benzene and toluene (R^2 = 0.955, 0.988). Besides, recorded maximum adsorption capacity was about 218.2 and 257.7 μg/g for benzene and toluene. Hence, a heterogeneous process involved with physisorption between sorbate molecule in aqueous media and sorbent surface can be suggested as the benzene and toluene removal mechanism.

4 Conclusions

The segregated cellulose fraction from MSW is rather effective for production of BC. Moreover, the MSW-BC properties seemed to be well fitted on pollution removal through different mechanisms such as ion exchange, complication and precipitation. Therefore, the production of BC from MSW through slow pyrolysis process is a sustainable approach for landfill improvement due to the net benefit towards greener environment through pollution removal.

By the case studies and literature, possible conclusions can be made for MSW-BC as a potential sorbent for the removal of HA, dye, heavy metals, pesticides and BTEX like benzene and toluene from the contaminated water sources. However, more research should be undertaken to understand the change in the environmental conditions on the desorption of pollutants. Hence, the MSW-BC can be considered as a value-added material from MSW which has a wide range of applications for environmental remediation. At the same time, MSW can be recycled and reused to remediate its own pollutants while reducing the volume of waste by producing MSW-BC.

Acknowledgements Equipment and financial supports from the JICA-JST SATREPS grant and NRC 15-24 are gratefully acknowledged.

References

Agarwal M, Tardio J, Mohan SV (2015) Pyrolysis biochar from cellulosic municipal solid waste as adsorbent for azo dye removal: equilibrium isotherms and kinetics analysis. Int J Environ Sci Dev 6:67

Agrafioti E, Kalderis D, Diamadopoulos E (2014) Arsenic and chromium removal from water using biochars derived from rice husk, organic solid wastes and sewage sludge. J Environ Manag 133:309–314

Ateş F, Miskolczi N, Borsodi N (2013) Comparison of real waste (MSW and MPW) pyrolysis in batch reactor over different catalysts. Part I: product yields, gas and pyrolysis oil properties. Bioresour Technol 133:443–454

Barrett A, Lawlor J (1995) The economics of solid waste management in Ireland, Economic and Social Research Institute (ESRI)

Bernardo M, Lapa N, Gonçalves M, Mendes B, Pinto F, Fonseca I, Lopes H (2012) Physico-chemical properties of chars obtained in the co-pyrolysis of waste mixtures. J Hazard Mater 219:196–202

Bornemann LC, Kookana RS, Welp G (2007) Differential sorption behaviour of aromatic hydrocarbons on charcoals prepared at different temperatures from grass and wood. Chemosphere 67:1033–1042

Buah W, Cunliffe A, Williams P (2007) Characterization of products from the pyrolysis of municipal solid waste. Process Saf Environ Prot 85:450–457

Chen T, Zhang Y, Wang H, Lu W, Zhou Z, Zhang Y, Ren L (2014) Influence of pyrolysis temperature on characteristics and heavy metal adsorptive performance of biochar derived from municipal sewage sludge. Bioresour Technol 164:47–54

Enders A, Hanley K, Whitman T, Joseph S, Lehmann J (2012) Characterization of biochars to evaluate recalcitrance and agronomic performance. Bioresour Technol 114:644–653

Essandoh M, Kunwar B, Pittman CU, Mohan D, Mlsna T (2015) Sorptive removal of salicylic acid and ibuprofen from aqueous solutions using pine wood fast pyrolysis biochar. Chem Eng J 265:219–227

Fagbemi L, Khezami L, Capart R (2001) Pyrolysis products from different biomasses: application to the thermal cracking of tar. Appl Energy 69:293–306

He M, Xiao B, Liu S, Hu Z, Guo X, Luo S, Yang F (2010) Syngas production from pyrolysis of municipal solid waste (MSW) with dolomite as downstream catalysts. J Anal Appl Pyrolysis 87:181–187

Jin H, Capareda S, Chang Z, Gao J, Xu Y, Zhang J (2014) Biochar pyrolytically produced from municipal solid wastes for aqueous as (V) removal: adsorption property and its improvement with KOH activation. Bioresour Technol 169:622–629

Kabir MJ, Chowdhury AA, Rasul MG (2015) Pyrolysis of municipal green waste: a modelling, simulation and experimental analysis. Energies 8:7522–7541

Li G, Shen B, Wang Y, Yue S, Xi Y, An M, Ren K (2015) Comparative study of element mercury removal by three bio-chars from various solid wastes. Fuel 145:189–195

Liu T, Liu B, Zhang W (2014a) Nutrients and heavy metals in biochar produced by sewage sludge pyrolysis: Its application in soil amendment. Polish J Environ Stud 23:271–276

Liu T, Liu B, Zhang W (2014b) Nutrients and heavy metals in biochar produced by sewage sludge pyrolysis: its application in soil amendment. Pol J Environ Stud 23:271–275

Liu G, Xie M, Zhang S (2015) Effect of organic fraction of municipal solid waste (OFMSW)-based biochar on organic carbon mineralization in a dry land soil. J Mater Cycles Waste Manag 1–10

Malkow T (2004) Novel and innovative pyrolysis and gasification technologies for energy efficient and environmentally sound MSW disposal. Waste Manag 24:53–79

Mukherjee A, Zimmerman A, Harris W (2011) Surface chemistry variations among a series of laboratory-produced biochars. Geoderma 163:247–255

Omri A, Benzina M, Trabelsi W, Ammar N (2014) Adsorptive removal of humic acid on activated carbon prepared from almond shell: approach for the treatment of industrial phosphoric acid solution. Desalin Water Treat 52:2241–2252

Onyango D (2014) Pyrolysis: an alternative technology for municipal solid waste management. Scientific conference proceedings

Phan AN, Ryu C, Sharifi VN, Swithenbank J (2008) Characterisation of slow pyrolysis products from segregated wastes for energy production. J Anal Appl Pyrolysis 81:65–71

Roy G (1988) Municipal solid waste recycle-an economic proposition for a developing nation

Sadasivam BY, Reddy KR (2015) Adsorption and transport of methane in landfill cover soil amended with waste-wood biochars. J Environ Manag 158:11–23

Serio MA, Kroo E, Bassilakis R, Wójtowicz MA, Suuberg EM (2001) A prototype pyrolyzer for solid waste resource recovery in space. SAE Technical Paper

Sharypov V, Marin N, Beregovtsova N, Baryshnikov S, Kuznetsov B, Cebolla V, Weber J (2002) Co-pyrolysis of wood biomass and synthetic polymer mixtures. Part I: influence of experimental conditions on the evolution of solids, liquids and gases. J Anal Appl Pyrolysis 64:15–28

Sørum L, Grønli MG, Hustad JE (2001) Pyrolysis characteristics and kinetics of municipal solid wastes. Fuel 80:1217–1227

Van Zwieten L, Kimber S, Morris S, Chan K, Downie A, Rust J, Joseph S, Cowie A (2010) Effects of biochar from slow pyrolysis of papermill waste on agronomic performance and soil fertility. Plant Soil 327:235–246

Velghe I, Carleer R, Yperman J, Schreurs S (2011) Study of the pyrolysis of municipal solid waste for the production of valuable products. J Anal Appl Pyrolysis 92:366–375

Venderbosch R, Prins W (2010) Fast pyrolysis technology development. Biofuels Bioprod Biorefin 4:178

Yang H, Yan R, Chen H, Lee DH, Zheng C (2007) Characteristics of hemicellulose, cellulose and lignin pyrolysis. Fuel 86:1781–1788

Yuan H, Wang Y, Kobayashi N, Zhao D, Xing S (2015) Study of fuel properties of torrefied municipal solid waste. Energy Fuel 29:4976–4980
Zai-Ming C, Bao-Liang C, Dan-Dan Z (2013) Composition and sorption properties of rice-straw derived biochars. Acta Sci Circumst 33:9–19
Zhang J, Liu J, Liu R (2015) Effects of pyrolysis temperature and heating time on biochar obtained from the pyrolysis of straw and lignosulfonate. Bioresour Technology 176:288–291
Zhou H, Long Y, Meng A, Li Q, Zhang Y (2013) The pyrolysis simulation of five biomass species by hemi-cellulose, cellulose and lignin based on thermogravimetric curves. Thermochim Acta 566:36–43
Zornoza R, Moreno-Barriga F, Acosta J, Muñoz M, Faz A (2016) Stability, nutrient availability and hydrophobicity of biochars derived from manure, crop residues, and municipal solid waste for their use as soil amendments. Chemosphere 144:122–130

Farm/Industrial/Municipal Waste: Prospects of Nutrient (Phosphorus) Recovery

Sampriti Kataki, Helen West, Michèle Clarke, and D.C. Baruah

Abstract External supply of plant nutrients is a must to achieve intensification of agricultural sector, thereby meeting food demand of growing global population. However, non-substitutable nutrient like phosphorus (P) with finite natural reserve makes recovery of it a progressively attractive option from alternative waste sources. P recovery method identifies production of P-enriched mineral as mainstream option through technically and economically viable recovery processes. In this context, recovery of struvite ($MgNH_4PO_4.6H_2O$), an alternative P fertiliser from waste streams of farm, municipal and industrial origin, remains a research focus, as considerable fraction of P ends up in these P sinks. The recovery process creates added benefits of quality control of nutrient-laden waste before environmental disposal and reduction of waste volume. In this study, we evaluate some potential P wastes for struvite recovery, in terms of some compositional parameters (orthophosphate, ammonium, calcium). Among these sources, anaerobically digested waste represents suitable considerations for struvite recovery because of retention of nutrient with increased plant availability. Study assessed the potential of anaerobic digestate of different climatic zones (India and the UK) as prospective P source by determining their chemical composition. Through a range of treatments (acidification and chelating agent), P availability was shown to be elevated, indicating its further enhanced potential recovery.

Waste source composition and the need for their management strategy with a focus to their optimal utilisation are identified as the driving forces for the selection of a waste source for P recovery. A viable struvite recovery process is anticipated to develop a product with enriched value to retain P into the nutrient cycle and can be useful as waste treatment/management method.

Keywords Solid waste • Nutrient recovery • Phosphorus • Digestate • Struvite

S. Kataki • D.C. Baruah (✉)
Department of Energy, Tezpur University, Napaam 784028, Assam, India
e-mail: baruahd@tezu.ernet.in

H. West
School of Biosciences, University of Nottingham, Sutton Bonington Campus, Leicestershire LE12 5RD, UK

M. Clarke
School of Geography, University of Nottingham, University Park, Nottingham NG7 2RD, UK

S.K. Ghosh (ed.), *Utilization and Management of Bioresources*,
DOI 10.1007/978-981-10-5349-8_21

1 Introduction

At the current population growth rate, the associated food consumption is seen to grow at an annual growth rate of 3.1% (Heffer and Prud'homme 2014). Food and Agriculture Organization of the United Nations estimated a required growth of 60% in current agricultural production to meet the food demand (FAO 2013). For agricultural intensification, nutrient supplementation is a must, which is given through chemical fertiliser. Phosphorus (P) is one plant macronutrient which is supplied through chemical fertiliser. This is an agricultural input, without which the escalating food demand for growing population cannot be expected to meet, as it is non-substitutable in agriculture. P being a non-substitutable input in agriculture, agro-ecosystem stands for 80–90% of the world total P consumption (Childers et al. 2011). Presently the only source of commercial P fertiliser is natural phosphate rock. Considering the indispensable role of P in agriculture, its limiting nature in soil and limited availability as P rock, sustainable use of P has drawn growing attention. This necessitates trapping or conservation of P from non-conventional P-rich sources with the help of efficient and economically viable processes.

One of the most investigated methods for the recovery of P is through production of solid recoverable phosphate mineral or as a salt precipitate from different P-rich sources. In this context, 'struvite' or magnesium ammonium phosphate (MAP/ $MgNH_4PO_4$ $6H_2O$) is a prospective value added P source. Struvite precipitation requires the presence of Mg^{2+}, NH_4^+ and PO_4^{3-} in alkaline medium in an optimum molar ratio of 1:1:1 (Rahaman et al. 2008). A pH range of 8–11 is considered favourable (Kabdasli et al. 2009). Precipitation of P as struvite is governed by many factors such as pH, presence of impurities, solution strength, mixing energy, temperature, residence time of suspension during crystallisation and effect of seeding material (Doyle et al. 2003; Nelson et al. 2003; Capdevielle et al. 2013).

There are a number of P sinks in nature, where P ends up depositing during its circulation through biogeochemical cycle such as farm waste, municipal waste and industrial effluents. It is important to identify and investigate the suitability of these sources as a source of P recovery. Anaerobic digestate or biogas slurry is such an inevitable by-product of bioenergy conversion technology, 'anaerobic digestion' that generates biogas as a main product and which needs immediate managerial attention. The physico-chemical characteristics of digestate are influenced by the type of feedstock, combined with the effectiveness of the digestion process. Plant essential nutrients like nitrogen, phosphorus and potassium are conserved in digestate from the original feedstock but with enhanced availability because of the increased rate of mineralisation of the organic bound nutrients. Due to these reasons anaerobic digestate holds good potential as prospective struvite recovery source. Further, though the fertiliser value of digestate is realised, there are some limitations of its direct use such as *potential unintended introduction of contaminants* originating from feedstock, *handling and transportation problem* due to enormous quantity of water (about 90%) and *nutrient leaching* through improper method of application and storage leading to eutrophication of aquatic ecosystems. In country like India, where the most common method of utilisation of digestate is to store it in earthen pit and then to transport it to crop area or to directly release into crop area through unlined earthen drain, the nutrient value of digestate is compromised through ground leaching or volatilisation. Therefore, extraction of the only desirable nutrient from digestate seems to be an effective methodology for optimum environmental and economic benefit from biogas residue application.

Up to 94% of total P could be recovered in the form of struvite using anaerobic sludge liquor of wastewater treatment plant as shown by Munch and Barr, 2001. Similarly, 89% of total P contained in digestate from swine waste-based biogas plant could be recovered in the form of struvite (Perera et al. 2007). However, fixation of P with other constituents decreases P in soluble form required for struvite formation, which is the limitation of biogas residue to recover the P (Marti et al. 2008). Potential of digestate for P recovery is likely to be increased if the P concentration in the solution can be enhanced or maximum P can be liberated into available form from the fixed fraction. Recovery potential could be enhanced by pretreatment methods like acidification and chelating agent treatment (Zhang et al. 2010; Stark 2005) that increases available P in solution. In anaerobically digested dairy manure, lowering of pH to 3.8 has been reported to increase available P level by 500% (Zhang et al. 2010). Use of EDTA has been reported as another method to increase dissolved P by up to 93% in digested dairy manure, after which struvite recovery has been enhanced (Zhang et al. 2012).

In addition to fertiliser production, this technology increases suitability of disposal of waste source preventing environmental nuisance, reducing waste volume to be handled. Keeping in view of the above discussion, as a prerequisite for the formulation of appropriate management strategy, the present study has been adopted to assess prospects of three waste streams (farm/municipal/industrial) as struvite source. Present work also investigated prospects of anaerobic digestate as struvite source through characterisation and P availability enhancement study.

2 Materials and Methods

2.1 Suitability of Potential Waste Sources for Recovery of Struvite

To assess the struvite recovery potential from a particular waste source, the study categorises all the potential sources into three types based on their origin of occurrence. Through extensive literature study, the concentration ranges of three selected parameters, viz. PO_4^{3-}, NH_4^+ and Ca, within these sources have been found out to understand their suitability for struvite recovery and other source-specific requirements (need of pretreatment and chemical supplement). To assess the suitability of a source, higher concentration of PO_4^{3-} and NH_4^+ ions and lower concentration of Ca were considered favourable.

2.2 Description of Digestate Samples Considered for Struvite Recovery Potential Study

Study considers two anaerobic digestate samples collected from biogas units of two climatic zones, India and the UK. Prior to analysis the samples were stored at 4 °C. Moisture content (wb) (ASTM D442-92, 2007), ash content (wb) (ASTM E1755-01, 2007) and total solid (% db) of the biogas residue samples were determined using standard method. pH of the raw samples was measured using digital pH meter (Systronics digital pH meter 802). Determination of Ca, Mg and total P was done by ICP-OES (Perkin Elmer Optima 2100 DV). Available P (orthophosphate) was determined using Olsen P method. Total NH_4^+ concentration in the raw samples was determined colorimetrically. Total solid contents of the samples remaining after moisture removal were analysed using EDX analysis for compositional analysis.

2.3 P Enhancement with Acidification and EDTA

A series of tests consisting of acidification and chelating agent (EDTA) addition were used as treatment method to investigate their effect on availability of P (ortho P) in anaerobic digestate. To see the effect of acidification, each sample was adjusted at pH 2, 3, 4 and 5 using HCl (35% w/w). After pH adjustment, the samples were mixed in rotary shaker for 15 min. Following mixing samples were centrifuged at 2000 rpm for 15 min and supernatants were analysed for available P using Olsen P method. To investigate the effect of the chelating agents on available P, each sample was treated with a final concentration of 0, 20, 40, 60, 80, and 100 mmol EDTA. Following addition, and 15 min shaking in rotary shaker, samples were centrifuged at 2000 rpm for 15 min and supernatants were analysed for available P using Olsen P method. In both the set of experiments (acidification, EDTA treatment), experiments were done in triplicate and tenfold diluted samples were used for high total solid content of the samples.

3 Results and Discussion

3.1 Suitability of Potential Waste Sources for Struvite Recovery

After extensive literature survey, we have listed out the range of suitable sources that have been investigated for the recovery of struvite (Table 1). We have categorised these sources into three categories depending upon their occurrence or origin, viz. **farm waste, municipal waste** and **industrial waste**. To consider a waste as prospective struvite source, two main factors of concern are its abundance and need

Table 1 Description of the digestate samples

Sample	Location	Plant type	Feedstock	Feeding rate (kg/day)	Feed used for animals
Ind1	Assam, India	Fixed dome (household plant)	Cow dung	35–40	Straw, vegetable waste, rice bran
UK1	Nottinghamshire, UK	Fixed dome (commercial plant)	Maize	100,000	Not applicable

for its treatment before being released to the environment. However, there are some source-specific issues that need to be taken care off to increase the efficiency of the recovery process such as pretreatments to release bound P into available form (Shen et al. 2011). In case of sources with relatively lower concentration of participating ion (NH_4^+ and PO_4^{3-}), supplementation of necessary chemicals may be required. A brief description on each type of sources category is provided below.

Farm wastes: Farm-based wastes are the most inexpensive sources for struvite recovery with reliable availability. Cattle manure (Shen et al. 2011), swine manure (Zhang et al. 2012), poultry manure (Yetilmezsoy and Sapci-Zengin 2009) and cattle urine (Prabhu and Mutnuri 2014) are some of the sources where successful struvite recovery has been reported. It has been reported that, in dairy manure, total P concentration varies in the range of 100–460 mg/l, whereas in poultry and in swine manure, these vary in the of 370–600 mg/l and 90–200 mg/l, respectively. However, in case of farm-based sources, incorporation of pretreatment method is a prerequisite to enhance the recovery process due to the presence of P in particulate form. Use of acid leaching in dairy manure and poultry litter, chelating agent, microwave treatment and anaerobic digestion in dairy manure are the pretreatment methods reported in struvite recovery from farm waste (Szogi et al. 2008; Qureshi et al. 2008; Moody et al. 2009).

Municipal waste: Spontaneous struvite precipitation happened to be a nuisance in municipal sewer system. With gradual understanding over the process condition through R and D, the scope of controlled struvite recovery using other waste also was realised. Therefore, anaerobic sludge effluent of municipal waste water is the most widely investigated struvite source (Turker and Celen 2007; Uysal et al. 2010; Pastor et al. 2010). It contains PO_4^{3-} concentration in the range of 21–270 mg/l and NH_4^+ concentration in the range of 168–1400 mg/l. In case of sources with PO_4^{3-} concentration below 50 mg/l, external addition of P salt such as H_3PO_4, KH_2PO_4 was reported to be added for struvite precipitation (Turker and Celen 2007; Uysal et al. 2010). Precipitation of struvite has also been recommended as a treatment method to bring down the high NH_4^+content of waste such as landfill leachate (average 2430 mg/l NH_4^+) (Kim et al. 2006; Iaconi et al. 2010) and human urine (average 3000 mg/l NH_4^+) (Ganrot et al. 2007; Morales et al. 2013). However, because of low ortho P content in leachate (~11 mg/l) (Kim et al. 2006), P salt has to be supplemented for struvite precipitation (Iaconi et al. 2010).

Table 2 Physical characterisation of biogas residue samples

Sample	pH	Total solid, % wet basis	Ash, % of total solid
Ind1	8.8	5.41	28.57
UK1	8.3	9.73	19.28

Industrial wastes: Struvite recovery has been mentioned as an alternative way to reduce both P and NH_4^+ concentrations in potentially polluting industrial effluent to meet the environmental standards set, as the industrial effluent contains substantial amount of nutrients from input material. Recovery of struvite at laboratory scale has been reported in waste water from tannery industry (Tunay et al. 1997); textile industry (Huang et al. 2012); carmine dye (Chimenos et al. 2003), semiconductor (Kim et al. 2009) and slaughterhouse industry (Kabdasli et al. 2009); potato processing and molasses-based industry (Turker and Celen 2010); rare earth (Huang et al. 2011), coking (Kumar and Pal 2013), 7-aminocephalosporanic acid (Li et al. 2012) and yeast (Uysal and Demir 2013) industry. The main target to employ struvite recovery in industrial waste water is to capture the ammonia-N fraction. However, their lower orthophosphate concentration (10–50 mg/l) compared to farm waste makes P salt supplementation a necessity to make the crystallisation effective.

From compositional status, human urine can be considered as a suitable waste source for struvite recovery. Though the PO_4^{3-} concentration of urine is relatively lower compared to other potential wastes, high concentration of other desirable ion NH_4^+ (average 3000 mg/l) and lower concentration of inhibiting ion Ca (16–234 mg/l) make it a favourable source. Further, no pretreatment method has been reported prior to struvite recovery using human urine in literature.

3.2 *Characterisation of Biogas Residue*

pH in both the samples showed moderate alkalinity due to ammoniacal nitrogen formation during the digestion process. The total solid concentration of the samples was in the range of 5.41–9.73%. These two parameters are directly related to feedstock quality and operational parameters. In digestate sample from India with cow dung as feedstock, ash content is higher (28.57%) compared to the ash content of digestate from maize (Table 2).

The chemical characterisations (total P, available P, Ca, NH_4^+, Mg) of the samples are shown in Table 3. The total P concentration of the samples varied between 64 (IND1) and 87.14 mg/l (UK1). However, the samples had lower orthophosphate concentration with a value of 22.87 mg/l (IND1) and 9.71 mg/l (UK1). These concentration values were found to be lower compared to other anaerobically digested wastes (digested dairy manure, swine manure, anaerobic sludge from wastewater treatment) where P recovery was successful. Though both the samples have relatively higher NH_4^+ concentration after anaerobic digestion, its concentration was

Table 3 Concentrations of different ionic constituents of digestate

Parameters	IND1	UK1
Total P (mg/l)	64.36	87.14
Available P (mg/l)	22.87	9.71
Total Ca (mg/l)	460.28	486.09
Total NH_4^+ (mg/l)	210.00	3250.00
Total Mg (mg/l)	396.86	71.96

Table 4 EDX analysis of the samples

	Element (mg/g)										
Sample	C	O	Mg	Na	Si	P	S	K	Ca	P:Ca	Mg:P
IND1	495.90	438.67	2.5	3.1	45.3	3.8	1.5	3.2	2.6	1: 0.68	1: 1.52
UK1	516.38	321.71	13	NA	24.8	9.6	5.9	35.6	25.5	1:2.65	1: 0.73

found to be higher in UK sample (3250 mg/l) compared to Indian samples (210 mg/l), which may be attributed to ammonia volatilisation loss from Indian digestate samples because of the method of storage in uncovered and unlined earthen pit.

The Ca concentration in both the samples was significantly higher with marginal variation. Presence of Ca can lengthen the induction time preceding the precipitation of P mineral, since it will bind for phosphate ions (Le Corre et al. 2005). With high Ca content, available P would be limited in forming particulate P, and it was hypothesised that crystal formation would be inhibited. Reason of low availability of available P in the samples could be related to the high Ca content of the samples. The low P availability within the samples may be further explained by the composition of the solid fraction of the samples determined using EDX (Table 4). P:Ca ratio showed almost equal or double the concentration of Ca than that of P in all the samples. This possibly indicated the presence of P in combination with Ca as particulate form within the samples.

The results obtained from the characterisation study of the samples indicated possible hindrance on P recovery by the presence of P in particulate form. Therefore, P enhancement to increase available P fraction in solution was investigated using the same digestate samples.

3.3 *Effect of Acidification and EDTA Addition*

Acidification has been suggested as an effective method of dissolving particulate P into solution (Zhang et al. 2010). Acidification causes the protonation of particulate phosphate lowering their ionic product below their equilibrium solubility product which shifts the equilibrium causing higher dissolution of particulate mineral phosphate and increasing the availability of P in solution (Zhang et al. 2010). Reduced pH has been shown to release from P from both the samples (Fig. 1) which proves

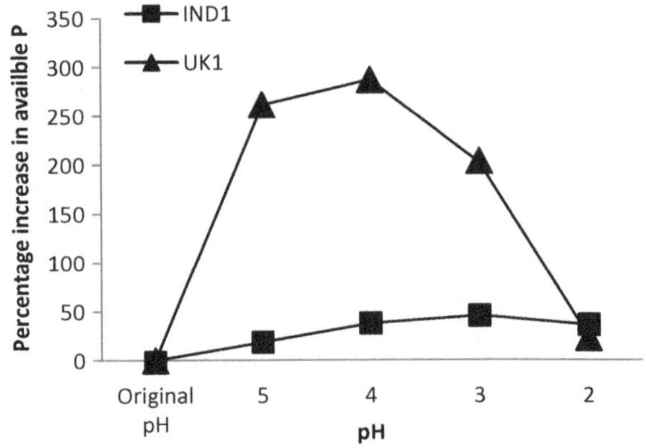

Fig. 1 Percentage increase in available P after acidification treatment in digestates

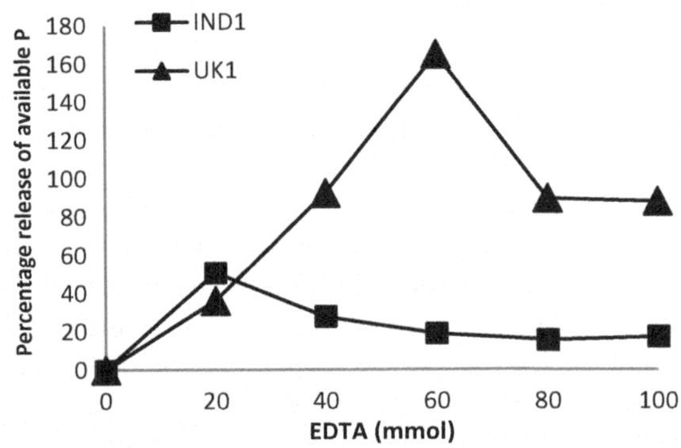

Fig. 2 Percentage increase in available P after chelating agent treatment in digestates

the proof of concept. It has been shown that, by acidification of up to 287%, more P can be liberated from the fixed P fraction. At further higher acidification, no enhancement of P release could be seen, which may be due to the presence of remaining P as a part of nucleation particles in centre matrix of digestate (Hong et al. 2005).

Use of chelating agent has been suggested as an effective way to control inhibitory effect of Ca on struvite precipitation and enhance recovery (Zhang et al. 2010). It reacts with Ca of particulate [Ca-PO_4] to form soluble [Ca-EDTA], and eventually PO_4^{3-} is released as shown in reaction (1). It has been seen that in both the digestate samples, P was released into solution after the addition of EDTA (Fig. 2), making it available for further recovery with maximum release of 267% in UK digestate sample:

$$\left[\mathrm{Ca}-\mathrm{PO}_4\right]^{-}_{(s)} + \left[\mathrm{EDTA}\right]^{4-}_{(aq)} \leftrightarrow \left[\mathrm{EDTA}-\mathrm{Ca}\right]^{2-}_{(aq)} + \mathrm{PO}_4{}^{3-}{}_{(aq)} \quad (1)$$

It was found that with increase in EDTA addition, concentration of available P simultaneously increases. However, after certain optimum concentration of EDTA, no further increase in release of PO_4^{3}- could be seen. The chelating agent quantity again needs to be optimised for the reason that, once all the calcium becomes bound to EDTA, the chelating agent may chelate other minerals such as Mg based on its affinity to chelate metal ions. This will subsequently limit the potential for P recovery as struvite. Therefore, the quantity of EDTA should be such that only Ca fraction becomes bound and not the Mg.

Enhancement of P availability after acidification and EDTA treatment indicated its enhanced potential recovery as struvite using digestate and established the effectiveness of these pretreatments in biogas digestate also.

4 Conclusions

The present work investigates the potential of three types of waste streams for struvite production through their nutritional status to understand their suitability for struvite recovery. Since biogas digestate varies in chemical composition based on feedstock, digestion conditions and digester type, the method for extracting P will be dependent on numerous factors and inherent concentration of participating as well as non-participating ions needs to be monitored and controlled. Through the use of chelating agent acidification, we have shown that up 267% more P can be made available into solution; by restricting the binding effect of calcium, however, the process again needs to be optimised as the effect is sample specific.

In addition to producing a supplement to diminishing phosphate fertiliser, successful execution of recovery process would produce ways to deal with the nutrient-laden waste for effective environmental disposal through cost-effective relocation of excess nutrients.

References

Capdevielle A, Sykorova E, Biscans B, Beline F, Daumer ML (2013) Optimization of struvite precipitation in synthetic biologically treated swine wastewater—determination of the optimal process parameters. J Hazard Mater 244–245:357–369

Childers DL, Corman J, Edwards M, Elser JJ (2011) Sustainability challenges of phosphorus and food: solutions from closing the human phosphorus cycle. Bioscience 61:117–124

Chimenos JM, Fernandez AI, Villalba G, Segarra M, Urruticoechea A (2003) Removal of ammonium and phosphates from waste-water resulting from the process of cochineal extraction using MgO-containing by-product. Water Res 37:1601–1607

Doyle JD, Oldring K, Churchley J, Price C, Parsons SA (2003) Chemical control of struvite precipitation. J Environ Eng 129:419–426

FAO statistical yearbook (2013) World food and agriculture, food and agriculture Organization of the United. Nations, Rome. Available at: http://www.fao.org/docrep/018/i3107e/i3107e.PDF

Ganrot Z, Dave G, Nilsson E (2007) Recovery of N and P from human urine by freezing, struvite precipitation and adsorption to zeolite and active carbon. Bioresour Technol 98:3112–3121

Heffer P, Prud'homme M (2014) Fertilizer outlook 2014–2018. In: 82nd annual conference. International Fertilizer Industry Association, Sydney, Australia

Hong KJ, Tarutani N, Shinya Y, Kajiuchi T (2005) Study on the recovery of phosphorus from waste-activated sludge incinerator ash. J Environ Sci Health J Environ Sci Health 40(3):617–631

Huang HM, Xiao XM, Yang LP, Yan B (2011) Removal of ammonium from rare-earth wastewater using natural brucite as a magnesium source of struvite precipitation. Water Sci Technol 63:468–474

Huang H, Song Q, Wang W, Wu S, Dai J (2012) Treatment of anaerobic digester effluents of nylon wastewater through chemical precipitation and a sequencing batch reactor process. J Environ Manag 30:68–74

Iaconi CD, Pagano M, Ramadori R, Lopez A (2010) Nitrogen recovery from a stabilized municipal landfill leachate. Bioresour Technol 101:1732–1736

Kabdasli I, Tünay O, Özcan P (2009) Application of struvite precipitation coupled with biological treatment to slaughterhouse wastewaters. Environ Technol 30:1095–1101

Kim D, Ryu HD, Kim MS, Kim J, Lee SI (2006) Enhancing struvite precipitation potential for ammonia nitrogen removal in municipal landfill leachate. J Hazard Mater 146:81–85

Kim D, Kim J, Ryu HD, Lee SI (2009) Effect of mixing on spontaneous struvite precipitation from semiconductor wastewater. Bioresour Technol 100:74–78

Kumar R, Pal P (2013) Turning hazardous waste into value-added products: production and characterization of struvite from ammoniacal waste with new approaches. J Clean Prod 43:59–70

Le Corre KS, Valsami EJ, Hobbs P, Parsons SA (2005) Impact of calcium on struvite crystal size, shape and purity. J Cryst Growth 283:514–522

Li Z, Ren X, Zuo J, Liu Y, Duan E, Yang J, Chen P, Wang Y (2012) Struvite precipitation for ammonia nitrogen removal in 7-Aminocephalosporanic acid wastewater. Molecules 17:2126–2139

Marti N, Bouzas A, Seco A, Ferrer J (2008) Struvite precipitation assessment in anaerobic digestion processes. Chem Eng J 141:67–74

Moody LB, Burns RT, Stalder KJ (2009) Effect of anaerobic digestion on manure characteristics for phosphorus precipitation from swine waste. Appl Eng Agric 25:97–102

Morales N, Boehler MA, Buettner S, Liebi C, Siegrist H (2013) Recovery of N and P from urine by struvite precipitation followed by combined stripping with digester sludge liquid at full scale. Water 5:1262–1278

Nelson NO, Mikkelsen RL, Hesterberg DL (2003) Struvite precipitation in anaerobic swine lagoon liquid: effect of pH and Mg:P ratio and determination of rate constant. Bioresour Technol 89:229–236

Pastor L, Mangin D, Ferrer J, Seco A (2010) Struvite formation from the supernatants of an anaerobic digestion pilot plant. Bioresour Technol 101:118–125

Perera PWA, Han ZY, Chen YX, Wu WX (2007) Recovery of nitrogen and phosphorous as struvite from swine waste biogas digester effluent. Biomed Environ Sci 20:343–350

Prabhu M, Mutnuri S (2014) Cow urine as a potential source for struvite production. Int J Recycl Org Waste Agric 3:49

Qureshi A, Lo KV, Liao PH (2008) Microwave treatment and struvite recovery potential of dairy manure. J Environ Sci Health B 43:350–357

Rahaman MS, Ellis N, Mavinic DS (2008) Effects of various process parameters on struvite precipitation kinetics and subsequent determination of rate constants. Water Sci Technol 57:647–654

Shen Y, Ogejo JA, Bowers KE (2011) Abating the effects of calcium on struvite precipitation in liquid dairy manure. Trans ASABE 54:325–336

Stark K (2005) Phosphorus release and recovery from treated sewage sludge. Ph.D thesis, Land and Water Resources Engineering, KTH Architecture and the Built Environment, Stockholm, Sweden

Szogi AA, Vanotti MB, Hunt PG (2008) Phosphorus recovery from poultry litter. Trans ASAE 51:1727–1734

Tunay O, Kabdasli I, Orhon D, Kolçak S (1997) Ammonia removal by magnesium ammonium phosphate precipitation in industrial wastewaters. Water Sci Technol 36:225–228

Turker M, Celen I (2007) Removal of ammonia as struvite from anaerobic digester effluents and Recycling of magnesium and phosphate. Bioresour Technol 98:1529–1534

Turker M, Celen I (2010) Chemical equilibrium model of struvite precipitation from anaerobic digester effluents. Turk J Eng Environ Sci 34:39–48

Uysal A, Demir S (2013) Struvite Pyrolysate recycling for removing ammonium from Baker's yeast industry wastewater. In: Proceedings of international conference on Environment Science and Technology, Turkey

Uysal A, Yilmazel YD, Demirer GN (2010) The determination of fertilizer quality of the formed struvite from effluent of a sewage sludge anaerobic digester. J Hazard Mater 181:248–254

Yetilmezsoy K, Sapci-Zengin Z (2009) Recovery of ammonium nitrogen from the effluent of UASB treating poultry manure wastewater by MAP precipitation as a slow release fertilizer. J Hazard Mater 166:260–269

Zhang T, Bowers KE, Harrison JH, Chen S (2010) Releasing phosphorus from calcium from struvite fertilizer production from anaerobically digested dairy effluent. Water Environ Res 82:34–42

Zhang DM, Chen YX, Jilani G, Wu WX, Liu W, Hen J (2012) Optimization of struvite crystallization protocol for pretreating the swine wastewater and its impact on subsequent anaerobic biodegradation of pollutants. Bioresour Technol 116:386–395

Removal of Copper from Bioleachate of Electronic Waste Using Banana-Activated Carbon (BAC) and Comparison with Commercial-Activated Carbon (CAC)

J. Senophiyah Mary and T. Meenambal

Abstract Electronic waste is the waste electrical and electronic equipment which is not fit for its original intended use. Out of which printed circuit board (PCB) has approximately ten different metals with it, while the percentage of metals present in PCB differ according to the instruments. The extraction of these metals play a major role because of the depleting natural resources. The lixiviant pool of bioleaching was chosen for the study for the extraction of metal. Copper has been chosen among all metals because of its increased usage and high content of about 30% on the PCB. Adsorption is used as a key tool in the extraction process. An attempt has been made to minimise waste and convert them into activated carbon. The activated carbon synthesised with banana peel (BAC) could replace commercial-activated carbon (CAC) which could reduce the depletion of natural resources. The copper adsorption capacity for BAC was more than 80%. The results of atomic absorption spectroscopy (AAS) indicated that banana peel is an effective adsorbent of copper from electronic waste leachate which results in waste minimisation.

Keywords E-waste • Bioleachate • Metal extraction • Commercial activated carbon • Banana activated carbon

1 Introduction

Electronic waste is the waste electrical and electronic equipment which is not fit for its original intended use. They are tomorrow's source and today's alarm as there is a depletion of mines and increase in e-waste due to lack of standard operating procedures to treat e-waste. It could be treated by various treatment processes like pyrometallurgy, hydrometallurgy, bioleaching and plasma treatment. The processes stated have various pros and cons.

J. Senophiyah Mary (✉) • T. Meenambal
Department of Environmental Engineering, Government College of Technology, Coimbatore, India
e-mail: senophiyah.mary8@gmail.com

S.K. Ghosh (ed.), *Utilization and Management of Bioresources*,
DOI 10.1007/978-981-10-5349-8_22

Bioleaching is a process in which microorganisms play an important role in the conversion of insoluble metals into its solubilised form. As bioleaching is a time consuming process it has not been industrialised in various recycling units while comparing to other treatment methods. Soon it will be industrialised by using microbes that are genetically modified. Fungal strains are highly involved in this process like *Aspergillus niger* and *Penicillium chrysogenum* (Aharoni and Ungarish 1977). The lixiviant pool that is produced by the microorganisms after the bioleaching process contains a mixture of various metals like copper, lead, nickel, tin and precious metals like gold, silver, platinum, etc. The metal concentration differs with different circuit boards and different microorganisms involved in the process. There lies a serious problem in the extraction of the specified metals out of the lixiviant pool (Hosseini et al. 2003).

Copper, where the need is increasing day by day, as there is a growth in electronic devices, being the major component in various electronic devices, is chosen because of its high conductivity. While metals like lead, nickel and tin might get extracted during the process. This work is focussed only on copper (Maldhure and Ekhe 2011).

Though there are various methods for the extraction of metals out of the lixiviant pool, adsorption is chosen because of its simplicity and low cost and low time consumption (Bai and Mardina 2002). The advantage is that it can adsorb nanoparticles which could be desorbed and can be used directly for various medical, industrial and wastewater treatment processes (Bernard and Jimoh 2013; APHA 2005). The adsorbent used is prepared from peels of banana which is easily degradable (Bansal et al. 1988; Zheng and Wang 2013). In general banana peels are used as fodder for cattle. Still, the waste generated is high because of the increasing number of hot chips shop, as it is a common snack all over the world, and it is famous in parts of Tamil Nadu and Kerala.

Preparation of activated carbon itself is an art. There are various processes like carbonisation and activation processes like thermal activation, chemical activation, etc. Annadurai et al. (2002). The main objective of this work was to convert waste into wealth in a healthy way. Various pretreatments were done to improve the adsorption process (Brunauer et al. 1938; Arancon Norman et al. 2008). The adsorbate obtained after the copper adsorption process was characterised by atomic absorption spectroscopy.

2 Materials and Methods

2.1 *Bioleachate*

Electronic wastes were treated with various effective microorganisms for 20–25 days and the leachate was collected, and henceforth it would be called as bioleachate. They are autoclaved to kill the liable microorganisms and centrifuged at 10,000 rpm for 20 min at 5 °C. It is then used as a main source for the process. The centrifugation is done to remove some heavy weight particles like cell debris. That cell debris is then lysed to extract the metals out of it. Initially, the metal concentration of bioleachate was analysed using Agilent atomic absorption spectroscopy (AAS) and marked as I_i. It was then diluted to 10 ppm. This was used as a source for the adsorption process. In order to prepare the standards, 1 g of copper turnings was immersed in concentrated nitric acid which forms copper nitrate. This is called as the stock solution. This sample was then diluted from 1 to 10 ppm for further analysis in atomic absorption spectroscopy (AAS).

2.2 *Activated Carbon*

As stated by Hema and AVVS (2011), Holden (1982), Hosseini et al. (2003) the activated carbon was prepared. The activated carbon was prepared from the peel of banana. The peel was collected from local chips shop. It was washed with distilled water and NaCl to remove contaminants. They are then chopped in different sizes. It is then sun-dried for 24 h to remove moisture. They are dried at 100 °C in a hot air oven for 24 h for thermal carbonisation. The samples were again heated at higher temperature at 800 °C for about 10 min in a muffle furnace for thermal activation (Ho and McKay 1999). The formed activated carbon was crushed in a ball mill apparatus to get a uniform size. This activated carbon is named as banana-activated carbon (BAC).

(a) **pH**

The pH value was determined using an Orion pH meter, after overnight shaking of 0.2 g activated carbon in 25 ml distilled water.

(b) **FTIR (Fourier transform infrared spectroscopy)**

PerkinElmer Fourier transform infrared spectroscopy was used for this process. It was done to determine the functional groups which are very useful in adsorption process.

(c) **Scanning electron microscope**

This SEM imaging is done to know the topography of the particle. It also evaluates the particle size.

2.3 *Experimental Procedure*

Fifty millilitre of 100 ppm sample is taken in an Erlenmeyer flask into which 0.5 g adsorbents like BAC and the same was done for CAC. It was then given agitation in an orbital shaker for 1 h at 150 rpm at 35 °C. The sample is allowed to settle and it was taken for further analysis using AAS (Cooney 1998) (Cheung 2000).

The following studies were done to optimise the adsorption process.

2.3.1 Optimisation of Contact Time

All chemicals and reagents used in this study were of analytical grade. Adsorbent dosages of 0.1 g were added to a sample of 500 ppm. It was kept in an orbital shaker at 150 rpm for various contact times like 20 min, 40 min and 1–5 h, respectively.

2.3.2 Optimisation of the Sample

The ability of 0.1 g of BAC and CAC should be optimised to avoid wastage of activated carbon. In order to optimise the sample quantity to be used, a series of copper nitrate solution in the dilution rate of 10, 50, 100, 150 and 200 ppm should be used to find the optimum metal concentration with which the adsorbent could act on adsorbate. The amount of copper samples was tested using atomic absorption spectrometer.

2.3.3 Optimum Dosage of Adsorbents

The amount of adsorbent should be optimised before use. A sample size of 0.1, 0.2, 0.3, 0.4 and 0.5 g were analysed for the sample of 500 ppm.

2.3.4 Optimisation of Time

Adsorption is a simple process within which time plays a key role. The sample was mixed with an optimum value of adsorbent and kept in an orbital shaker at 180 rpm in 20 min, 40 min and 1, 2, 3, 4 and 5 h, respectively.

The same experiment is done for commercial-activated carbon too.

3 Results and Discussion

3.1 E-Waste Bioleachate

The electronic waste bioleachate is a pregnant solution which has so many metals in the form of oxides, carbonates, nitrates, etc. Characteristics of e-waste bioleachate mainly pH were found to be 4.5. The amount of copper metal present in the sample is determined using Agilent atomic absorption spectroscopy (AAS) using copper flame. The amount of copper metal present in the supernatant was approximately 500 ppm.

3.2 *Activated Carbon*

Characteristics of activated carbon pH, presence of functional group by using Fourier transform infrared spectroscopy and SEM imaging were determined.

(a) **pH of activated carbon**

pH of banana activated carbon was determined using a pH meter. It shows that thermally activated carbon has higher pH value of 8.05 and highly suitable for adsorption process.

(b) **FTIR**

FTIR (Fourier transform infrared spectroscopy) PerkinElmer Grade equipment was used for this process. It was done to determine the functional groups which are very useful in adsorption process.

On comparing the IR spectra of the activated carbon prepared from banana peel, presence of various functional groups could be predicted. The displayed bands in the region 3400–3500 cm^{-1} could be attributed to O–H stretching, and bands around 2900–2800 cm^{-1} were due to C–H stretching vibrations in aromatic rings. Peaks in the range 1620 cm^{-1} suggested the presence of C=O carbonyl group quinonic from the stretching of C=C. Peaks in the range 600–700 cm^{-1} were due to out-of-plane C–H bending with different degrees of substitution. It is evident from Fig. 1 that most of the peaks are common in the two activated carbons. Based on the collective FT-IR data, the commercial-activated carbon has high heterogeneity which is useful for more adsorption of metals. The surface heterogeneity is due to the presence of the functional groups like C=O, aromatic C=C, aromatic C–H, hydrogen-bonded O–H groups.

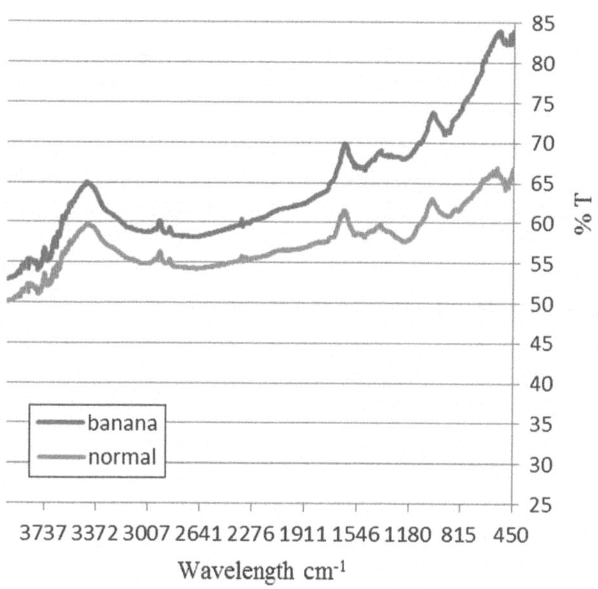

Fig. 1 Results of FTIR for BAC and CAC

Fig. 2 SEM image of BAC

3.3 *Scanning Electron Microscope*

The SEM image of banana-activated carbon is shown in Fig. 2. It shows that the homogenity of the sample (BAC) was low while comparing to CAC. The size of the particle is less than 1 μm. The activation can be improved by addition of some chemicals during activation as well as effective ball milling process (Dubinin 1960, 1965).

3.4 *Adsorption*

Adsorption process is carried out by taking 50 ml of 100 ppm sample in a 100 ml Erlenmeyer flask within which 0.5 g of ACs was added and kept in orbital shaker at 150 rpm in 35 °C for 4 h. After the process, the sample is filtered and percentage removal of copper is determined by testing for residual amount of copper in the sample by using AAS, and the result shows that banana-activated carbon and commercial-activated carbon has comparatively same adsorbing capacity.

3.4.1 Optimisation of the Sample

In order to optimise the sample quantity to be used, a series of copper nitrate solution in the dilution rate of 10, 50, 100, 150 and 200 ppm were used to find the optimum metal concentration. In many cases, BAC has more capacity in adsorbing metal while comparing to CAC as shown in Fig. 3. The amount of copper content before and after adsorption was tested using atomic absorption spectrometer. An optimum amount of 100 ppm sample can be used to recover copper out of the leachate which showed an effective result.

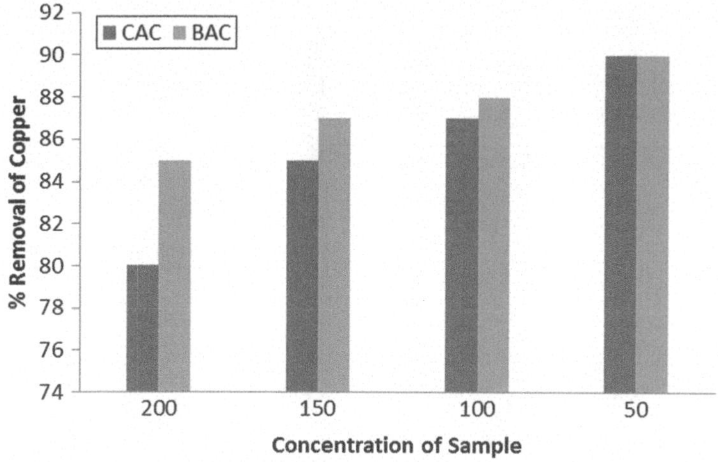

Fig. 3 % removal of copper due to change in concentration of the sample

3.4.2 Optimum Dosage of Adsorbents

The amount of adsorbent was optimised. The sample size of 0.05, 0.1, 0.25 and 0.5 g was analysed for sample concentration of 500 ppm. The results showed that an optimum dosage of 0.1 g would be effective in treating the copper sample. Figure 4 would explain the removal efficiency which was clear that BAC could be an effective replacement to CAC, though BAC doesnot have a proper homogenity.

3.4.3 Optimisation of Time

The sample was mixed with an adsorbent size of 0.1 g and kept in an orbital shaker at 150 rpm in 20 min, 40 min and 1, 2, 3, 4 and 5 h, respectively. It showed that an optimum time of 1 h was efficient for the removal of copper, and it showed more than 90% adsorption (Fig. 5).

An optimum amount of 0.1 g of BAC was treated to various bioleachates which was given agitation at 10 rpm for 1 h that gave a result greater than 90%. The effect of optimised result on the adsorption process on various treated bioleachates (Fig. 6) showed that this could be an effective and cheap process and environmentally sound management in disposing the bioleachate which is very easy that does not require any skilled labour.

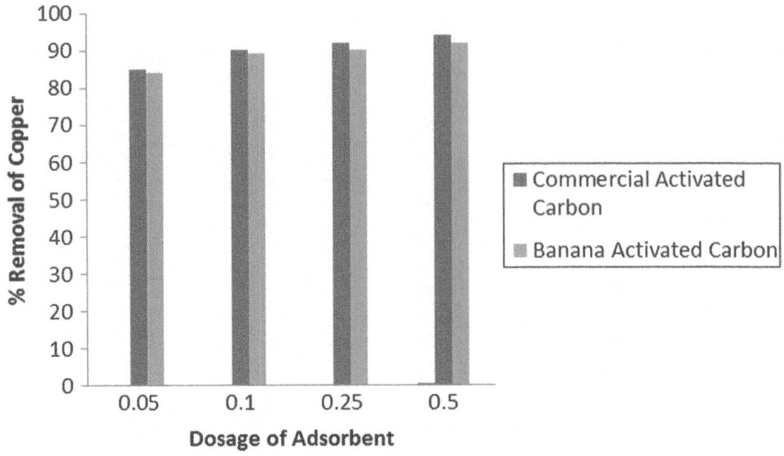

Fig. 4 % removal of copper due to change in dosage of adsorbent

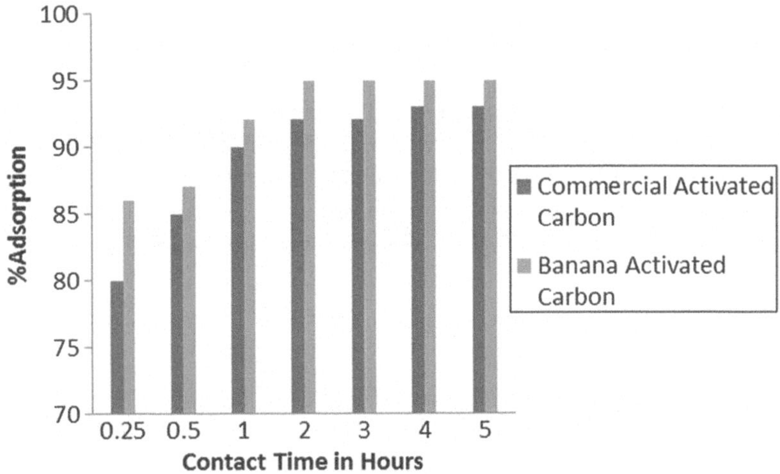

Fig. 5 Percentage adsorption of copper due to effect of time

4 Conclusions

Banana peel has been used as a fodder for animals but not all are used efficiently. The left out were only concentrated for the entire process. The waste banana peel was converted into BAC whose characteristics were similar to commercial-activated carbon which could effectively and efficiently replace commercial-activated carbon. It was also found that they could adsorb more than 90% of copper, out of which no cost or low cost is involved which could save people's health and welfare.

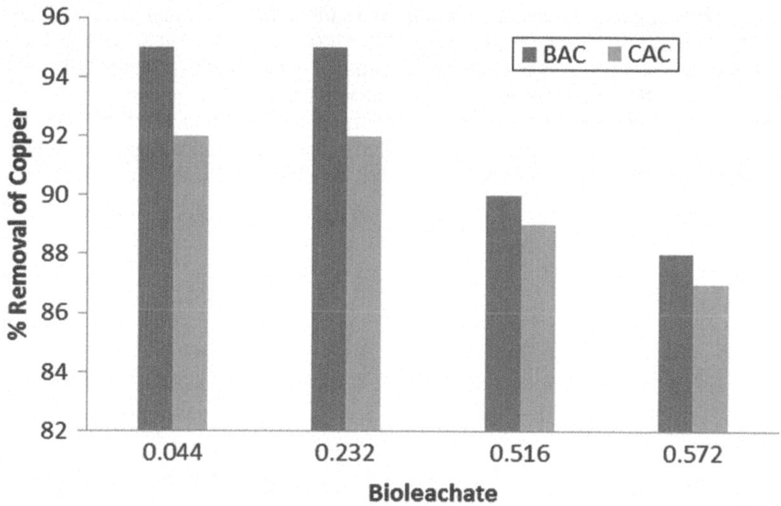

Fig. 6 Percentage removal of copper from bioleachate

References

Aharoni C, Ungarish M (1977) Kinetics of activated chemisorption part 2. Theoretical models. J Chem Soc Faraday Trans 73:456–464

Annadurai G, Juang RS, Lee DJ (2002) Adsorption of heavy metals from water using banana and orange peels. Water Sci Technol 47(1):185–190

APHA (2005) Standard methods for examination of water and waste water, 21st edn. American Public Health Association, Washington

Arancon Norman Q, Clive A, Babenko A, Cannon J, Galvis P, James Metzger D (2008) Influences of vermicomposts, produced by earthworms and microorganisms from cattle manure, food waste and paper waste, on the germination, growth and flowering of petunias in the greenhouse. Appl Soil Ecol 39:91–99

Bai R, Mardina S (2002) The practice and challenges of solid waste management in Singapore. Waste Manag 22(5):557–567

Bansal RC, Donnet JB, Stoeckli HF (1988) Active carbon. Marcel Dekker, New York

Bernard E, Jimoh A (2013) Adsorption of Pb, Fe, Cu, and Zn from industrial electroplating wastewater by orange peels activated carbon. Int J Eng Appl Sci 4(2) ISSN2305

Brunauer S, Emmett PH, Teller E (1938) Adsorption of gases in multimolecular layers. J Am Chem Soc 60:309–319

Cheung W, Porter JF, McKay G (2000) Sorption kinetics for the removal of copper and zinc from effluents using bone char. Sep Purif Technol 19:55–64

Cooney D (1998) Adsorption design for waste water treatment. Lavis Publishers, Bola Raton

Dubinin MM (1960) The potential theory of adsorption of gases and vapours for adsorbents with energetically non-uniform surface. Chem Rev 60:235–266

Dubinin MM (1965) Modern state of the theory of volume filling of micropore adsorbents during adsorption of gases and steams on carbon adsorbents. Zh Fiz Khimii 39:1305–1317

Hema KR, AVVS S (2011) Studies on the removal of Ni (II) from aqueous solutions using powder of Mosambi fruit peelings as a low cost sorbent. Chem Sci J 2011:CSJ-31 1

Ho YS, McKay G (1999) Pseudo second order model for sorption processes. Adsorption 7:451–465

Holden MJ (1982) Manufacture and uses of activated carbon. Effluent Water Treat J 22:27
Hosseini M, Mertens SFL, Ghorbani M, Arshadi MR (2003) Asymmetrical Schiff bases as inhibitors of mild steel corrosion in sulphuric acid media. Mater Chem Phys 78:800–807
Maldhure AV, Ekhe JD (2011) Preparation and characterization of microwave assisted activated carbons from industrial waste lignin for Cu(II) sorption. Chem Eng J 168:1103–1111
Zheng H, Wang L. 12013(2013) Banana peel carbon containing functional groups applied to the selective adsorption of Au(III) from waste printed circuit boards. Soft Nanosci Lett 3:29

Effect of Tannery Effluent on Germination and Early Seedling Growth of *Oryza sativa* Var. IET-4786

S. Biswas, A. Bhattacharya, and P. Basak

Abstract The present investigation focuses on the effect of a tannery effluent on germination of early seedling growth and of *Oryza sativa* var. IET-4786. Seedling growth is indicated by root length, shoot length, vigour index, tolerance index and photosynthetic pigment content including chlorophyll (Chla, Chlb and total Chl) and carotenoids. Paddy seeds were procured from the Rice Research Institute, Chinsurah, West Bengal, and selected to grow in Petri plates irrigated with various effluent concentrations [control (0%), 20%, 25%, 50%, 75% and 100%]. Imbibition was higher in the lower effluent concentrations and gradually decreased with increasing concentrations during the soaking process before they were placed in Petri plates lined with effluent-soaked filter papers. At dilutions up to 50%, the effluent imparted growth-promoting traits, while at higher concentrations (75% onwards), inhibition of germination and reduction in root and shoot length, vigour index, and tolerance index were recorded; photosynthetic pigment (both chlorophylls and carotenoids) content was also found to decline.

Keywords Effluent • *Oryza sativa* var. IET-4786 • Germination • Early growth • Photosynthetic pigments

1 Introduction

Rapid urbanization has led to the rise of a plethora of industries, which actively contribute not only to the economic growth but also to the growing levels of pollution. Water pollution is hailed as one of the most serious concerns throughout the world. Effluents from the industries are released in the surrounding water bodies relentlessly, rendering these water resources highly polluted. The leather industry is a water-based industry and generates huge amount of waste water during processes like curing, soaking, liming, detaining, bating, picking, degreasing and tanning

S. Biswas (✉) • A. Bhattacharya • P. Basak
School of Bioscience and Engineering, Jadavpur University, Kolkata, India
e-mail: shreyabiswas2011@gmail.com

S.K. Ghosh (ed.), *Utilization and Management of Bioresources*,
DOI 10.1007/978-981-10-5349-8_23

(Rao et al. 2014). Tannery waste liquors (TWL) poses as an important environmental problem due to their high organic load, and they are also particularly difficult to treat (Vlyssides and Israilides 1997). Agriculture plays a pivotal role in shaping the world economy. Paddy fields account for more than 11% of the world's area of agricultural lands (Maclean 2002). Cultivated rice, *Oryza sativa* L., represents the world's most economically important crop species serving as the staple food for more than half of the global population (Londo et al. 2006). The present study was carried out to investigate the effect of a tannery effluent on rice in terms of seed germination; early seedling growth including root length, shoot length, vigour index and tolerance index; and photosynthetic pigment content.

2 Review of Literature

Tannery effluents characteristically contain high amounts of salts and chromium (Babyshakila and Usha 2009). The poisonous heavy metal chromium, which is let out through these effluents, affects plant and animal life, as well as causes environmental damage (Belay 2010). Like tanneries, fertilizer industries also use huge amounts of water and at the same time produces a lot of effluents. They can thus be held responsible for water and soil pollution to a considerable extent (Saravanan et al. 2001). However, since these effluents are also rich in other salts, they may serve as source of nutrient to crops irrigated with them and may reduce the requirement of chemical fertilizers (Kumar and Chopra 2012). Waste water has been reported to impart growth promotion in rice and wheat (Dash 2012; Kohli and Malaviya 2013). Other than these cereal crops, the impact of effluents has been studied in pea, lentil and gram (Khan et al. 2011); in *Vigna angularis*, *Vigna cylindrica* and *Sorghum cernuum* (Doke et al. 2011); in *Brassica napus* (Malaviya and Sharma 2011); in fenugreek (Kumar and Chopra 2012); etc. Different varieties of a single crop species might have different levels of tolerance towards the same effluent; for example, wheat varieties PBW-343 and HS-365 were more tolerant than HS-295 towards a tannery effluent (Kohli and Malaviya 2013). The effect of presoaking of sesame seeds in either water or gibberellic acid for different time intervals on their germination and root and shoot length was studied by Kyauk et al. (1995). Seed germination has been considered as acritical test for probable crop productivity since it ensures reproduction (Dash 2012). Shoot length and root length have been considered as very significant growth parameters for early seedling growth by many authors (Rao et al. 2014; Dash 2012; Peralta et al. 2001; Sundaramoorthy and Kunjithapatham 2000). Effects of industrial effluents on vigour index and tolerance index have been studied in *Vigna radiata* and *Cicer arietinum* by Mehta and Bhardwaj (2012). Chlorophylls and carotenoids are the major pigments associated with photosynthesis in higher plants, including rice. Photosynthetic pigments can give a direct indication about the photosynthetic capacity of a plant and plant nutrition. The impact of industrial effluents on

Table 1 Physico-chemical analysis of the effluent

Sl. No.	Parameters	Raw effluent	Tolerance limit as per CPCB
1	Colour	Brownish	Should be absent
2	Odour	Unpleasant	Should be absent
3	pH	6.1	5.5–9
4	Total solids	12,360	200
5	Biological oxygen demand	96.4	100
6	Chemical oxygen demand	164	250

All parameters were expressed in mg/L except colour, odour and pH
CPCB Central Pollution Control Board

chlorophyll (Chla, Chlb and Chl total) has been studied in several plants (Sundaramoorthy and Kunjithapatham 2000; Sangannavar and Kalshetty 2011; Singh et al. 2006).

3 Materials and Methods

Seeds of *Oryza sativa* var. IET-4786 were procured from the Rice Research Institute, Chinsurah, West Bengal. The tannery effluent was collected from a tannery unit in Park Circus area, Kolkata, in sterile glass containers and kept under refrigeration at 4 °C. The physico-chemical parameters of the effluents were analysed, results of which are shown in Table 1. Various dilutions (20, 25, 50, 75 and 100%) of the effluent sample were made. Tap water was taken as control for the experiments. Paddy seeds were surface-sterilized by washing them in 0.5% sodium hypochlorite solution for 2 min followed by 70% ethanol for another 2 min and then given three consecutive washes in sterile distilled water and dried to remove surface moisture. Dry weights of each of these groups of 20 seeds were taken. These seeds were soaked in 20 ml of the different effluent concentrations for 4 h. Imbibition percentages were calculated henceforth using the formula:

(weight after soaking − weight before soaking) × 100/weight after soaking

The soaked seeds were then placed equidistantly in sterile Petri plates lined with two Whatman filter papers (No. 1), soaked in the corresponding effluent concentrations. Twenty seeds were kept per Petri plate for each concentration and all experiments were done in triplicate. The Petri plates were maintained in a B.O.D incubator at normal physiological conditions (25 ± 2 °C). Plates were irrigated with different concentrations of effluents uniformly at regular intervals to prevent drying up of the seeds. The number of seeds germinated was counted each day until further germinations stopped. Seeds which showed emergence of radical piercing through the seed coat were considered to be germinated. All seeds that were germinated and grown in tap water served as control. Data for germination were recorded on the third, fifth and seventh day. Percentage of germination was calculated separately for each treatment using the formula (Czabator 1962):

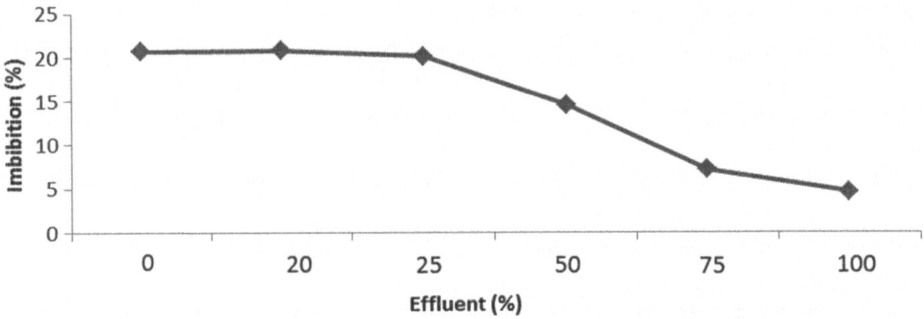

Fig. 1 Percentage of imbibition (during soaking) of different effluent concentration by the seeds of *Oryza sativa* var. IET-4786

Percentage germination = (No . of seeds germinated/No . of seeds sown) × 100.

Root length and shoot length were measured and data was recorded for 14th and 21st day. Seedling vigour index was calculated using the formula (Orchard 1977):

Vigor index = Mean germination percentage × Mean seedling length

The tolerance index was calculated using the formula (Turner and Marshall 1972):

Tolerance index = Mean length of longest root in treatment/Mean length oflongest root in control

The chlorophyll contents of each treatment were estimated using a Cary 60 UV-Vis spectrophotometer from Agilent Technologies following Wellburn and Lichtenthaler (1984).

4 Result

The physico-chemical characteristics of the effluent are represented in Table 1. It was brownish in colour with pH 6.1 and contained suspended and dissolved solids in large amounts (12,360 mg/L). It was observed during the 2-h soaking phase of the surface-sterilized, dried seeds that they could imbibe effluents of lower concentrations more than those of the higher concentrations (Fig. 1). Highest imbibition was shown for tap water and 20% effluent concentration, whereas 75% and 100% effluent concentrations were least imbibed by the seeds.

The highest rate of germination was recorded in tap water, followed by 25% effluent concentration, but germination was severely inhibited at higher effluent concentrations; no seed germinated at 100% till day 3. Germination rates followed a similar trend in day 5 and day 7 with highest germination rates in tap water which progressively decreased to the lowest in the raw effluent (100%) (Fig. 2). Difference in the reduction of seed germination in the 75% and the 100% effluent concentrations was highly significant ($p < 0.001$).

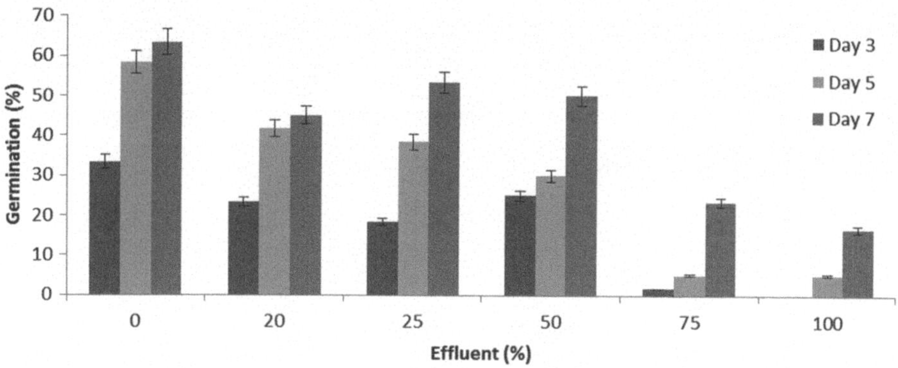

Fig. 2 Percentage germination of *Oryza sativa* var. IET-4786 in different concentrations of the effluent at different time intervals

Table 2 Effect of the effluent on root and shoot lengths of *Oryza sativa* var. IET-4786

Effluent concentration (%)	Days of exposure			
	14 days		21 days	
	Root	Shoot	Root	Shoot
0	6.037 ± 1.972	6.945 ± 1.735	7.2 ± 2.199	8.037 ± 1.751
20	6.722 ± 1.083	7.07 ± 1.273	7.515 ± 1.09	8.007 ± 1.29
25	7.141 ± 1.153	7.338 ± 1.233	8.112 ± 1.078	8.259 ± 1.099
50	7.19 ± 1.027	7.27 ± 1.171	8.253 ± 1.069	8.37 ± 1.141
75	5.25 ± 0.992	5.929 ± 1.085	6.514 ± 0.571	6.892 ± 0.823
100	3.97 ± 1.051	4.02 ± 0.881	4.91 ± 0.877	5.06 ± 0.842

Values are arithmetic means ± standard deviations of three replicates

Both root length and shoot length were found to increase progressively in the lower concentrations of effluent. At 20, 25 and 50% effluent concentrations, increase in root length was statistically significant ($p < 0.01$). However, decrease in root length was recorded at higher effluent concentrations. Reduction in root length was highly significant ($p < 0.001$) in 75 and 100% effluent concentrations. At 25 and 50% effluent concentration, increase in shoot length was statistically significant ($p < 0.01$), but at higher concentrations (75 and 100%), highly significant ($p < 0.001$) decrease in shoot length was recorded on day 14 and day 21 (Table 2).

Vigour index was maximum in the control, followed by 25 and 50% effluent concentrations, but decreased thereafter from 75%, and minimum vigour index was estimated for raw effluent (100%) (Fig. 3). Highest tolerance index was recorded for 25% followed by 50% effluent concentrations. Tolerance index decreased at 75% concentration and was its lowest for the raw effluent (100%) (Fig. 4).

Figure 5 shows effect of different concentrations of effluent on photosynthetic pigments including chlorophyll (Chla, Chlb and Chl total) and total carotenoids on the 21st day. At 25% concentration, increase in all the pigments was recorded when compared to the control. Pigment contents further increased and reached their

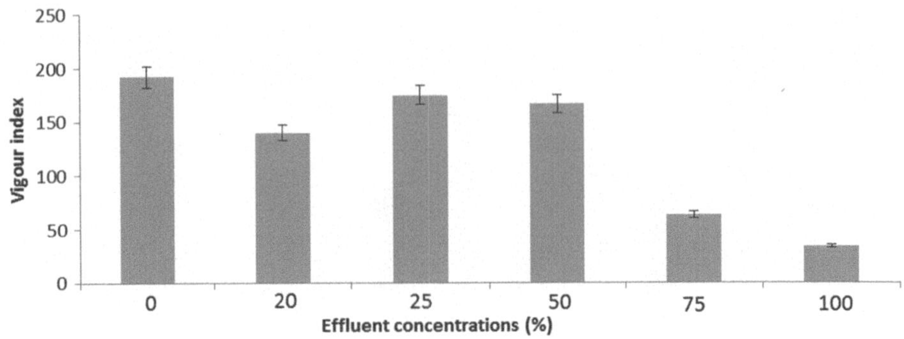

Fig. 3 Effect of different concentrations of the effluent on vigour index of *Oryza sativa* var. IET-4786

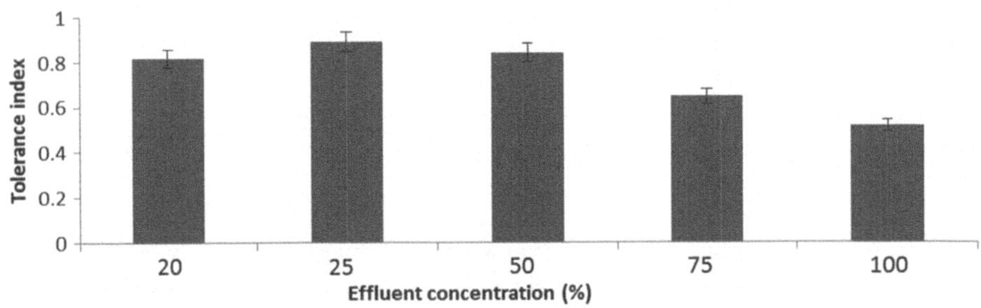

Fig. 4 Effect of different concentrations of the effluent on tolerance index of *Oryza sativa* var. IET-4786

maximum at 50% effluent treatment. Decrease in pigment concentration was visible from higher concentrations. Minimum concentration of photosynthetic pigments was recorded for the treatment with raw effluent (100%).

5 Discussion

The present study was performed with a tannery effluent which contained high levels of dissolved and suspended solid particles. The rates of imbibitions decreased with the corresponding increase in effluent concentration during the soaking period, which can be attributed to the higher amount of solid particles present. Malaviya and Sharma (2011) reported that a correlation exists between effluent concentration and seed germination (Malaviya and Sharma 2011). Highest germination was seen in tap water in the present study. Significant reduction in germination was visible from 75% effluent concentration. According Srivastava and Sahai (1987), this

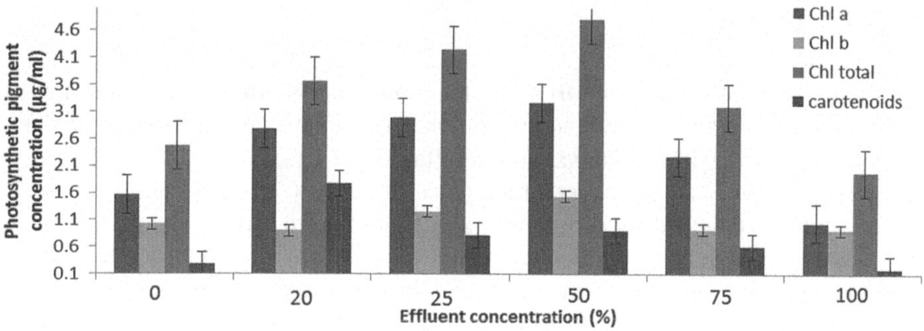

Fig. 5 Effect of different concentrations of the effluent on photosynthetic pigment content of *Oryza sativa* var. IET-4786

reduction can be correlated to the high amounts of organic matter, sulphates, calcium, magnesium, etc. At high effluent concentrations, germinating seeds get limited oxygen which restricts their energy supply (Hadas 1976). Rates of rice and wheat cultivars started decreasing after 50% concentration. Khan and Sheikh (1976) reported significant reduction and delay in germination of *Capsicum annuum* seeds (Khan and Sheikh 1976) during sewage treatment. Elevated salinity and conductivity of solutes absorbed by the seeds are direct effects of high levels of dissolved solids which may correspond to the inhibition of seed germination at higher concentration (Sundaramoorthy and Kunjithapatham 2000). In the present study, both root and shoot length increased progressively up to 50% effluent concentration and then started decreasing from 75% and was lowest for raw effluent (100%). According to Mishra and Bera (1996), lower concentration of tannery effluent promoted seedling growth, while higher concentrations showed considerable reduction in maize (Mishra and Bera 1996). On sewage sludge application, increase in shoot length was observed (Ramamoorthy et al. 1992; Kalavathi and Ramamoorthy 1992). In our study, vigour index was maximum for control, followed by 25% concentration. At 75% concentration, vigour index was reduced and it was minimum for raw effluent (100%). Tolerance index was maximum for 25% effluent concentration and minimum for 100% effluent treatment. Vigour index and tolerance index were found to be minimum in untreated effluent for *Vigna radiata* and *Cicer arietinum* (Mehta and Bhardwaj 2012). All the photosynthetic pigments including chlorophylls and carotenoids were found to increase significantly ($p < 0.01$) with increasing concentration of the effluents, which clearly indicated healthy plant growth. Highest concentrations of all the pigments were observed for the 50% effluent treatment. Thereafter, pigment concentrations declined, and finally lowest amounts of pigments were found to occur for the plants grown in raw effluent.

6 Conclusion

Imbibition gradually decreased from the maximum at 0% effluent concentration to the minimum in 100% effluent concentration. The highest germination percentage of *Oryza sativa* var. IET-4786 was found in tap water (0% effluent, control) which progressively decreased to lowest for the raw effluent (100%) treatment. Early seedling growth as indicated by root length and shoot length, vigour index, tolerance index and photosynthetic pigment concentrations (Chla, Chlb, Chl total, xanthophylls) followed a similar pattern: effluents in lower concentrations (up to 50%) promoted early seedling growth and that at higher concentrations (75% onwards) hindered overall seedling growth.

7 Future Scope

Response of different varieties of a plant to the same amount and type of stress can be different as previously shown by Dash (2012) for wheat. Different varieties of rice can be tested for various amounts and types of effluents to see their responses. The most tolerant varieties might be selected thereafter for better crop yield which can be expected to be more adapted to grow in polluted soil or effluent irrigated agricultural lands. Since effluents at lower concentrations have shown to promote growth, whether they can be used in lieu of costly chemical fertilizers still remains to be determined.

Acknowledgements The authors express their gratitude to the Department of Biotechnology, Government of West Bengal, for financial assistance to carry out the research work.

References

Babyshakila P, Usha K (2009) Effect of diluted effluent on soil properties and plant growth. Adv Stud Biol 1(8):391–398

Belay AA (2010) Impacts of chromium from tannery effluent and evaluation of alternative treatment options. J Environ Prot 1(01):53

Czabator FJ (1962) Germination value: an index combining speed and completeness of pine seed germination. For Sci 8(4):386–396

Dash AK (2012) Impact of domestic waste water on seed germination and physiological parameters of rice and wheat. Int J Res Rev Appl Sci 12(2)

Doke KM, Khan EM, Rapolu J, Shaikh A (2011) Physico-chemical analysis of sugar industry effluent and its effect on seed germination of vigna angularis, vigna cylindrica

Hadas A (1976) Water uptake and germination of leguminous seeds under changing external water potential in osmotic solutions. J Exp Bot 27(3):480–489

Kalavathi D, Ramamoorthy K (1992) A note on the effect of seed size on viability and vigour of seed in cluster beans Cyamopsis tetragonoloba L. cultivar PusaNavbhagar. Madras Agric J 79(9):530–532

Khan SS, Sheikh KH (1976) Effects of different levels of salinity on seed germination and growth of Capsicum annuum L. Biologia

Khan MG, Daniel G, Konjit M, Thomas A, Eyasu SS, Awoke G (2011) Impact of textile waste water on seed germination and some physiological parameters in pea (Pisum sativum L.), Lentil (Lens esculentum L.) and gram (Cicer arietinum L.) Asian J Plant Sci 10(4):269

Kohli R, Malaviya P (2013) Impact of tannery effluent on germination of various varieties of wheat (Triticum aestivum L). J Appl Nat Sci 5(2):302–305

Kumar V, Chopra AK (2012) Fertigation effect of distillery effluent on agronomical practices of Trigonella foenum-graecum L.(Fenugreek). Environ Monit Assess 184(3):1207–1219

Kyauk H, Hopper NW, Brigham RD (1995) Effects of temperature and presoaking on germination, root length and shoot length of sesame (Sesamum indicum L.) Environ Exp Bot 35(3):345–351

Londo JP, Chiang YC, Hung KH, Chiang TY, Schaal BA (2006) Phylogeography of Asian wild rice, Oryza rufipogon, reveals multiple independent domestications of cultivated rice, Oryza sativa. Proc Natl Acad Sci 103(25):9578–9583

Maclean JL (2002) Rice almanac: source book for the most important economic activity on earth. Int Rice Res Inst

Malaviya P, Sharma A (2011) Impact of distillery effluent on germination behaviour of Brassica napus L

Mehta A, Bhardwaj N (2012) Phytotoxic effect of industrial effluents on seed germination and seedling growth of Vigna radiata and Cicer arietinum. Glob J Biosci Biotechnol 1:1–5

Mishra P, Bera AK (1996) Effect of tannery effluent on seed germination and seedling growth in maize (Zea mays L. CvVikram). Environment and ecology. Kalyani 14(4):752–754

Orchard TJ (1977) Estimating the parameters of plant seedling emergence. Seed Sci Technol

Peralta JR, Gardea-Torresdey JL, Tiemann KJ, Gomez E, Arteaga S, Rascon E, Parsons JG (2001) Uptake and effects of five heavy metals on seed germination and plant growth in alfalfa (Medicago sativa L.) Bull Environ Contam Toxicol 66(6):727–734

Rao S, Kumar PN, Reddy PH, Inayat M (2014) Study on effect of raw and diluted tannery effluent on the seed germination and seedling growth of selected crops. Eur J BiotechnolBiosci 2(3):1–4

Sangannavar MC, Kalshetty BM (2011) Morphological parameters and chlorophyll content of green vegetable plants available around Jamkhandi. Curr World Environ 6(1):67–70

Saravanan S, Subramani A, Sundaramoorthy P, Selvaraju M, Lakshmanachary AS (2001) Influence of copper sulphate on germination, growth and yield of soybean [Glycine max (L.) Merr.] Ecol Environ Conserv 7:101–104

Singh PP, Mall M, Singh J (2006) Impact of fertilizer factory effluent on seed germination, seedling growth and chlorophyll content of gram (Cicer arietinum). Magnesium 31:2–06

Srivastava N, Sahai R (1987) Effects of distillery waste on the performance of Cicer arietinum L. Environ Pollut 43(2):91–102

Sundaramoorthy P, Kunjithapatham J (2000) Effect of paper mill effluent on seed germination and seedling growth of six varieties of groundnut (Arachis hypogaea). J Ecotoxicol Environ Monit 10(1):53–57

Turner RG, Marshall C (1972) The accumulation of zinc by subcellular fractions of roots of AgrostistenuisSibth.in relation to zinc tolerance. New Phytol 71(4):671–676

Vlyssides AG, Israilides CJ (1997) Detoxification of tannery waste liquors with an electrolysis system. Environ Pollut 97(1):147–152

Wellburn AR, Lichtenthaler H (1984) Formulae and program to determine total carotenoids and chlorophylls a and b of leaf extracts in different solvents. In: Advances in photosynthesis research. Springer, Netherlands, pp 9–12

Characterization of Lipase-Producing Bacteria Isolated from Degrading Oil Cakes

Sutripta Sarkar and Anubrati Chatterji

Abstract Lipases occupy a prominent place amongst biocatalysts as they can hydrolyse fats into fatty acids and glycerol at the water-lipid interface. Lipase production is a prime area of interest for microbiologists, process engineers and biochemists. Recent advances in green energy have led to renewed interest in the ability of microbial lipases to catalyse transesterification reaction for biodiesel production. Research carried out in this field has revealed that microbes, especially fungi and bacteria, are the tools of choices for commercial production.

The primary objective of the study was to isolate lipase-producing microbes from the waste product of oil industries and to assess their ability to grow in different oil medium. Samples (oil cakes) were collected from a mustard oil factory in Barrackpore, Kolkata. Lipid-degrading microbes were trapped in specialized medium containing various types of oils. All isolated strains were morphologically, physiologically and biochemically characterized. Strain L_J was found to be the most efficient strain as it showed highest lipase activity (25 μmoles/ml/min) after 24 h of incubation at 37 °C amongst the strains.

Though the lipase activity obtained was lower compared to reported fungal lipases, the study is significant because of its exploratory nature and has the potential of identifying new organisms having lipolytic capability.

Keywords Waste management • Oil industry waste • Lipolytic bacteria • Lipases

1 Introduction

Lipases (triacylglycerol acylhydrolases, EC 3.1.1.3) catalyse the hydrolysis and the synthesis of esters formed from glycerol and long-chain fatty acids (Mobarak-Qamsari et al. 2011). They are biotechnologically important enzymes and find

S. Sarkar (✉)
Department of Food and Nutrition, Barrackpore Rastraguru Surendranath College, Kolkata, West Bengal, India
e-mail: sarkar.sutripta@gmail.com

A. Chatterji
Department of Microbiology, Barrackpore Rastraguru Surendranath College, Kolkata, West Bengal, India

S.K. Ghosh (ed.), *Utilization and Management of Bioresources*,
DOI 10.1007/978-981-10-5349-8_24

tremendous application in food, leather, detergents and pharmaceutical industries. Most of the commercial lipases are of microbial origin and have been isolated from soil contaminated with oil, industrial wastes, dairy industries, oil industries, decaying food, organic wastes, etc. (Haki and Rakshit 2003). Some lipolytic bacteria have been isolated from niche environment like the Amazonian soils (Willerding et al. 2011) and hypersaline environment (Ghasemi et al. 2011). These microbes primarily belong to the genera *Bacillus*, *Pseudomonas* and *Burkholderia* (Gupta et al. 2004). The recent interest in green and renewable energy has led to a lot of research in biodiesel production (Boonmahome and Mongolthanaruk 2013). Lipases can catalyse transesterification reaction of fats to produce biodiesel (Zaks and Klibanov 1985; Macrae 1989). Lipases immobilized on electrodes along with glucose oxidase functions as lipid biosensor and are used in blood cholesterol determination (Imammura et al. 1989). They also play an important role in the management of industrial wastes (Sharma et al. 2001). Lipases are used in aerobic degradation of activated sludge where fats form a thin film and block aeration (Bailey and Ollis 1986). Waste water from food, dairy and poultry industries is also laden with fats which can be treated by lipases.

The current study attempts to isolate and identify lipolytic enzyme-producing bacteria from degrading oil cakes (an oil industry waste). Microbes were screened based on their ability to break down different types of oils so that they can be used as inoculums for treatment of waste contaminated with oils. The work is expected to add to the existing knowledge about lipase-producing organisms and also open up new avenues for research.

2 Materials and Methods

Sample Collection: The degraded oil cake sample was collected from a mustard oil factory in Barrackpore (North 24 Parganas District, West Bengal, India) in sterile containers and was stored at −20 °C till use. For colony count, 10 g of sample was mixed with 90 ml of sterile distilled water in a 250 ml Erlenmeyer flask and kept in a shaker at room temperature for 30 min. Serial dilution was prepared and plated in specific medium. All plates were incubated at 37 °C for 24 h. Cell density was measured at 660 nm.

The following medium was used for growth and isolation of lipolytic bacteria: LB broth/agar (0.5% yeast extract, 1% tryptone, 1% NaCl, 2% agar, pH 7). Minimal medium (starch-olive oil medium) containing peptone 2%, starch 2%, KH_2PO_4 0.2%, $(NH_4)NO_3$ 0.5%, Na_2HPO_4, $12H_2O$ 0.8%, $MgSO_4$, $7H_2O$ 0.001%, $CaCl_2$, $2H_2O$ 0.01%, olive oil (1%) (V/V), agar 2% and pH was maintained at 7.5. Tributyrin medium was prepared by adding peptone 5 g, yeast extract 3 g, and tributyrin 10 ml per 1000 ml of distilled water (pH 7.0). For Tween 20 medium, Tween 80 medium, olive oil medium, coconut oil medium and castor oil medium, 4.7 ml of Tween 20, Tween 80, olive oil, coconut oil and castor oil were added to 1,000 ml of distilled water along with 10 g peptone and 2 g $CaCl_2$. pH of the medium that was maintained at 7. 2% agar was added to the broths for making plates.

Reagents Used for Biochemical Tests: Barritt's reagent (Voges-Proskauer test); methyl red indicator (for indicating acid production); Kovac's reagent (indole production test); crystal violet, Gram's iodine, safranin and 95% ethanol (for Gram staining); H_2O_2 solution (for catalase test); and sugar utilization kit (KB009 HiCarbohydrate Kit, Himedia, India).

For lipase assay, the method described by Kanwar et al. (2005) was followed. PNP (para-nitrophenol) stock of 60 μg/ml concentration was prepared for the standard curve. Twenty millimetre stock solution of PNPP (para-nitrophenyl palmitate) was prepared in isopropanol and 0.05 M Tris-HCl buffer of pH 8.5. Twenty-four-hour old cultures of different isolated strains and two standard strains, *Pseudomonas aeruginosa* and *Escherichia coli*, which were grown in starch-olive oil medium were centrifuged at 10,000 rpm for 20 min at 4 °C, and the supernatant was collected. PNPP and the buffer were added to the supernatant and incubated at 37 °C for 10 min.

Ethanol/acetone (1:1) mixture was added to stop the reaction. The optical density was measured at 410 nm. Blank for each tube was made by adding heat-inactivated (boiling in water bath for 10 min) crude enzyme; everything else was remained same. The absorbance of blank tubes was subtracted from the respective test sample tubes, and actual absorbance was recorded for each strain.

One unit (1U) was defined as that amount of enzyme that liberated 1 μmol of *p*NP per minute per ml under the test conditions (Karadzic et al. 2006).

3 Results and Discussion

Thirteen lipolytic bacterial strains (L_A, L_B, L_C, L_D, L_E, L_F, L_G, L_H, L_I, L_J, L_K, L_L, L_M) were isolated from degrading oil cakes by dilution method. All 13 strains were characterized morphologically and biochemically (Tables 1 and 3), and their Gram characters (Table 2) were also noted. Most of the cells where coccus baring strain L_E,

Table 1 Morphological characteristics of different isolated strains

Different isolated strains	Morphological characters
L_A	White in colour
L_B	White in colour
L_C	White in colour
L_D	White in colour
L_E	Creamish colonies
L_F	Creamish colonies
L_G	Creamish smooth
L_H	Creamish in colour
L_I	Whitish smooth
L_J	Whitish smooth
L_K	Creamish smooth
L_L	Creamish smooth
L_M	Whitish in colour

Table 2 Gram characters of different isolated strains

Name of strains	Gram characters	Cellular character
L_A	Gram +ve	Coccus
L_B	Gram +ve	Small coccus
L_C	Gram +ve	Coccus
L_D	Gram +ve	Coccus
L_E	Gram +ve	Short rods
L_F	Gram +ve	Coccus
L_G	Gram +ve	Coccus
L_H	Gram +ve	Coccus
L_I	Gram +ve	Coccus
L_J	Gram +ve	Coccus
L_K	Gram +ve	Small coccus
L_L	Gram +ve	Small coccus sometimes cluster of two to three cells
L_M	Gram +ve	Small coccus sometimes cluster of two cells

Table 3 Results of biochemical tests

Name of strains	Catalase test	Indole test	MR test	VP test
L_A	++	–	+	–
L_B	++	–	+	–
L_C	++	–	+	–
L_D	++	–	+	–
L_E	++	++	++	+
L_F	++	++	++	+
L_G	++	–	–	–
L_H	++	–	–	++
L_I	++	–	–	+
L_J	+	–	–	++
L_K	++	–	–	–
L_L	++	–	–	–
L_M	++	–	–	–

Positive result =++, Negative result = –, Feebly positive result = +

which was rod shaped. All the strains were Gram +ve and also gave positive catalase test. Strains L_E and L_F gave positive results for indole test. Results of the biochemical tests are shown in Table 3. Strains L_A, L_B,L_C, L_D, L_E and L_F showed positive methyl red (MR) test, and VP (Voges-Proskauer) test was positive in case of L_E, L_F, L_H, L_I and L_J. Most of the strains could utilize a range of sugars as evident from Table 4. All isolated strains were studied for the growth in seven different growth mediums (Kathiravan et al. 2012) (Figs. 1 and 2).

In case of LB medium, all strains showed high growth after 24 h of incubation in 37 °C. Strains L_I and L_K showed very high growth and strains L_G and L_M showed moderate growth in Tween 80 medium, but all of these strains showed higher growth

Table 4 Result for sugar utilization tests

Name of the sugars	Name of the different bacterial strains	
	Name of strains with positive results	Name of strains with negative result
Lactose	L_A,L_B,L_E,L_M,L_F, *E. coli*, *Pseudomonas* sp.	L_C, L_D, L_G, L_H, L_I, L_J, L_K, L_L
Xylose	All strain	–
Maltose	All strains	–
Fructose	All Strains	–
Dextrose	All Strains	–
Galactose	All strains	–
Raffinose	All strains	–
Trehalose	All strains	–
Melibiose	L_A,L_B,L_E,L_M, *E. coli*, L_C, L_D, L_G, L_H, L_I, L_J, L_K, L_L *Pseudomonas* sp.	L_F
Sucrose	All strains	–
L-Arabinose	All strains	–
Mannose	All strains	
Insulin	–	All strains
Sodium gluconate	L_A, L_B, *E. coli*	L_E,L_M, L_C, L_D, L_G, L_H, L_I, L_J, L_K, L_L,L_F *Pseudomonas* sp.
Glycerol	All strains	–
Salicin	All Strain	–
Dulcitol	All strains	–
Inositol	L_H, L_I, L_J, L_K, L_L,	L_A, L_B, L_C, L_D, L_E, L_F, L_G, L_M, *E. coli*, *Pseudomonas* sp.
Sorbitol	L_A, L_B, L_C, L_D, L_G, L_H, L_I, L_J, L_K, L_L,	L_E, L_F, L_M, *E. coli*, *Pseudomonas* sp.
Mannitol	All strains	–
Adonitol	–	All strains
Arabitol	–	All strains
Erythritol	–	All strains
α-Methyl-D-glucoside	L_H	All other strains
Rhamnose	All strains (L_F feebly positive)	–
Cellobiose	All strains (*Pseudomonas* sp. feebly positive)	–
Melezitose	L_M	Rest of the strains
O-Methyl-D-mannoside	L_A,L_B	Rest of the strains
Xylitol	Rest of the strains	L_F,L_G, *Pseudomonas* sp.
ONPG	All strains	–
Esculin	All strains	–
D-Arabinose	Rest of the strains	L_F, L_G, L_I, L_J, L_K, L_L,
Citrate	All positive	–
Malonate	All strains	–
Sorbose	–	All strains

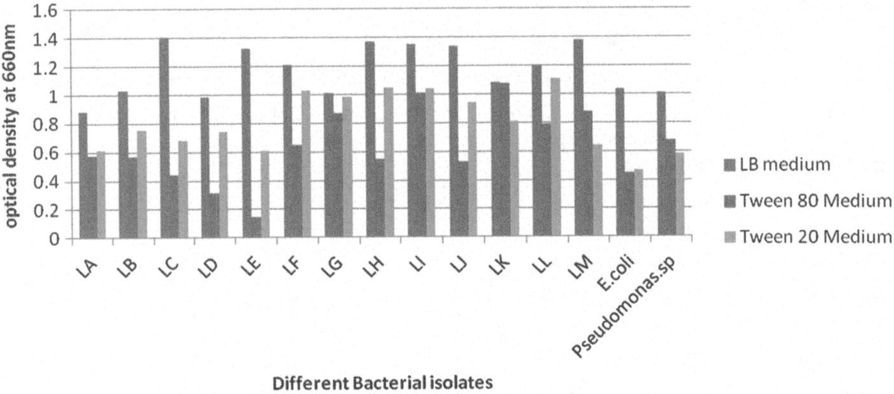

Fig. 1 Growth in LB, Tween 80 & Tween 20 medium after 24 hrs. incubation at 37 °C

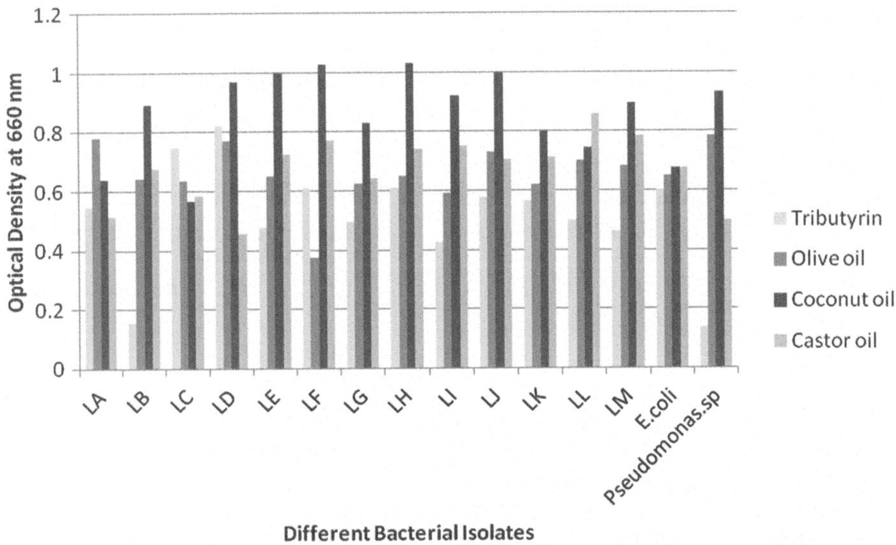

Fig. 2 Growth in tributyrin, olive oil, coconut oil & castor oil after 24 hrs. at 37 °C

than standard strain *Pseudomonas aeruginosa* which is supposed to be a good lipolytic strain (Kathiravan et al. 2012). Strain L_E showed very low growth as the O.D at 660 nm was 0.146. Strains L_F, L_H, L_I and L_L showed very high growth in Tween 20 medium, and strains L_G and L_K showed moderate growth after 24 h of incubation at 37 °C, but the *Pseudomonas aeruginosa* and *E. coli* showed low growth as the O.D were 0.580 and 0.465, respectively, at 660 nm.

In case of tributyrin medium without strain L_D which showed moderate growth, all other strains including *E. coli* and *Pseudomonas aeruginosa* showed low growth after incubation at 37 °C for 24 h.

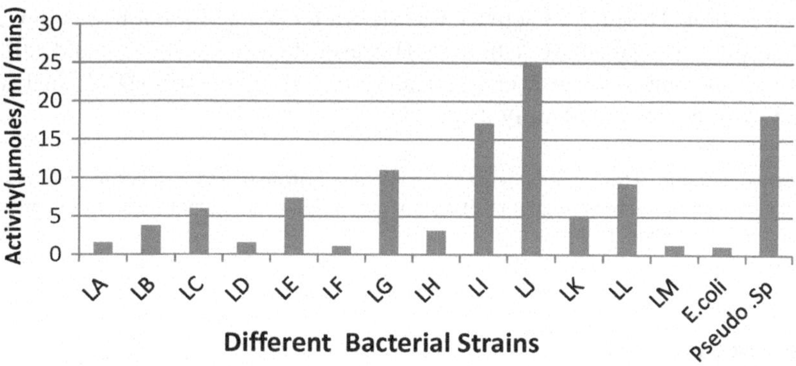

Fig. 3 Activity of different lipolytic strains

In the olive oil medium, all strains including the standard strains showed moderate growth after 24 h of incubation at 37 °C.

In coconut oil medium, all strains showed good growth; strains L_A, L_C, L_L, L_F and L_H and *E. coli* showed very good growth after 24 h of incubation at 37 °C.

However, in castor oil medium, strain L_A showed higher growth than other strains. All other strains showed moderate growth including *E. coli* and strains L_A, L_C and L_D; *Pseudomonas aeruginosa* showed low growth after 24 h of incubation at 37 °C (Fig. 2).

Colorimetric assay was done to determine the lipase activities of the isolates. The unknown amount of product formed was calculated from the standard curve of PNP, and then the activity of different strains including the standard strains was calculated as product formed/ml of enzyme/min or µmoles/ml/mins (Fig. 3). *Pseudomonas aeruginosa* has been reported as good lipolytic organisms (Kathiravan et al. 2012; Amara and Salem 2009). They showed maximum activity of 1121.00 µmoles/mins/ml at 37 °C in colorimetric assay using para-nitrophenyl palmitate. But we got 1.10 µmoles/min/ml activity for the *Pseudomonas aeruginosa* in our experiment that was much less. The difference may be due to the different experimental conditions and the strain used in our experiment. Although strain L_J showed moderate growth in oil medium, it gave high activity (25.0 µmoles/min/ml) which was even higher than the strain *Pseudomonas aeruginosa* after 24 h of incubation at 37 °C. So, according to experimental conditioned, strain L_J was found to be most efficient lipase producer amongst all the isolates.

4 Conclusions

Lipase is a commercially important enzyme but very few bacterial lipases are actually used in industries. The present study was an attempt to isolate and identify a few lipase-producing bacterial strains from oil industry waste which can be commercially viable and can be effectively used for waste management and bioremediation. The highest activity was shown by isolate L_J, but it was low compared to other fungal lipase activities. However, that the isolates could utilize a variety of oils and fats was a unique feature of this study. 16S rDNA analysis of strain L_J will be done to identify its genera.

Acknowledgements The authors wish to thank the principal and teacher-in-charge of the Barrackpore Rastraguru Surendranath College for encouraging and supporting the work and to the UGC, India, for funding the project.

References

Amara AA, Salem SR (2009) Degradation of castor oil and lipase production by *Pseudomonas aeruginosa*. Am-Eurasian J Agric Environ Sci 5(4):556–563

Bailey JE, Ollis DF (1986) Applied enzyme catalysis. In: Biochemical engineering fundamentals, 2nd edn. McGraw-Hill, New York, pp 157–227

Boonmahome P, Mongkolthanaruk W (2013) Lipase-producing bacterium and its enzyme characterization. J Life Sci Technol 1(4):196–200

Ghasemi Y, Rasoul Amini S, Kazemi A, Zarrinic G, Morowvat MH, Kargar M (2011) Isolation and characterization of some moderately halophilic bacteria with lipase activity. Microbiology 80(4):483–487

Gupta R, Gupta N, Rathi P (2004) Bacterial lipases: an overview of production, purification and biochemical properties. Appl Microbial Biotechnol 64:763–781

Haki GD, Rakshit SK (2003) Developments in industrially important thermostable enzymes: a review. Bioresour Technol 89(1):17–34

Imamura S, Takahashi M, Misaki H, Matsuura K (1989) Method and reagent containing lipases for enzymatic determination of triglycerides. West Germany Patent 3:912,226

Kanwar S, Kaushal R et al (2005) Methods for inhibition of residual lipase activity in colorimetric assay: a comparative study. Indian J Biochem Biophys 42:233–237

Karadzic I, Masui A, Lidija Izrael Zivkovic LI, Fujiwara N (2006) Purification and characterization of an alkaline lipase from *Pseudomonas aeruginosa* isolated from putrid mineral cutting oil as component of metal working fluid. J Biosci Bioeng 102:82–89

Kathiravan T, Marykala J et al (2012) Studies on nutritional requirements of *Pseudomonas aeruginosa* for lipase production. Adv Appl Sci Res 3(1):591–598

Macrae AR (1989) Biochem Soc Trans 17:1146–1148

Mobarak-Qamsari E, Kasra-Kermanshahi R, Moosavi-nejad Z (2011) Isolation and identification of a novel, lipase-producing bacterium, *Pseudomonas aeruginosa* KM110. Iranian J Microbiol 3(2):92–98

Sharma R, Chisti Y, Banerjee UC (2001) Production, purification, characterization, and applications of lipases. Biotechnol Adv 19:627–662

Willerding AL, Luiz A, Moreira FW, Germano MG, Chagas AF Jr (2011) Lipase activity among bacteria isolated from Amazonian soils. Enzyme Res. doi:10.4061/2011/720194

Zaks A, Klibanov AM (1985) Proc Natl Acad Sci U S A 82:3192–3196

By-Products of Bioenergy Systems (Anaerobic Digestion and Gasification): Generation and Prospects of Utilization

Kangkana Saikia, Pallavi Roy, Dipam Patowary, Sampriti Kataki, Hemantajeet Medhi, Pranjal Das, Helen West, Michèle Clarke, and D.C. Baruah

Abstract The producer gas and biogas are biomass-based renewable energy sources drawing increased attention worldwide. Substantial quantities of by-products are also generated during the production of these fuels from biomass. While primary emphasis is on main energy form, management and utilization of by-products are also of concern, both from economic and environmental points of view, as these are the inevitable commodities. There have been some reports on utilization of these by-products, mostly traditionally, to supplement soil nutrients. However, lack of proper management could lead to their nonoptimal use through loss of nutrients. The present investigation considers the by-products of two bioenergy systems, viz. digested slurry from anaerobic digestion unit and tar from bio-gasification unit, available in food industry through case studies to understand the rate of by-product generation, characterization and prospective utilization. From the total slurry generated using cow dung as feedstock, an estimated amount of 3.68 kg of dry mass (solid digestate) and 36.28 l of liquid (liquid digestate) per cubic metre of biogas were obtained. While analysing the digested slurry in terms of thermal and nutritional characteristics, it is known that recovery of second-stage thermal energy without losing essential nutrients could be possible if appropriate management strategy is taken. It has been estimated that 54.56 MJ energy could be made available per cubic metre of biogas, considering second-stage energy recovery route through briquetting technology using solid fraction of digestate. While estimating gasification tar, for generating 1 m^3 of producer gas using bamboo, an average of

K. Saikia • P. Roy • H. Medhi • P. Das
Tezpur University, Napaam, Assam, India

D. Patowary • S. Kataki • D.C. Baruah (✉)
Department of Energy, Tezpur University, Napaam 784028, Assam, India
e-mail: baruahd@tezu.ernet.in

H. West
School of Biosciences, University of Nottingham,
Sutton Bonington Campus, Leicestershire LE12 5RD, UK

M. Clarke
School of Geography, University of Nottingham, University Park, Nottingham NG7 2RD, UK

S.K. Ghosh (ed.), *Utilization and Management of Bioresources*,
DOI 10.1007/978-981-10-5349-8_25

0.041 g tar could be obtained. The prospects of utilization of gasification tar could be possible from the presence of a number of useful chemicals as indicated by characterization study.

Keywords Bioenergy by-product • Anaerobic digestion • Gasification • Tar • Biogas digested slurry

1 Introduction

In India, nearly 72% of the total energy consumption is based on fossil fuels which are gradually at a depleting stage. Hence, scarcity of local resources and financial constraint on importing the resources along with increasing level of emission associated with their use has become a serious concern for all the nations including India with a little degree of variation. Presently, emphasis is given on identification of new and reliable energy source so as to meet the ever-increasing energy demand of the society. To face these challenges, people have attempted to substitute the energy system by the extensive use of renewable energy sources comprising of mostly biomass, solar, wind and tidal.

Biomass is almost equitably a distributed resource in India as compared to other renewables. The current availability of biomass in India is estimated at about 500 million metric tonnes per year (MNRE, India). Since India is an agricultural-based country and about 69% of Indian population resides in rural areas, biomass to energy conversion can play an important role for fulfilling the energy demand of rural India (Galvez et al. 2012). Biomass which is the traditional source of energy has a rapid growth in its demand with an annual average rate of 1.4% during the period 2002–2009 (MNES Annual Report 2003–2004). Energy conversion from biomass has been achieved with greater extent of success through different routes at different places. Along with the desired form of energy, the bioenergy conversion routes result in a wide range of side products also known as bioenergy by-products (Sawin 2012; Taheripour et al. 2010) or bioenergy residue (Galvez et al. 2012). Such growth in the bioenergy sector raises questions about the end use of the bioenergy residues, as these are the inevitable commodities of conversion process.

The two main bioenergy conversion routes are thermochemical and biochemical conversion. Anaerobic digestion under biochemical route and gasification under thermochemical route are the two different established technologies which produce biogas and producer gas, respectively. Anaerobic digestion generates anaerobic digestate, and gasification generates tar and char as their respective by-products. Biogas technology is gaining momentum as a fuel for cooking in the rural areas due to its low cost and easy availability of cattle manure. Depending upon feedstock, in a biogas plant, 20–95% of organic content of feedstock is converted into biogas, and the rest is generated as digestate. For a biogas plant fed with cow dung, approximately 48% of the input cow dung is generated as digested slurry (Gell et al. 2011).

Gasification is another established technology to produce producer gas for electricity generation. With the increasing demand of these technologies, substantial amount of by-product is also generated which would demand special managerial attention. This management would benefit in the following ways by preventing environmental nuisance if remain unattended, i.e. waste reduction, compensating the loss of energy against its potential, increases in the overall economy of that system. Keeping in view of the above discussion, it is well understood that specific strategy should be adopted for proper management of bioenergy by-products. As a prerequisite for formulation of their appropriate management strategy, the present study has been adopted to study available by-products generated in a bioenergy process, their estimation and rate of generation and their characterization. The knowledge on these aspects will promote applications of these by-products as fertilizer or other use such as for chemical extraction.

2 Methodology

The present study aims to investigate the optimal utilization of by-products obtained from two bioenergy processes, i.e. anaerobic digestate from anaerobic digestion plant (AD) and tar from thermal gasification plant (TG). The methodologies concerning the estimation of rate of generation, characterization and the potential applications of these by-products are discussed below.

2.1 Estimation of By-Products Generated by AD and Gasification System

There are many variable factors which govern the rate of generation of by-products. In the present investigation, we have considered two specific experimental systems, viz. *anaerobic digestion system* and *gasification system*, to estimate the rate of by-product generation under a given specific working condition. The details of the conditions and methods of estimation are discussed below.

2.1.1 Estimation of Anaerobic Digestate

A lab-sized commercial anaerobic digester (*Shakti Surabhi-type biogas digester, VK Nardep, Kanyakumari, India*, 0.25 m^3) installed in the Department of Energy, Tezpur University (Fig. 1), is used to estimate the rate of digestate generation. The feedstock used for the study was cow dung. The quantity of feedstock used, gas production and digestate generated were monitored daily. Anaerobic digestate obtained as a by-product was collected in pot through slurry collection outlet of the

Fig. 1 Anaerobic digester unit with digestate collection outlet

digester. The outcome of this experiment is used to estimate the digestate produced per kg of dry mass of feedstock. 150 μ filter bags were used to determine the solid and liquid fraction of digestate by manual separation.

2.1.2 Estimation of Gasification Tar

To estimate the rate of generation of tar in a gasification system, a 10 kW downdraft gasification unit was considered present in bamboo-based gasification system of Nezone Biscuits factory in Tezpur. The gasification system was made to run twice using bamboo chips as feedstock for 2 h each to collect the data, and average values were used to estimate the biomass tar generated. Tar and fine particles were produced in the gasifier along with the producer gas. Substantial portion of tar along with the fine particles are trapped in water scrubbing system and sawdust filter. Tar was collected from these two sections of the gasifier for each run of the system. Tar-mixed fine particles coming out of the gasifier form a foamy layer while passing through the scrubbing system. Considering that tar-mixed fine particle is insoluble in water, the samples of foamy layers were collected to estimate the amount of tar in water scrubbing system using the following relationship:

$$W_T = \frac{f \times m \times R}{vM} \tag{1}$$

where

W_T = amount of tar scrubbed by per unit volume of producer gas generation, (g m^{-3})
m = mass of fine particle-mixed tar in sample water of volume v, (g l^{-1})

Table 1 Parameters investigated for characterization of AD digestate and tar

Parameters	Analysis/methods
Moisture content, volatile matter, ash and fixed carbon content	Proximate analysis, ASTM E870–82 (2006)
Calorific value	Auto bomb calorimeter (SE-1AC/ML, automatic calorimeter)
Elemental analysis	EDX
Ultimate analysis (carbon, hydrogen, nitrogen)	CHN analyser

R = rate of foam production in scrubbing system ($l\ h^{-1}$)
M = rate of gas production, $m^3\ h^{-1}$
f = fraction of tar in the tar and fine particle mixture

In the present investigation, 50 ml of sample water (v) was collected. It was considered that tar and fine particulate mixtures were uniformly distributed in the water tank of water scrubbing system, and fraction of tar is 50% ($f = 0.5$) in the mixture of tar and fine particulate matter. For the estimation, the average value of M for the gasifier is considered. The rate of water circulation is assumed as the rate of foam (R) production.

The producer gas coming out of water scrubber (tar laden) was then passed through a sawdust filter. The tar-laden gas gets purified in the sawdust filter. The mass of tar trapped in sawdust was estimated from the difference of masses of fresh gas sample and treated gas sample of a unit volume collected during a specified duration of time. The difference of weight provided an estimate of tar trapped in the sawdust filter which was then expressed as the rate of tar generated per kg of biomass consumption. Neglecting the tar stuck to the interior wall of the pipelines, the total amount of tar per kg of biomass was assessed from the sum total of the above two estimates. The process of tar sampling was initiated only after the temperatures in the gasifier approached steady state, and there was evidence of combustible gases produced, indicated by a self-sustaining flame in the flare.

2.2 *Characterization of By-Product*

The collected anaerobic digestate was characterized to understand its optimal use in bioenergy applications through its combustible property as well as fertilizer application. The tar was characterized and investigated for its potential uses as chemicals. Raw bamboo feedstock was also characterized to understand the change in property after its conversion into tar through gasification route. Table 1 lists the different parameters determined to characterize digestate and tar along with the standard protocols used.

2.3 *Assessment of Bioenergy Potential from By-Products*

Availabilities of digestate and bamboo tar per unit biogas and producer gas produced have been calculated as discussed (Sect. 2.2) above. Taking the calorific value (CV) of the primary products (biogas from AD and producer gas from TG) and their respective by-products (anaerobic digestate and tar), gross energy yields from the AD, and TG processes were finally determined. Enhancement in net energy yield considering by-product utilization was estimated. The digested slurry is composed of both liquid and solid fraction. Therefore, in case of anaerobic digestate, only the solid fraction was considered for energy recovery through a secondary energy route, i.e. briquetting technology and gross energy yield from both biogas and briquettes has been determined.

3 Results and Discussions

3.1 *Estimation of Bioenergy By-Product*

3.1.1 Estimation of Anaerobic Digestate

The average biogas production rate of the system was found to be 24.7 l d^{-1} which is equivalent to 29.64 g d^{-1} (considering density of biogas as 1.2 kg m^{-3}). On an average, 1,010 g d^{-1} of digestate was generated by the setup. An amount of 3.68 kg solid digestate and 36.28 l of liquid digestate per cubic metre of biogas have been estimated from the experimental observation.

3.1.2 Estimation of Tar Generated

As discussed in Sect. 3.2.2, gravimetric tar and fine particulate matters at the two sections, viz. water scrubbing system and sawdust filter of the gasifier system for generating 1 m^3 of producer gas using bamboo, were found to be about 0.151 g and 0.182 g, respectively. Therefore, considering the two systems, a total of tar 0.333 g was produced.

3.2 *Characteristics of By-Product*

3.2.1 Characteristics of Digestate

Proximate and ultimate analysis results of anaerobic digestate are shown in Table 2. The total solid content of the digestate was found to be 9%, which signifies the semisolid nature of it. The pH of the digestate was found to be basic (8.2) because

Table 2 Characterization of digestate

Total solid (% of dry matter)	9
Ash (%) (dry basis)	27.31
pH	8.2

Table 3 EDX analysis of cow dung digestate

Element	Weight %	Element	Weight %
C	41.18	Cl	0.04
O	53.8	K	0.14
Mg	0.08	Ca	0.79
Si	1.19	Mn	0.03
P	0.93	Zn	0.14
S	0.03	–	–

Table 4 Proximate analysis of bamboo tar in comparison with raw bamboo

Parameter	Raw bamboo	Bamboo tar
Moisture content (%)	5.80	2.35
Volatile matter (%)	67.83	21.55
Fixed carbon (%)	21.87	67.75
Ash content (%)	4.5	8.35
Calorific value (MJ/Kg)	16.83	26.85

of ammonia-N formation during anaerobic digestion (ADAS 2007). Ash content (27.31%) of digestate was found to be high enough, which indicates presence of considerable amount of mineral. EDX analysis of solid fraction of digestate also supports this point which is shown in Table 3. It shows that the solid content of digestate contains different plant macronutrients like P and K and micronutrients like Ca, Mg, Mn, Cl, etc. (Table 3).

3.2.2 Characteristics of Tar

Ultimate analysis and proximate analysis of bamboo tar collected from gasifier unit are shown in Tables 4 and 5. These results are compared with raw bamboor There is an increase in fixed carbon content and ash when bamboo is converted to with a significant increase in calorific value because of increase in fixed carbon content. Elemental analysis of bamboo tar shows substantial concentration of carbon and oxygen which indicated presence of hydrocarbons. EDX analysis of tar is shown in Table 6.

Table 5 Ultimate analysis of bamboo tar in comparison with raw bamboo

Parameters	Bamboo tar	Raw bamboo
C (%)	55.51	57.27
H (%)	6.25	8.24
N (%)	4.13	5.52
O (%)	34.11	28.97
C/H ratio	8.38	6.95

Table 6 EDX analysis of bamboo tar

Element	Weight %
C	73.96
O	25.16
Si	0.03
Cl	0.12
K	0.12
Fe	0.61

4 Potential Applications of Bioenergy By-Products

While estimating the enhancement of energy output of the gasification system considering energy recovery from tar, only a marginal increase (3%) in energy output is observed. However, substantial increase in energy output would be resulted if we consider digestate of AD system as a solid fuel. It has been estimated that overall 54.56 MJ of energy could be made available per cubic metre of biogas including briquetting as a secondary route of energy.

In general, physico-chemical characteristics of the digestate are determined by the composition and quality of the feedstock, combined with the effectiveness of the anaerobic digestion process (Makádi et al. 2012). In digestate plant, essential nutrients like nitrogen, phosphorus and potassium are conserved from the original feedstock with minimum variation (Bachmann and Eichler-Löbermann 2009; Zhang et al. 2010), which makes this a suitable soil quality enhancer. Digestate has been already reported to replace 50–75% of recommended dose of inorganic fertilizer without significantly affecting the grain yield (Singh et al. 2007; Garfi et al. 2011). The present study shows that the overall efficiency of the biogas technology can be increased by using the thermal output of the prepared briquettes from the digestate along with the thermal output of the biogas. There is also a scope that the nutrient content of the digestate need not to be compromised as the leftover ash, after burning of briquettes, could be circulated back as fertilizer. This technique is feasible to carry out in rural areas and briquettes can be used as a solid fuel in rural household applications.

The gasification by-product tar is a mixture of various hydrocarbons, mostly oxygen-containing compounds as found in the chemical characterization. Previous investigation reported presence of derivatives of phenol, free fatty acids and esters of fatty acids in tar which could be precursor to other chemicals (Steve et al. 2007).

Therefore, tar could be considered as a source of chemicals based on chemical analysis, while digestate could be a prospective source of fuel source as well as fertilizer.

5 Conclusions

The objective of by-product management is to reduce the waste stream volume as well as to extract optimum benefit by recovery of materials and energy. Since with increasing market of bioenergy, mismanagement of simultaneous by-product generation could possess some nuisance; hence, it is very essential to step in effective by-product management system. The characterization of the by-products from anaerobic digestion and gasification shows scope for their utilization for various purposes. Apart from the soil fertilizing property of the digestate by virtue of their nutrient retention from feedstock, it is shown that by using briquetting technology, digestate can be utilized as a source of solid fuel during the season of fuel crisis. The overall efficiency of biogas technology can be increased by utilizing the thermal output of the biogas as well as the thermal output of the prepared briquettes from the same feedstock given, where the nutritional value of digestate remains unaltered if the ash produced from briquette could be circulated back. The higher calorific value of tar compared to raw feedstock also suggests its potential application as fuel. But due to the marginal increase in energy output, its chemical application could be a better option. Therefore, optimum utilization of the bioenergy by-products for maximum economic and environmental benefit should be ensured while planning a bioenergy programme to make this an asset rather than a liability.

References

ADAS U (2007) Nutrient value of digestate from farm-based biogas plants in Scotland. Report for Scottish executive environment and rural affairs department-ADA/009/06

Bachmann S, Eichler-Löbermann B (2009) Fertilizing effect of biogas slurries. In: More sustainability in agriculture – new fertilizers and fertilization management, 18th Symposium CIEC, Rome, Nov 2009

Galvez A, Sinicco T, Cayuela ML, Mingorance MD, Fornasier F, Mondini C (2012) Short term effects of bioenergy by-products on soil C and N dynamics, nutrient availability and biochemical properties. Agric Ecosyst Environ 160:3–14

Garfi M, Gelman P, Comas J, Carrasco W, Fermer I (2011) Agricultural reuse of the digestate from low- cost tubular digestates in rural Andean communities. Waste Manag 31:2584–2589

Gell K, de Ruijter FJ, Kuntke P, De Graaff M, Smit AL (2011) Safety and effectiveness of struvite from black water and urine as a phosphorus fertilizer. J Agric Sci 3(3):67. http://data.worldbank.org/indicator/EG.USE.COMM.FO.ZS

Makádi M, Tomócsik A, Orosz V (2012) Digestate: a new nutrient source–review. Energy 4(7.5):8–7

MNES annual report (2003–04) Govt. of India

Sawin JL (2012) Renewables 2012–global status report, Renewables Energy Police Network of the 21st Century (REN21). Zuletzt eingesehen am, 22

Singh KP, Suman A, Singh PN, Srivastava TK (2007) Improving quality of sugarcane-growing soils by organic amendments under subtropical climatic conditions of India. Biol Ferti Soils 44(2):367–376

Steve PD, Daniel L, Carpenter R, French J (2007) Quantitative measurement of biomass gasifier tars using a molecular-beam mass spectrometer: comparison with traditional impinger sampling. Energy Fuel 21(5):3036–3043

Taheripour F, Hertel TW, Tyner WE, Beckman JF, Birur DK (2010) Biofuels and their by-products: global economic and environmental implications. Biomass Bioenergy 34(3):278–289. www.mnre.org

Zhang T, Bowers KE, Harrison JH, Chen S (2010) Releasing phosphorus from calcium for struvite fertilizer production from anaerobically digested dairy effluent. Water Environ Res 82(1):34–42

Evaluation of Blending of Lowest Emission Biodiesel with Jet A for Producing Aviation Biofuels

N.K. Attia, E.A. Abdel Kader, G. ElDiwani, H.S. Hussein, and R. El-Araby

Abstract Today, a huge volume of jet fuel is consumed through air transportation. Petroleum-based jet fuels, including Jet A-1, produce considerable amount of particulate and gaseous pollutions which affect the world climate change. To overcome these pollutions, portion of jet fuel should be replaced by jet biofuels. This can be achieved by blending Jet A with efficient energy source with lowest emission such as biofuels from vegetable oils. Biodiesel from palm, *Jatropha curcas*, and waste cooking oils show promising potential in producing aviation biofuels when blending with Jet A. Various volumetric ratios of biodiesel (5–25%) were blended with high-grade kerosene. The binary blends of jet biofuels were characterized and compared with Jet A fuels. The comparison indicated that biofuels with 5% ester content have almost similar characteristics with Jet A aviation. The physicochemical properties of 5% binary blends of palm, *Jatropha curca*s, and waste cooking oils were kinematic viscosity (at −20 °C) of 8.6, 5.1, and 5.1; high heating value (HHV) around 43 Mj/kg and freezing point of −14.5, −15.5, and −25.5 °C for binary blends of the biodiesel from the three types of oils with Jet A, respectively. Additives may be added to binary blends to reach exact Jet A physicochemical characteristics. The yield % | C8–C15 were determined for the 5% ester content for each binary blends using GC-mass spectrometry.

Keywords Biodiesel • Blending • Aviation biofuels • Evaluation

1 Introduction

Development of aviation fuels has continued for many years. Several studies have been conducted on aviation fuels in order to diminish production costs, ease supplies, reduce the dependence on fossil fuels, and hence reduce its carbon footprint (Solmaz et al. 2014). The aviation industry contributes to air pollution by the greenhouse effect due to increase of carbon emission (about 5.1 global GHG emission by

N.K. Attia (✉) • E.A. Abdel Kader • G. ElDiwani • H.S. Hussein • R. El-Araby
Chemical Engineering and Pilot Plant Department, National Research Center, Cairo, Egypt
e-mail: nahed_attia@yahoo.com

S.K. Ghosh (ed.), *Utilization and Management of Bioresources*,
DOI 10.1007/978-981-10-5349-8_26

2050) (Reksowadojo et al. 2014). World consumption of jet fuel has increased from 189.1 to 198.3 million gallons per day within the period of 2001–2008. Moreover, world jet fuel demand was predicted to increase by almost 38% from 2008 to 2025 (Reksowadojo et al. 2014; Intergovernmental Panel on Climate Change (IPCC) 1999). Consequently, the price of jet fuel will be increased rapidly in the next 30 years as the average price in the year 2013 was \$2.82/gallon (KandaramathHari et al. 2015). So, the need for developing renewable, sustainable and environmentally friendly fuel is a must. Jet fuels are often classified as kerosene or naphtha type including the combination of a large number of different hydrocarbons. Kerosene-type fuels include Jet A, Jet A-1, JP-5, and JP-8, while naphtha-type jet fuels, sometimes referred to as "wide-cut" jet fuel, include Jet B and JP-4. Jet fuel kerosene-type Jet A-1 is the predominant fuel in the world and has a wide range of fractions between about 8 and 16 carbon numbers (Aviation biofuel 2016; Baroutian et al. 2013).

The current aviation fuel, kerosene (Jet A/A-1), is derived from oil and is a middle distillate between gasoline and diesel, and the only relevant difference between them is the freezing point (−40 °C for Jet A and −47 °C for Jet A-1) (Rosillo-Calle et al. 2012). The use of biofuel in aviation must consider the high-quality standard requirements, where biodiesel must be compatible with aviation fuel properties. It must be thermally stable to avoid freezing or gelling at low temperatures and to satisfy other requirements in terms of viscosity, surface tension, and ignition properties in agreement with the materials typically used in aviation (Sandquist and MatasGüell 2012).

Blending of biodiesel with jet fuel must improve the cold flow properties which make it more suitable for jet engines (Baroutian et al. 2013). More than 95% of the world production of biofuel was from edible oil, which causes negative impact on human food supplies and increases the risk of food security (Baroutian et al. 2013; Veny et al. 2009). Therefore, scientists pay attention for producing biofuel from nonedible oils like oil from *Jatropha curcas* seeds and the waste cooking oil. Biodiesel can be blended with jet fuel having the same of its characteristics. However, the main drawbacks of vegetable oil are the high viscosity and low volatility which may cause poor combustion in engines (Ong et al. 2011; Antony Raja et al. 2011)

Currently, different methods are used in biodiesel preparation, for example, pyrolysis, hydroprocessing, and transesterification (Baroutian et al. 2013). Transesterification is a more efficient method and the most promising process to convert vegetable oil into methyl ester. Also, it was successfully employed to reduce the viscosity of biodiesel and improve its properties (Ong et al. 2011; Antony Raja et al. 2011; Balat and Balat 2008). It is a simple process with high production capacity and low-cost operation due to moderate reaction temperature and short-time and low-cost catalysis (Baroutian et al. 2013; Antony Raja et al. 2011).

Blending of biodiesel with jet fuel will decrease the rate of exhaust emission and then decrease the greenhouse effects, global warming, and weakening of ozone layer (Ganjehkaviri et al. 2016). In this study, blending of biodiesel with JetA-1 fuel (kerosene) at different proportions has been proceeded. The produced blended biodiesel with JetA-1 fuels is characterized and evaluated by measuring their viscosity at −20, calorific values, and freezing points.

2 Materials and Methods

2.1 *Materials*

Aviation Jet A-1 fuel was obtained from Misr for Petroleum Co.; waste cooking oil was prepared, and *Jatropha* oil and palm oil were obtained from the agriculture ministry in Egypt. Potassium hydroxide was used as a catalyst. Methanol (99.5%) was supplied by Sigma-Aldrich Co.

2.2 *Biodiesel Production*

Jatropha curcas, palm oil, and waste oil were converted into methyl ester by methylation with potassium hydroxide using homogenous catalyst method. It was necessary to purify waste cooking oil before using as biofuel feedstock by filtration. The acid value, for the oils, must be determined before using as biofuel. The acid values are listed in Table 1. Conversion of free fatty acid was conducted in case of *J. curcas* (AV = 14.1 mg KOH/ gm) as increasing of free fatty acid values will lead to fatty acid salt formation (soap) during alkali transesterification (ref).

2.2.1 Esterification of *J. curcas*

Reducing free fatty acid of *J. curcas* oil was conducted by using methanol 40% (v/v) and sulfuric acid as catalyst (v/v) at 60 °C 2 h. The mixture is transferred to a separating funnel to separate the esterified *J. curcas*, and then evaporate the excess alcohol.

2.2.2 Transesterification Process

The free fatty acid value is reduced by esterification step, and then the oils were ready for transesterification process which was carried out as the following:

Table 1 Acid value of *J. curcas*, waste cooking and palm oil

Item	*J. curcas*	Waste cooking oil	Palm oil	Esterified *J. curcas*
Acid value, mg KOH/g sample	14.1	0.3	0.1	1.05

About 200 ml of oil was mixed with KOH (0.7% w/v) and 20% v/v methanol at 65 °C for 2 h. The product methyl esters were separated from glycerol (by-product), and then the prepared biodiesel was washed with warm water and acetic acid. Methyl ester was dried and characterized by gas chromatography. The process of production of biofuel from *J. curcas*, waste cooking, and palm oil is represented in Fig. 1

2.3 Biojet Fuel Preparation

Biofuels were prepared by mixing biodiesels with Jet A-1 at different fraction volumes (5–25%) at room temperature and kept in glass container for characterization. The prepared biofuels were analyzed for their chemical and physical characteristics.

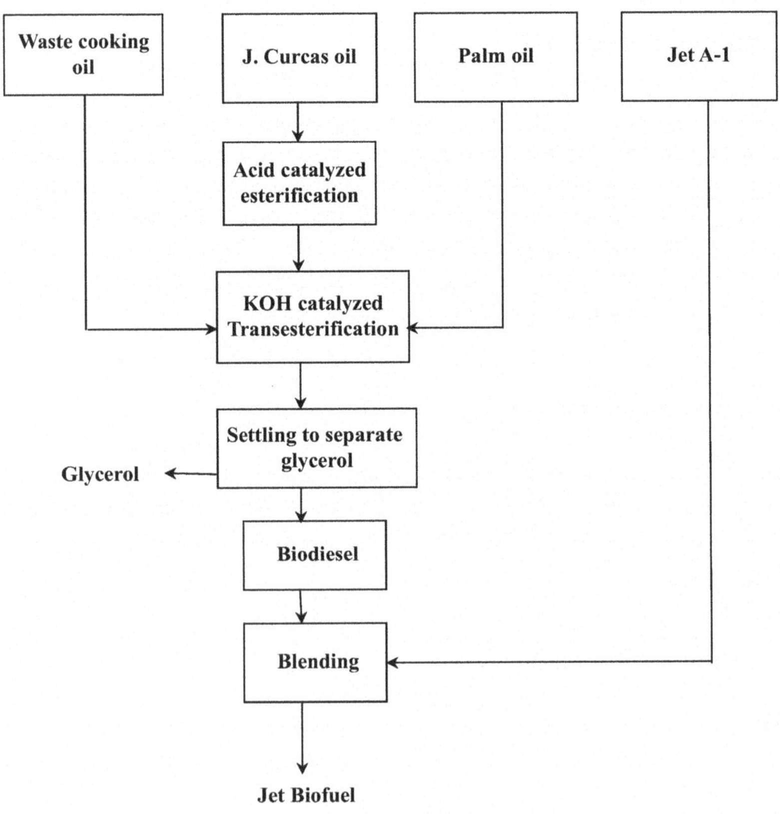

Fig. 1 Jet biofuel production from used cooking, *Jatropha*, and palm oil biodiesel

2.4 Characterizations of the Biojet Fuels

The produced biojet fuel standard characteristics such as viscosity at −20 °C, calorific value, and freezing point were measured for each blending ratio of biodiesel with Jet A-1 according to test methods ASTMD341, ASTMD4529, and ASTMD2386, respectively.

Elemental analysis was performed for the 5% ratio blending of biodiesels with Jet A-1 for the used cooking, *Jatropha*, and palm oils to measure the effect of present elements in the produced biojet fuels on their suitability as aviation fuels.

2.5 GC-Mass Spectrometry Characterization

GC-mass spectrometry characterization of the 5% blending biofuels was performed to measure fuel range hydrocarbon percentage (wt.%) using GC Ultra-ISQ mass spectrometer (Thermo Scientific, Austin, TX, USA) with a direct capillary column TG-5MS (30 m × 0.25 mm × 0.25 μm film thickness). The column oven temperature was initially held at 60 °C and then increased by 5 °C/min to reach 200 °C and hold 2 min then increased to 280, 10 °C/min.

3 Results and Discussion

To reduce the emission of petroleum aviation fuel and meet the jet fuel specifications, Jet A-1 should be blended with aromatic renewable source. The major natural resources are from transesterification of oils to produce biodiesel. Various volumetric ratios of biodiesel (5–25%) were blended with aviation kerosene to produce biojet fuels. The results of measured biojet fuels' specifications are shown in Table 2.

Table 2 Slandered characteristics of the produced biojet fuels

	Biodiesel ratio % with Jet A-1											
	Kinematic viscosity at −20 °C (Cst.)				Calorific value (Mj/Kg)				Freezing point (− °C)			
Type of oil	5	10	20	25	5	10	20	25	5	10	20	25
Used cooking	4.1	6.4	8.7	19.8	43.145	42.893	42.601	42.348	14.5	0.5	+6.5	+2
Jatropha	8.6	5.1	6.0	7.8	42.970	42.543	42.068	41.963	15.5	12	1.5	+1
Palm	8.6	10.1	11.3	14.1	43.124	42.915	42.610	42.418	25.5	12.5	8.5	6.5

3.1 Viscosity

Highly viscous aviation fuels lead to poor combustion and blocking of fuel injectors. The produced biofuels at 25% blend of biodiesel with Jet A-1 lie outside the required specification. This indicates that biodiesel of 25% blends are not suitable as an aviation fuel. Low level of 5% biodiesel blends is more suitable as jet fuels, as they are within the required viscosity specification. It is potentially beneficial for better atomization and pumping. Table 2 shows the kinematic viscosities ($mm^2\ s^{-1}$) of the produced biofuel blends at −20 °C.

3.2 Calorific Value

Calorific value is the amount of energy produced by the complete combustion of a fuel. Calorific values of all types of produced biodiesel oils lied in the acceptable range for substitute biojet fuels as presented in Table 2. The produced biojet fuels with lower biodiesel to Jet A-1 ratios have higher calorific values than that of higher blending ratios. Calorific values for palm oil and waste cooking oil biodiesel blending biojet fuels are higher than those of *Jatropha* oil. The calorific values of 5% blending biojet fuels of used cooking, palm, and *Jatropha* biodiesel oils are about 43 Mj/Kg.

3.3 Freezing Point

Freezing point is the temperature below which solid hydrocarbon crystals of aviation fuels may form in turbine. Biofuels with lower freezing point are more suitable as jet fuel substitute. The freezing points of biofuels of higher blending ratio of biodiesel to JetA-1 are increasing with the increase of the blending ratio as presented in Table 2. This indicates that the freezing points of biofuels with lower blending ratios may reach the required values when increasing the additives which lead to the decrease of freezing points. The freezing points of the obtained biojet fuel depend on the source of oil used to produce biodiesel. As shown in Table 2, the freezing points of the 5% biodiesel blending with Jet A-1 of the used cooking oil, Jatropha, and palm oils are −14.5, −15.5, and −25.5, respectively.

3.4 Oxygen Percent in the Produced Biojet Fuels

Table 3 illustrates the results of the elemental analysis of produced biojet fuels of 5% biodiesel blending with Jet A-1. It is obvious that the biojet fuel produced from blending of 5% palm oil biodiesel has the less content of oxygen (0.95%) than other

Table 3 Elemental analysis of produced biojet fuels of 5% biodiesel blending with Jet A-1

	Elements present %				
Biodiesel oil type	Nitrogen	Carbon	Sulfur	Hydrogen	Oxygen
Used cooking	0.47	65.75	–	27.65	6.13
Jatropha	0.14	80.78	–	16.55	2.53
Palm	0.33	71.07	–	27.65	0.95

types. The less oxygen present indicates that the substitute biojet fuel can be successfully produced by blending of 5% palm oil biodiesel with Jet A-1.

It is also noticed that no sulfur content in all produced biofuels blends, which indicates safe emission of this aviation biofuels. The carbon percent present in all types of biodiesel blends shows that the biofuels will demonstrate the desired combustion which agrees with the measured calorific values.

3.5 GC-Mass Spectrometry Characterization

Paraffin and aromatic hydrocarbons are the major constituents of jet fuel range hydrocarbons (C8–C15). The results of GC-mass spectrometry characterization of the 5% blending biofuels show that the biofuels contain cyclo- and n-alkane (about 15 and 80%), respectively, which are the major constituents, while aromatic hydrocarbons was less than 5%. This indicates that the quality of 5% blending of biodiesel with Jet A-1 is the most suitable blending for producing biojet aviation.

4 Conclusions

Biodiesel is a renewable fuel derived from transesterification of palm, *Jatropha*, and used cooking oils considered as alternative to jet fuel range hydrocarbons. From elemental analysis, a small amount of oxygen content (0.95%) in 5% palm biodiesel blended with Jet A gave a promising aviation biofuel. Biofuel of 5% palm biodiesel blended with jet A is the most suitable substitute of traditional kerosene where no further upgrading process is required. All three fuel range hydrocarbons (blends 5% palm, *Jatropha*, and used cooking oil biodiesel) have excellent calorific values which would likely be similar to Jet A-1. Substituting part of petroleum source Jet A-1 fuel with biodiesel can reduce the undesired emissions which has a good impact on the environment.

Acknowledgments This project was supported financially by the Science and Technology Development Fund (STDF), Egypt, Grant No 4991.

References

Antony Raja S, Robinson Smart DS, Lindon Robert Lee C (2011) Biodiesel production from jatropha oil and its characterization. Res J Chem Sci 1(1):70–80

Aviation biofuel – Wikipedia (2016) The free encyclopedia.htm

Balat M, Balat HA (2008) Critical review of bio-diesel as a vehicular fuel. Energy Convers Manag 49(10):2727–2741

Baroutian S, Aroua MK, Abdul Raman A, Shafie A, Ismail RA, Hamdan H (2013) Blended aviation biofuel from esterified Jatropha curcas and waste vegetable oils. J Taiwan Inst Chem Eng 44:911–916

Ganjehkaviri A, Jaafar MNM, Hosseini SE, Musthafa AB (2016) Performance evaluation of palm oil-based biodiesel combustion in an oil burner. Energies 9:97. (1–10)

Intergovernmental Panel on Climate Change (IPCC) (1999) Aviation and the global atmosphere. In: Penner J, Lister D, Griggs D, Dokken D, McFarland M (eds) A special report of IPCC working group I and III. Cambrigde University Press, Cambridge

KandaramathHari T, Yaakob Z, Binitha NN (2015) Aviation biofuel from renewable resources: routes, opportunities and challenges. Renew Sust Energ Rev 42(C):1234–1244

Ong HC, Mahlia TMI, Masjuki HH, Norhasyima RS (2011) Comparison of palm oil, Jatropha curcas and Calophyllum inophyllum for biodiesel: a review. Renew Sustain Energy Rev 15:3501–3515

Reksowadojo IK, Duong LH, Ng D (2014) Pham, review of typical biofuel foraviation alternative fuel purposes. Engineering, Institut Teknologi Bandung, Bandung

Rosillo-Calle F, Thrän D, Seiffert M, Teelucksingh S (2012) The potential role of biofuels in commercial air transport – biojetfuel, Commissioned by: IEA Bioenergy Task 40 Sustainable International Bioenergy Trade. Available at: www.bioenergytrade.org

Sandquist J, Güell M (2012) Overview of biofuels for aviation. Chem Eng Trans 29:1147–1152

Solmaz H, Yamık H, İçingür Y, Calam A (2014) Investigation of the effects of civil aviation fuel jet A1 blends on diesel engine performance and emission characteristics. Indian J Eng Mat Sci 21:200–206

Veny H, Baroutian S, Aroua MK, Hasan M, Raman AA, Sulaiman NMN (2009) Density of Jatropha curcas seed oil and its methyl esters: measurement and estimations. Int J Thermophys 30:529–541. Epub 541

Value Addition to Horticultural Solid Waste by Applying It in Biosynthesis of Industrially Important Enzyme: Cellulase

Modhuleena Mandal and Uma Ghosh

Abstract Banana (*Musa* sp.) which is a popular fruit consumed worldwide has a yearly production of over 145 million tons approximately. The peel is generally discarded after consumption of the fruit, which eventually leads to the generation of a significant amount of organic waste causing environmental pollution. Meaningful utilization of this nutrient-rich waste has not drawn much attention. The peel can be biotechnologically converted to value-added products which can reduce environmental pollution to a great extent. Cellulases represent a major group of the industrially important enzyme. Its production has been studied widely by the technique of solid-state fermentation (SSF). The present study deals with the cost-effective biosynthesis of endoglucanase (CMCase) utilizing solid waste, banana peel. Optimization of nutritional parameters was done by the conventional one variable at a time (OVAT) method followed by chemical characterization of banana peel by energy-dispersive X-ray spectroscopy (EDX). By applying OVAT methodology for optimization of nutritional parameters for cellulase production, the specific activity of endoglucanase (CMCase) was found to be 54 U/mg. EDX analysis showed that, as the solid waste was found to be rich in carbon source, no additional carbon supplement was essential for the growth of cellulase-producing microorganism used for the production of secondary metabolite by SSF.

Keywords CMCase • FPase • EDX

1 Introduction

Biodegradable waste management has become an acute problem which includes collection and disposal of the wastes. Meaningful utilization of the wastes to prepare value-added products will not only reduce the pollution hazard associated with the waste but also will make industrial products cheaper as the products will be

M. Mandal • U. Ghosh (✉)
Department of Food Technology & Biochemical Engineering,
Jadavpur University, 700032 Kolkata, West Bengal, India
e-mail: ughoshftbe@yahoo.co.in

S.K. Ghosh (ed.), *Utilization and Management of Bioresources*,
DOI 10.1007/978-981-10-5349-8_27

synthesized in a cost-effective way. Banana being an extensively grown fruit in tropical and subtropical countries is consumed as a nutrient-rich fruit; besides that studies have also been carried out for the production of banana powder, chips, and wine (Krishna et al. 2012). Due to this, a large amount of peel is reported to be accumulated which can cause serious environmental pollution. Efforts have been made by the researchers to utilize the horticultural waste for large-scale production of alcohol and enzymes (Krishna et al. 2012) like amylase and laccases. Cellulose present in the renewable lignocellulose-containing material is considered to be the most abundant organic substrate on earth for the production of glucose, for biofuel, and for chemical feedstock. Research on cellulase has progressed very rapidly in the past few decades, the emphasis being on enzymatic hydrolysis of cellulose. The conversion of cellulose to glucose requires the use of hydrolytic enzyme cellulase which is a multiple enzyme system consisting of endo 1, 4,-*b*-D-glucanases [CMCase, EC 3.2.1.4] and exo 1, 4,-*b*-D-glucanases [FPA, EC 3.2.1.91] along with cellobiase [beta-D-glucoside glucanohydrolase, EC 3.2.1.21] (Krishna 1998). Commercial production of cellulases has been tried by different techniques like batch, fed-batch, continuous flow processes, and solid-state and submerged fermentation. Media used in cellulase fermentations usually contain pure cellulose (Domingues et al. 2000); the use of raw lignocellulosic substrates (Sukumaran et al. 2005) serves as an alternative to the expensive pure cellulose which is true especially in the case of solid-state fermentation (SSF) making it advantageous compared to submerged fermentation (SmF). Besides that cellulase yield is reported to be higher in SSF compared to SmF (Mandal and Ghosh 2015) when compared to their respective controls. Cellulases are inducible enzymes, and the most important challenge of industrial cellulase production is providing the appropriate inducer for cellulases without making the process too expensive. Therefore the present study is aimed at cellulase production in laboratory scale by utilizing banana peel as substrate by SSF. Optimization of the nutritional composition of the fermentation medium was also done by supplementing the media with various carbon, nitrogen, metal salt, surfactants, and amino acid sources. The study was carried out with isolated *Aspergillus niger* as a potent strain for cellulase production.

2 Materials and Methods

2.1 Chemicals and Substrate Required

Carboxymethyl cellulose (CMC) and amino acids were procured from LOBA chemicals. Whatman No. 1 filter paper was acquired from Sigma-Aldrich. 3,5-Dinitrosalicylic acid, Folin-Ciocalteu, carbon, nitrogen supplements, and surfactants were obtained from Merck. Solid waste banana peel was collected from local sources for SSF.

2.2 *Pretreatment of Substrate*

Ripe banana peels were collected from local sources, washed, dried at 70 °C, and were ground to fine powder by passing it through sieve with a of size 1.1 mm.

2.3 *Fungi*

The isolated fungal strain *Aspergillus niger* was maintained on potato dextrose agar and stored at 4 °C.

2.4 *SSF*

Solid-state fermentation was carried out in 100 ml Erlenmeyer flasks. Fermentation medium was inoculated with spore suspension made from the isolated fungi. Different nutritional supplements were added to figure out their positive or negative impact on production of cellulase. The flasks were incubated at 30 °C for 5 days.

2.4.1 Crude Enzyme Extraction

The enzyme was extracted from the solid moldy medium (Mandal and Ghosh 2015) and centrifuged. The clear supernatant was used as a source of crude extracellular enzyme.

2.4.2 Assay of Crude Enzyme

The endoglucanase (CMCase) activity was determined by the method of Ghose (Ghose 1987). Carboxymethyl cellulose (CMC) was used as substrate for the assay. The liberated reducing sugar was estimated by reading absorbance at 540 nm after addition of DNS. One unit of enzyme (CMCase) is the amount of enzyme releasing 1 μmol of reducing sugar from carboxymethyl cellulose/mL/min (Pradeep and Narasimha 2011). The enzyme activity was expressed as U/gds (i.e., unit per gram dry substrate). The dry weight of the samples was determined by drying them in a hot air oven at 80 °C to a constant weight.

2.4.3 Protein Estimation in Crude Enzyme

The same supernatant was used for extracellular protein estimation according to the method of Lowry (Lowry et al. 1951). Bovine serum albumin was used as the standard.

2.5 Parametric Optimization of SSF by the Isolated Strain

Cellulase production under SSF was optimized by adding different nutritional supplements in media used for SSF. The optimal level of one factor was determined by varying its level, keeping other factors in the medium constant (Sun et al. 2011). The environmental parameters maintained throughout the experiment were time of fermentation, 5 days; temperature, 30 °C; the amount of solid waste, 5 g; and hydration ratio, 3:1 (w/v). For optimization of nutritional parameters, the fermentation medium (banana peel) was supplemented with glucose, CMC, starch, sodium acetate, and xylose as the carbon source (1 %w/w) and glycine, sodium nitrate, gelatin, ammonium sulfate, and ammonium chloride as nitrogen sources (0.5 %w/w). Studies were also carried out with $MgSO_4$, $ZnSO_4$ $.7H_20$, $MnSO_4$. H_2O, $BaCl_2$, and $FeCl_3$ as metal additives (0.05 %w/w). Triton X-100, Tween 20 w/v, SDS, and EDTA were added as surfactant sources (0.3 %w/w) and glycine, serine, glutamic acid, tyrosine, and cysteine as amino acid sources (0.1 %w/w). After fermentation, the assay of endoglucanase (CMCase) was done for each case to optimize the enzyme production.

2.6 Specific Activity

The effect of the nutritional supplements added to the media was figured by calculating the specific activity which can be defined in terms of enzyme units per mg enzyme protein.

2.7 Statistical Analysis

Statistical interpretation of the data was done in origin 6.1 software; error bars with 1 % error were added for each value obtained for different nutritional parameters.

2.8 *Characterization of Banana Peel by Energy-Dispersive X-ray Spectroscopy (EDX)*

Energy-dispersive X-ray spectroscopy or energy-dispersive X-ray microanalysis (EDXMA) is an investigative technique used for the chemical characterization of a sample (Naidu et al. 2013). It is based on the primary principle that each element present in the test sample has an exclusive atomic structure allowing a unique set of peaks on its X-ray spectrum. EDX micrograph of banana peel was done to determine the elemental composition of the peel.

3 Results and Discussion

3.1 *Optimization of Nutritional Parameters for SSF by OVAT*

Extracellular enzyme production depends substantially on the chemical composition of the fermentation medium. In order to determine the effect of carbon sources on the cellulolytic activity of the isolated strain *Aspergillus niger* ISSFR-019, various carbon sources like CMC, xylose, starch, sodium acetate, and glucose were tested. The influence of the carbon sources on cellulase production is represented in Fig. 1, which denoted that none of them could impart positive effect on cellulase production as the medium used for fermentation is reported to be rich in the supplement (Sun et al. 2011; Naidu et al. 2013). Different nitrogen sources like sodium nitrate, gelatin, glycine, ammonium sulfate, and ammonium chloride have shown to

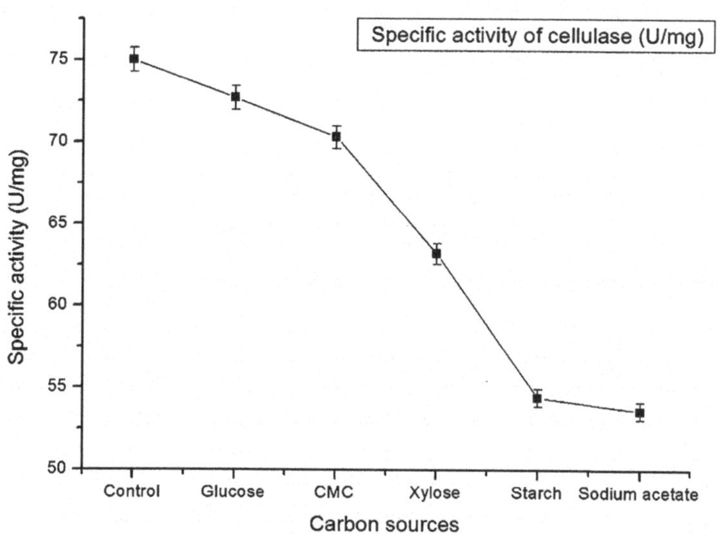

Fig. 1 Effect of carbon sources on cellulase production

Fig. 2a Effect of nitrogen sources on cellulase production

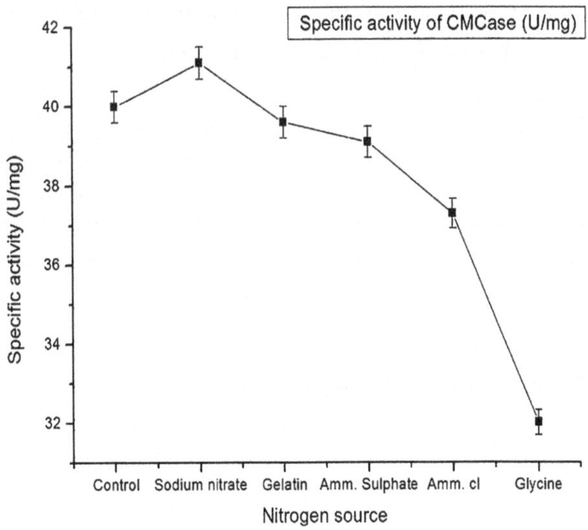

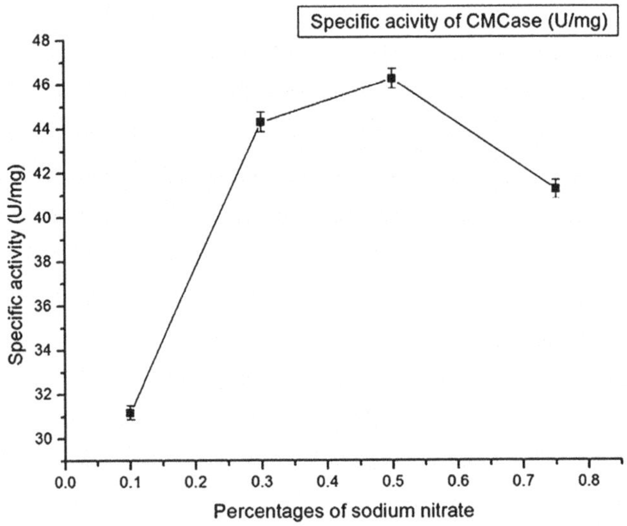

Fig. 2b Effect of different concentrations of $NaNO_3$ on cellulase production

affect the yield of cellulase (Fig. 2a). Inorganic nitrogen source sodium nitrate was found to produce maximum CMCase (specific activity 41.1 U/mg) compared to control. It also served as an effective replacement of organic nitrogen source as they are generally expensive (Sun et al. 1999). The optimization of the percentage of sodium nitrate added to the medium was also done which showed that 0.5 % (w/w) of $NaNO_3$ was ideal for maximum cellulase yield (Fig. 2b). Various metal salts like $MgSO_4$, $MnSO_4$, $BaCl_2$, $ZnSO_4$, and $FeSO_4$ were used as additives in the basal medium to establish their stimulatory or impeding effect on cellulase production

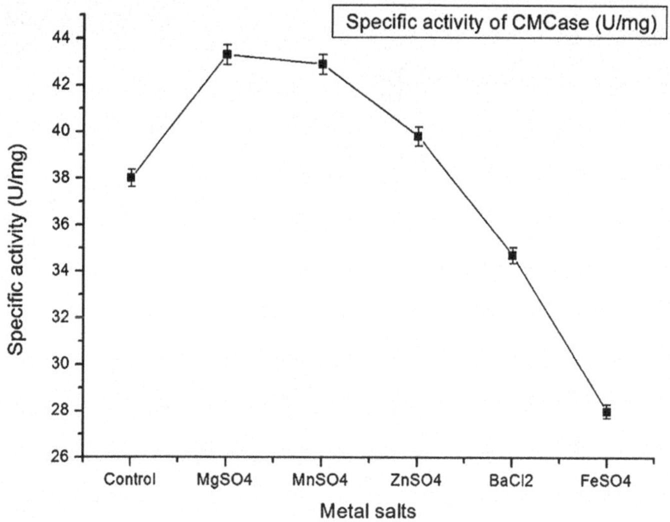

Fig. 3a Effect of metal salts on cellulase production

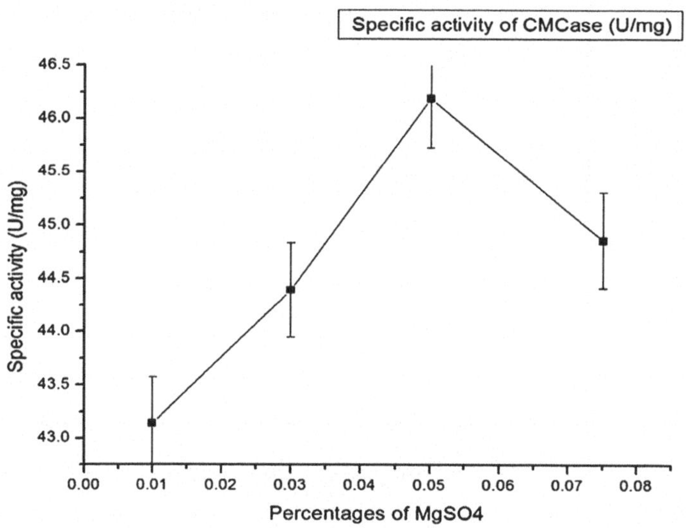

Fig. 3b Effect of concentration of $MgSO_4$ on cellulase production

(Fig. 3a and 3b). An increase in CMCase production was observed when the production medium was supplemented with 0.05 % $MgSO_4$ (specific activity 43.3 U/mg). The implementation of surfactants in the production medium of enzymes like cellulase is well recognized (Kapoor et al. 2008). Triton X-100 escalated cellulase yield (specific activity 53.8 U/mg) at the concentration of 0.3 % (w/v) (Fig. 4a and 4b). Surfactants possibly enhanced the permeability of the microorganisms' cell

Fig. 4a Effect of surfactants on cellulase production

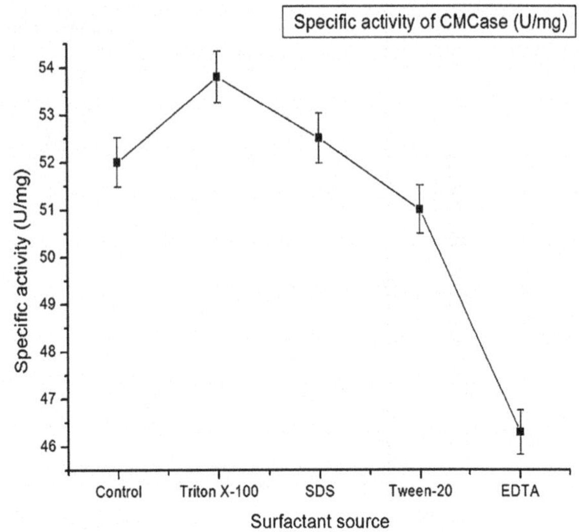

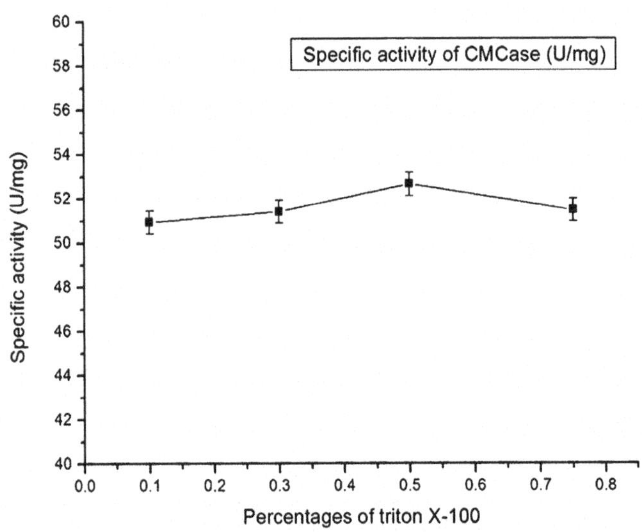

Fig. 4b Effect of percentage of triton X-100 on cellulase production

membrane resulting in rapid secretion of enzymes (Ahamed and Vermette 2008). The results were in accordance with studies done by Deswal et al. (Deswal et al. 2011). Different polar and nonpolar amino acids like serine, glutamic acid, glycine, tyrosine, and cysteine were assessed for their effect in cellulase production; among the amino acids tested, tyrosine (Fig. 5a and 5b) was found to have positive effect on endoglucanase production at the concentration of 0.1 % (w/w), whose specific activity was found to be 54 U/mg.

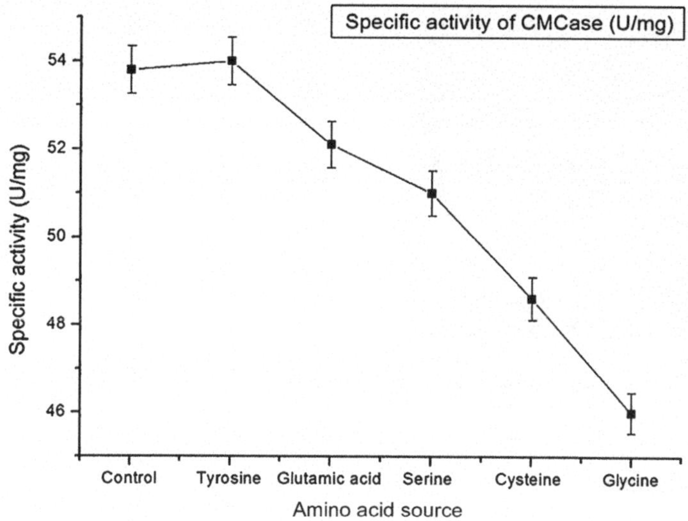

Fig. 5a Effect of amino acids on production of CMCase

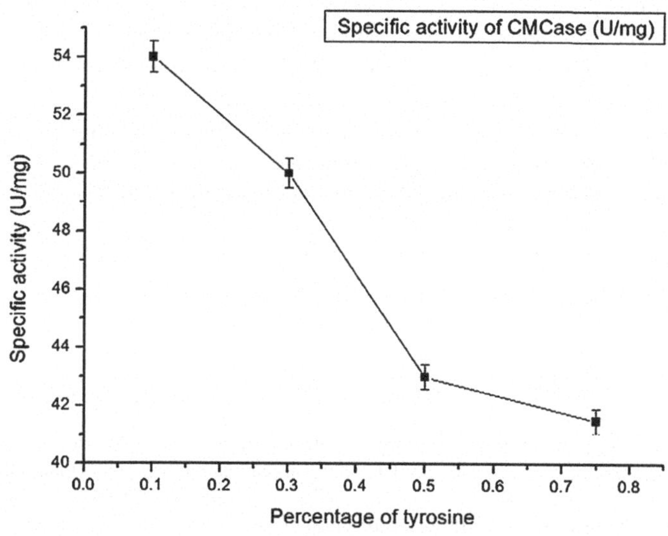

Fig. 5b Effect of percentage of tyrosine on cellulase production

3.2 *Energy-Dispersive X-ray Microanalysis*

Figure 6 showed the EDX analysis of banana peel. The chemical composition of the peel revealed that it is rich in carbon content. Besides that other elements like Mg, Al, P, K, Mn, Cu, and Zn were also present. The existence of C, Al, K, Mg, and Zn on the surface of the peel was previously reported by Kamsonlian et al. (Kamsonlian et al. 2001).

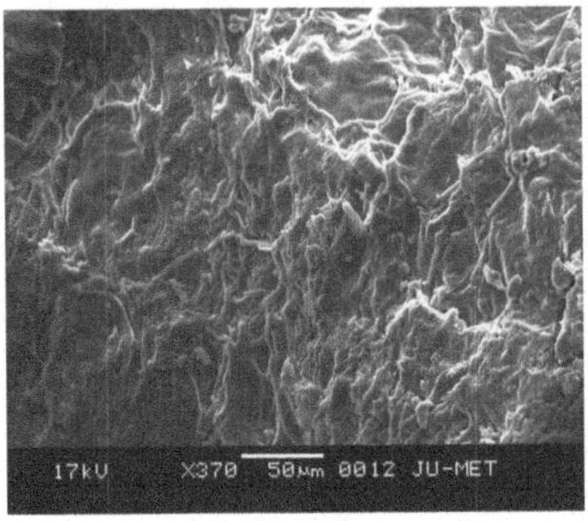

Element	Weight%	Atomic%
C K	41.86	52.29
Mg K	0.44	0.27
Al K	0.46	0.25
P K	0.26	0.13
S K	0.18	0.09
K K	6.59	2.53
Mn K	0.24	0.07
Fe K	0.32	0.09
Cu K	0.30	0.07
Zn K	0.15	0.04

Fig. 6 EDAX analysis of banana peel

4 Conclusion

The current investigation is based on the utilization of horticultural solid waste to biosynthesize secondary metabolite by SSF. The optimization of nutritional parameters for cellulase production by SSF was carried out employing isolated *Aspergillus niger* as cellulase producer. It was evident from the study that the solid waste selected can be used as an efficient substrate for fermentation as it supported the production of industrially important enzyme cellulase without any additional requirement of carbon source as the peel itself is rich in the supplement. Besides that the solid waste was found to contain ample amount of K, Mg, Al, Zn, Fe, and other metal ions which can be implemented in biosorption studies.

Acknowledgment The authors acknowledge UGC, New Delhi, India, for the financial assistance.

References

Ahamed A, Vermette P (2008) Culture-based strategies to enhance cellulase enzyme production from *Trichoderma reesei* RUT-C30 in bioreactor culture conditions. Biochem Eng J 140:399–407

Deswal D, Khasa YP, Kuhad RC (2011) Optimization of cellulase production by a brown rot fungus *Fomitopsis* sp. RCK2010 under solid state fermentation. Bioresour Technol 102:6065–6072

Domingues FC, Queiroz JA, Cabral JMS, Fonse (2000) The influence of culture conditions on mycelial structure and cellulase production by *Trichoderma reesei* RUT C-30. Enzym Microb Technol 26:394–401

Ghose TK (1987) Measurement of cellulase activities. Pure Appl Chem 59:257–268

Kamsonlian S, Suresh S, Majumder CB, Chand S (2001) Characterization of banana and orange peels: biosorption mechanism. IJSTM 2(4):1–7

Kapoor M, Nair LM, Kuhad RC (2008) Cost-effective xylanase production from free and immobilized *Bacillus pumilus* strain MK001 and its application in saccharification of *Prosopis juliflora*. Biochem Eng J 38:88–97

Krishna C (1998) Production of bacterial cellulases by solid state bioprocessing of banana wastes. Bioresour Technol 69:231–239

Krishna PR, Srivastava AK, Suprasana P, D'Souza SF (2012) Banana peel as substrate for α-amylase production using *Aspergillus niger* NCIM 616 and process optimization. IJBT 11:314–319

Lowry OH, Rosebrough NJ, Farr AL, Randall RJ (1951) Protein measurement with folin phenol reagent. J Biol Chem 193:265–275

Mandal M, Ghosh U (2015) Production of multi-enzyme system by isolated *Aspergillus* sp. under solid state fermentation and submerged fermentation. IJLTET 5(1):186–192

Naidu AL, Raghuveer D, Suman P (2013) Studies on characterization and mechanical behavior of banana peel reinforced epoxy composites. Int J Sci Eng Res 4(6):844

Pradeep MR, Narasimha G (2011) Utilization of pea seed husk as a substrate for cellulase production by mutant *Aspergillus niger*. Insight Biotechnol 1(2):17–22

Sukumaran RK, Singhania RR, Pandey A (2005) Microbial cellulases -production, applications and challenges. JSIR 64:832–844

Sun T, Liu BH, Li ZH, Liu DM (1999) Effect of air pressure amplitude on cellulase production by *Trichoderma viride* SL-1 in periodic pressure solid state fermentation. Process Biochem 34:25–29

Sun HY, Li J, Zhao P, Peng M (2011) Banana peel: a novel substrate for cellulase production under solid-state fermentation. Afr J Biotechnol 10(77):17887–17890

Bioconversion of Agro-waste to Value-Added Product Through Solid-State Fermentation by a Potent Fungal Strain *Aspergillus flavus* PUF5

P. Ghosh and U. Ghosh

Abstract Laccases are multicopper-containing oxidoreductase enzymes, which are commonly produced by plants and microbes, are involved in various biotechnological applications like bleaching of textile dyes, biofuels, biosensors, detoxification of effluent, bioremediation, etc. To compete efficiently with the chemical oxidizing agents, laccase must be produced in significant amounts up to industrial scale, and its production cost should also be minimized. In this concern, cheap waste products of agriculture and food industry can be used as inexpensive carbon source which can improve the economic feasibility of the enzyme production. Moreover, agro-wastes do not compete with the general food crops and cause environment pollution. So utilizing these renewable sources in solid-state fermentation (SSF) process can be considered very promising for the biotechnological production of laccase. Hence, the current study was focused on the optimization of process parameters for the higher production of laccase using a novel isolated fungi *Aspergillus flavus* PUF5 under SSF-utilizing waste ridge gourd peel as substrate. Fermentation parameters such as the amount of substrate, inoculum size, fermentation time, temperature, pH, and initial moisture content were optimized through one-variable-at-a-time (OVAT) method. Laccase production was further enhanced with additional carbon and nitrogen source supplementation. After 7 days of fermentation, maximum laccase yield (38.4 U/mg) was obtained using dried ridge gourd peel (5%) with 50% initial moisture content at 30 °C and pH 5.0 when 0.5% starch and peptone each were supplemented as carbon and nitrogen sources, respectively. Thus, the isolated fungal strain seems to be a potential candidate for laccase production through SSF. The process also supports economic valorization of agro-waste residues.

Keywords Agro-waste • Ridge gourd peel • Laccase • Solid-state fermentation

P. Ghosh • U. Ghosh (✉)
Department of Food Technology & Biochemical Engineering, Jadavpur University, 700032 Kolkata, West Bengal, India
e-mail: priyankaghsh28@gmail.com; ughoshftbe@yahoo.co.in

S.K. Ghosh (ed.), *Utilization and Management of Bioresources*,
DOI 10.1007/978-981-10-5349-8_28

1 Introduction

Laccases are multicopper-containing enzymes that are able to reduce molecular oxygen to water and instantaneously perform one-electron oxidation of various aromatic substrates such as diphenols, methoxy-substituted monophenols, and aromatic amines. For the last few decades, it has gained significant interests due to their potentiality for oxidizing phenolic and nonphenolic lignin-related compounds and different recalcitrant environmental pollutants. Currently in different fields like delignification of pulp, textile dye bleaching, effluent detoxification, biotransformation of antibiotics and steroids, development of biosensors, etc. (Chhaya and Gupte 2010). For better application biotechnologically, the yield of the enzyme should be high enough, and the production cost should be low. Considering this, solid-state fermentation (SSF) can be a better option, where different agro-wastes could be effectively utilized in enzyme production through microbial fermentation. In agro-industries, large quantity of agricultural residues accumulates. Disposal of these agro-wastes is one of the serious problems for industries. This condition could be partially resolved through bioconversion of these materials to value-added products through microbial fermentation. This is a promising technology because it has different beneficial aspects like lower-energy requirements, high product yield, and less infrastructural requirements and is eco-friendly. Therefore, SSF could be a promising alternate for valorization of agro-wastes to value-added products like laccases.

2 Materials and Methods

2.1 Microorganism and Inoculum Preparation

Aspergillus flavus PUF5, a pre-isolated fungal strain, was used in the study. It was grown on potato dextrose agar (composition: potato extract 20 g, dextrose 2 g, agar 1.75%) slants at 30 °C for 5 days and stored at 4 °C until further use.

About 5 ml of sterile distilled water was added to a fully sporulated culture slant. The spores were dislodged properly. A spore suspension of about to 5×10^9 spores/ml was considered as inoculum for further studies.

2.2 Laccase Production Through Solid-State Fermentation

Five grams of dried ridge gourd peel was used as substrates for laccase production by isolated fungal strain *Aspergillus flavus* PUF5. The buffer-moistened substrates were sterilized and incubated. Finally the crude enzyme was extracted, and the supernatants were assayed for laccase activity.

2.3 Optimization of Fermentation Parameters

For maximum laccase production by *Aspergillus flavus* PUF5-utilizing waste ridge gourd peel, different fermentation parameters such as amount of substrate (1–6 g), inoculum size (2–5%), fermentation time (3–9 days), temperature (25–45 °C), pH (Niladevi et al. 2007; Jang et al. 2006; Poojary and Mugeraya 2012; Soumya et al. 2016; Gochev and Krastanov 2007; Sivakumar et al. 2010), and substrate to initial moisture ratio were optimized through one-variable-at-a-time (OVAT) method. All experiments were done in triplicates. Results were shown as mean ± standard deviation.

2.4 Laccase Assay

Laccase activity was measured by the method of Desai et al. (2011). The laccase activity (U/ml) was calculated using the guaiacol at 450 nm. The specific laccase activity was determined as U/mg of protein, where the amount of soluble protein was determined by Lowry method (1951).

3 Results and Discussion

3.1 Effect of Substrate Amount and Inoculum Size for Laccase Production

In a fermentation process, the amount of substrate and the inoculum size are important factors that correspond to microbial growth and extracellular enzyme production. In the current experiment, fermentation was carried out with different amounts of ridge gourd peel (1–6 g) with varying inoculum sizes (2–5%), and the observation is summarized in Fig. 1. After 7 days of fermentation, a positive correlation was observed among the enzyme production, substrate concentration, and inoculum size. Laccase production was gradually increased (red color in Fig. 1) along with the increase of substrate amount (up to 5 g per flask) and inoculum size (5 ml). Below that it was insufficient for microbial growth and enzyme production. Niladevi et al. (2007) have reported a similar observation.

3.2 Study on Time Course for Production of Laccase

The influence of incubation time on laccase production by *Aspergillus flavus* PUF5 is summarized in Fig. 2. Under normal condition, maximum laccase secretion started on the third day of incubation, and the maximum activity (32.4 U/mg) was

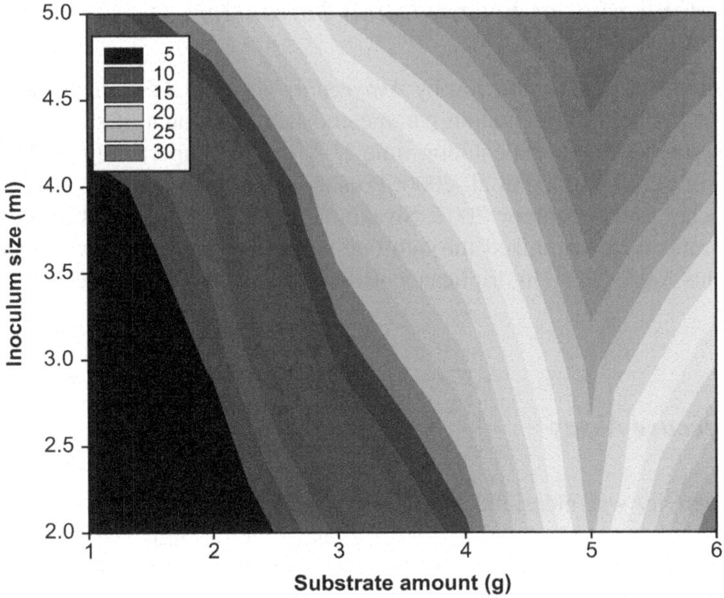

Fig. 1 Effect of substrate amount and inoculum size on laccase production by *A. flavus* PUF 5

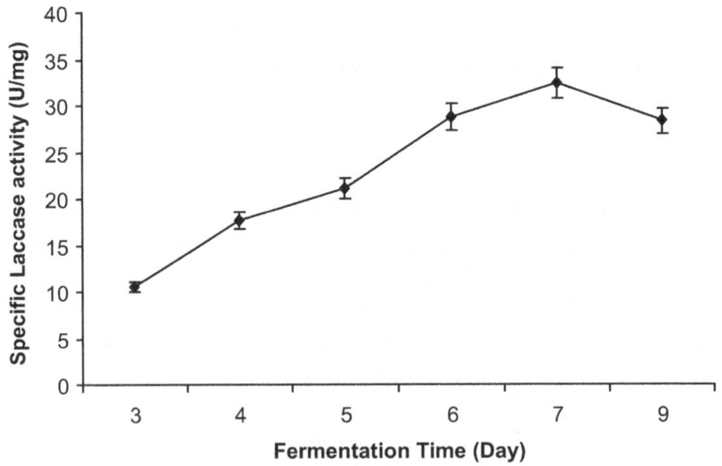

Fig. 2 Effect of incubation period on laccase production

achieved after 7 days of fermentation. A longer incubation time did not improve the level of laccase production. The incubation time depends on the microbial growth rate, and similar results were found by Jang et al. (2006) where *Trametes* sp. produced maximum extracellular laccase at the end of the stationary phase.

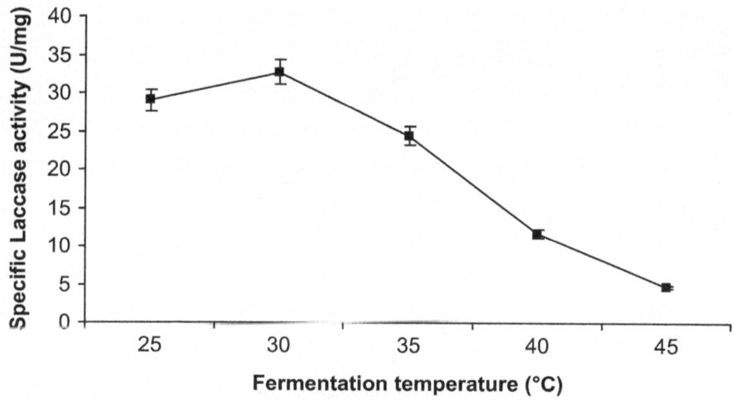

Fig. 3 Optimization of fermentation temperature for laccase production

3.3 *Effect of Incubation Temperature for Laccase Production*

Temperature influences the properties of aqueous environment as well as all metabolic processes of the organism including nutrient availability and uptake. During the optimization of culture condition for the highest yield of laccase from *A. flavus* PUF5, it was observed that temperature has a profound influence on enzyme activity (Fig. 3). Results indicated that *A. flavus* PUF5 produced maximum laccase (32.7 U/mg) activity at 30 °C and by increasing the incubation temperature, a gradual decrease in laccase production occurred. Above 40 °C, both the growth and laccase production minimized. This suggested that proper growth of *A. flavus* PUF5 is necessary for the optimal laccase production. Poojary and Mugeraya (2012) observed similar results during laccase production by *Phellinus noxius* hpF17. The reduction in the yield of laccase production may be that at lower temperature substrate was unable to move across the cells. At higher temperature, laccase production was lower due to the requirement of high maintenance energy and thermal denaturation of the enzyme.

3.4 *Influence of pH of the Medium*

The pH of the growth medium plays an important role in terms of inducing enzyme production and morphological changes in the microbes. Among different influencing parameters, medium pH affected laccase production by *Aspergillus flavus* PUF5. Figure 4 reveals that maximal specific activity (35.2 U/mg) of laccase was found at pH 5.0. A further increase in pH reduced the laccase activity. Similar results were found by Soumya et al. (2016) in different fungi. The reduced laccase production at higher pH may be due to alteration of the three-dimensional structure of the enzyme. The laccase production was thereby significantly affected by the pH of the medium.

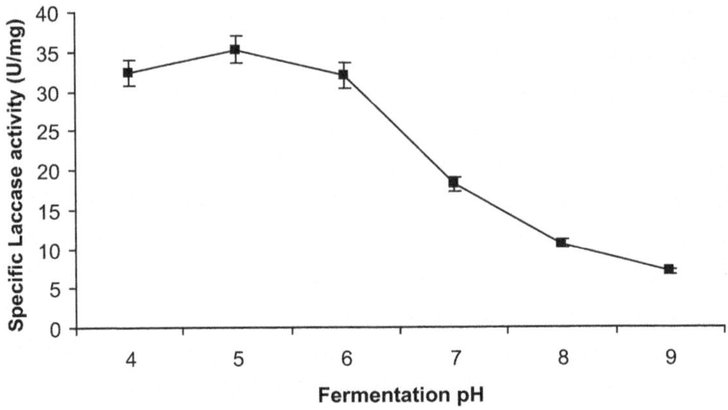

Fig. 4 Effect of fermentation pH on laccase production

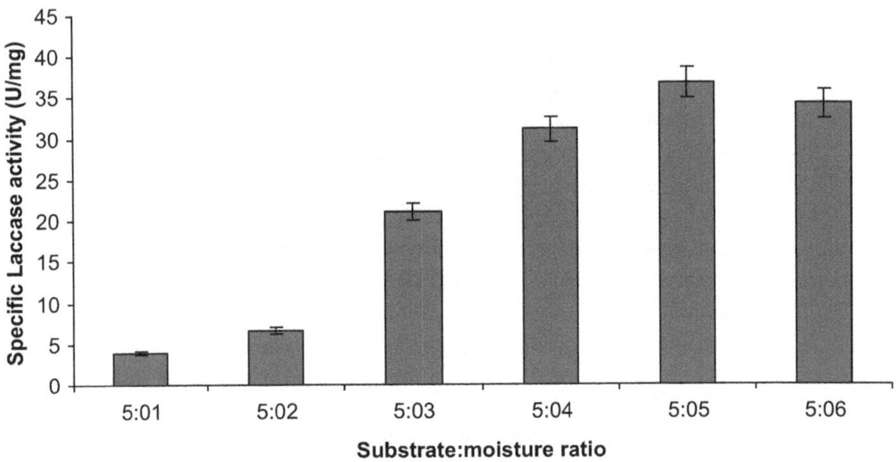

Fig. 5 Effect of hydration on laccase production

3.5 *Optimization of Initial Moisture*

In solid-state fermentation, it is crucial to provide sufficient moisture content for better microbial growth and product formation. Figure 5 indicated that specific laccase activity was maximum (36.8 U/mg) at 5:5 substrate to moisture ratio. At lower moisture contents, fungal growth as well as the rate of laccase production was decreased.

Table 1 Effect of additional carbon source on laccase production

Additional carbon source	Relative activity (%)
No additional carbon source	100
Glucose	60.5
Fructose	89.4
Sucrose	82.5
Starch	145.6
Carboxymethyl cellulose	143.2

Table 2 Effect of additional nitrogen source on laccase production

Additional nitrogen source	Relative activity (%)
No additional nitrogen source	100
Yeast extract	123.4
Peptone	135.6
Urea	120.3
Sodium nitrate	121.8
Ammonium chloride	128.3

3.6 *Effect of Additional Carbon Source on Laccase Production*

The concentration and nature of carbon source used in the medium showed a notable effect on enzyme production. The effect of different additional carbon sources (0.5% w/v) on the growth of *A. flavus* PUF5 and extracellular laccase activity was tested, and the results indicated that the extracellular laccase activity was greatly influenced by the exogenous addition of carbohydrates. Simple sugar like glucose supported the profuse fungal growth, but the yield of laccase production was significantly low. In laccase biosynthesis, catabolite repression by glucose and other easily metabolizable carbon sources is a commonly observed phenomenon (Gochev and Krastanov 2007). High extracellular laccase production was observed when starch followed by carboxymethyl cellulose was used as additional carbon sources (Table 1). Previous report also supports this phenomenon (Sivakumar et al. 2010) (Table 2).

3.7 *Effect of Additional Nitrogen Source on Laccase Production*

Nitrogen is involved in amino acid synthesis which makes up proteins and other value-added substances. In this study, the effects of different exogenous nitrogen source on the production of laccase by *A. flavus* PUF5 were investigated separately.

Results indicated that among the tested nitrogen sources, highest specific laccase activity was achieved with 0.5% peptone followed by yeast extract as compared to control. Nitrogen generally affects laccase production at the transcriptional level, and the expression of different laccase genes appears to be regulated by a process mediated by nitrilase family member 2-type proteins, which are involved in regulating nitrogen metabolism (Piscitelli et al. 2011).

4 Conclusion

The present study reveals that *Aspergillus flavus* PUF5 is a potent strain for the laccase production which showed maximum activity (38.4 U/mg) when solid-state fermentation was carried out using dried ridge gourd peel (5 g) moistened (50%) with initial moisture content buffer solution (pH 5.0) at 30 °C. Enzyme production was enhanced when starch (0.5%) and peptone (0.5%) were supplemented as additional carbon and nitrogen sources, respectively. The isolated strain PUF5 effectively utilized agro-waste for laccase production under ambient conditions without requiring any expensive additional growth factors. This indicates that this strain is suitable for cost-effective industrial productions.

Acknowledgment The authors are thankful to the University Grant Commission [Sanction no: F./2016-17/NFO201517OBCWES32372/(SAIII/Website)], Government of India, for the financial assistant in this study.

References

Chhaya U, Gupte A (2010) Optimization of media components for laccase production by litter dwelling fungal isolate *Fusarium incarnatum* LD-3. J Basic Microbiol 50(1):43–51

Desai SS, Tennali GB, Nityanand Channur N, Anup AC, Deshpande G, Murtuza BPA (2011) Isolation of laccase producing fungi and partial characterization of laccase. Biotechnol Bioinformatics Bioeng 1(4):543–549

Gochev VK, Krastanov AI (2007) Isolation of laccase producing Trichoderma Spp. Bulg J Agri Sci 13:171–176

Jang MY, Ryu WY, Cho MH (2006) Enhanced production of laccase from *Trametes* sp. by combination of various inducers. Biotechnol Bioprocess Eng 11(2):96–99

Lowry OH, Rosenbrough NJ, Farr AL, Randall RJ (1951) Protein measurement with the Folin Phenol Reagent. J Biol Chem 193:265–275

Niladevi KN, Sukumaran RK, Prema P (2007) Utilization of rice straw for laccase production by Streptomyces psammoticus in solid-state fermentation. J Ind Microbiol Biotechnol 34(10):665–674

Piscitelli A, Giardina P, Lettera V, Pezzella C (2011) Induction and transcriptional regulation of laccases in fungi. Curr Genomics 12:104–112

Poojary P, Mugeraya G (2012) Laccase production by Phellinus noxius hpF17: optimization of submerged culture conditions by response surface methodology. Res Biotechnol 3(1):09–20

Sivakumar R, Rajendran R, Balakumar C, Tamilvendan M (2010) Isolation, screening and optimization of production medium for thermostable laccase production from *Ganoderma* sp. Int J Eng Sci Technol 2(12):7133–7141

Soumya PS, Lakshmi MSK, Nambisan P (2016) Application of response surface methodology for the optimization of laccase production from Pleurotus ostreatus by solid state fermentation on pineapple leaf substrate. J Sci Ind Res 75:306–314

Toxic Metal Removal Using Biosorption Process and Inertization of Generated Hazardous Metal-Laden Biosorbent

L. Ramrakhiani, A. Halder, A.K. Mandal, S. Majumdar, and S. Ghosh

Abstract The tannery industry waste was used in this study as a low-cost biosorbent. The presence of large number of functional groups in the tannery waste resulting from both live and dead microbial fractions positively favours for higher and faster sequestration rate of metal ions. The batch mode of biosorption process were performed with dried activated tannery sludge for removal of Ni(II), Co(II), Zn(II) and Cd(II) ions in multi-metal system. Zn(II) and Cd(II) ions showed 99% removal efficiency within 10 min of contact time, while Ni(II) and Co(II) attained 98% removal at 20–24 h. The influence of various experimental variables like pH, contact time, biosorbent dosage and initial metal concentrations was studied. The biosorbent was characterized using FESEM-EDX and XPS analysis.

One of the major challenges in the field of biosorption is the management of the hazardous spent biosorbent even after metal recovery and effectively reuse over several cycles. The conventional ways of the disposal of such materials involve storage, landfill disposal or incineration which in turn again pollutes the environment. Landfill disposal are not suitable due to groundwater contamination prospects by leaching of metal ions. Again, the biomass burning processes create air pollution, generate hazardous volatiles and produce ash having high concentration of the desired metal. To solve this issue, the present investigation was designed for vitrification of toxic metals waste in a glass matrix.

Safe management of spent biosorbent has been developed as inertization in phosphate glass formulation. Control and multi-metal-loaded biosorbent up to 25% were successfully incorporated into phosphate glass matrix for safe disposal to the environment. Amorphous nature of good glass preparation was confirmed by XRD analysis. Results of dissolution test on the prepared phosphate glass showed no leaching of hazardous heavy metal ions up to 1 month of incubation under DI water with thermal cycle at 75 °C for 8 h/day.

L. Ramrakhiani • S. Majumdar • S. Ghosh (✉)
Ceramic Membrane Division, CSIR-Central Glass and Ceramic Research Institute, Kolkata, West Bengal, India
e-mail: sourja@cgcri.res.in

A. Halder • A.K. Mandal
Glass Division, CSIR-Central Glass and Ceramic Research Institute, Kolkata, West Bengal, India

S.K. Ghosh (ed.), *Utilization and Management of Bioresources*,
DOI 10.1007/978-981-10-5349-8_29

Keywords Heavy metals • Waste-derived biosorbent • Multi-metal removal • Spent biosorbent management

1 Introduction

Increased industrialization and human activities have created harmful impact on the environment through the disposal of waste containing heavy metals. Pb, Cr, Ni, Co, Zn, Cd and Cu are several of the general types of toxic metals originating from the waste water of various small- and large-scale industries such as mining, surface finishing, fertilizer, pesticide, batteries, metallurgy, iron and steel, electroplating, welding, electrolysis, electrical appliance manufacturing and aerospace and atomic energy installations, etc. The existence, exposure and mobility of such toxic heavy metals into the surroundings represent a significant and long-term environmental hazard. Even at low concentration, these metals can be lethal to organisms, including humans (Areco et al. 2012).

The biosorption process is an efficient and eco-friendly tool for removal of toxic metals from water and waste water. It has currently gained momentum for employing low-cost biological materials (biosorbents) with effective binding capacities towards different metal ions. Biosorbents for the removal of metals mainly come under the following categories: bacteria, fungi, algae, plants, industrial wastes, agricultural wastes and other polysaccharide materials. Moreover, the process occurs as a result of physicochemical interactions, mainly ion exchange or complex formation between metal ions and functional groups present on the cell surface (Lee et al. 2006).

Most of the scientific literatures have been focused on the use of different biosorbents for removal of single-metal ion (Li et al. 2010; Magalhães et al. 2004). However, single toxic metallic species rarely exist in natural and waste waters, and that the presence of multiple metal ions often causes an interactive effect, insufficient attention seems to have been given to this problem. Another contest in the biosorption field is the need of long-term solution for safe disposal of generated metal-laden biosorbent to the environment.

Generally, management of industrial sludge involves incineration, or landfill disposal. Incineration of sludge has been well recognized as the pretreatment technique before landfill disposal due to effectual volume reduction and detoxification. But worldwide incineration is criticized because it generates hazardous volatiles and ashes (Mandal et al. 2013). Thus it is very important to manage the metal-laden biosorbent in an environmentally sound and productive manner. Some of the novel insinuations have been signified in the literature for the management of hazardous industrial waste sludge which includes bricks, tile, concrete and cement materials, etc. (Park et al. 2010; Paulino et al. 2013; Ramrakhiani et al. 2016a). However, toxic characteristic leaching procedure (TCLP) tests on brick or tile showed unpredictable metal leaching for longer time periods (Park et al. 2010). Glasses have excellent preference for such immobilization due to higher environmental stability, high chemical durability, thermal stability, mechanical resistance and slow dissolution rates in water.

The present work has been formulated for removal of toxic metals as multi-metal system using industrial waste-derived biosorbent. The dried activated sludge from tannery industry was developed as a complex biomaterial because it contains both live and dead microbial fractions that have large number of binding sites for higher and faster sequestration rate of toxic metal ions.

The detail objectives were to establish adsorption characteristics both in single- and multi-metal removal system for the treatment of industrial waste water and inertization of generated hazardous metal-laden waste in phosphate glass matrix for its safe disposal. Different experimental approaches involving the influence of various environmental parameters such as pH, biosorbent dosage, initial metal concentrations and contact time on Ni(II), Co(II), Zn(II) and Cd(II) as single- and multi-metal removal system were investigated. Surface characterization of biosorbent was analysed by FESEM-EDX and XPS techniques. Finally, the possibility of accommodation of maximum amount of spent biosorbent in phosphate glass matrix was studied, and good glass structure was confirmed with the analysis of XRD patterns. Leaching experiments and dissolution test were also carried out with prepared glass.

2 Materials and Methods

2.1 *Biosorbent Collection and Characterization*

Activated sludge generated at tannery industry located at leather complex, Kolkata, India, was collected in ziplock bags. The samples were washed with generous amount of deionized water and dried at 50 °C for 48 h. Finally, the dried mass was powdered in a mortar and pestle and sieved through a 150 mesh sieve. The prepared biosorbent was stored in an airtight container for biosorption and safe disposal studies.

The biosorbent was characterized for initial pH, total organic matter, bulk density, porosity and ash content, zeta potential (Zetasizer of Malvern Instruments, UK) and Brunauer–Emmett–Teller (BET) surface area (Quantachrome Instruments, USA).

2.2 *Metal Solution Preparations and Analysis*

Stock solutions (1000 mg/L) of Ni(II), Co(II), Zn(II) and Cd(II) were prepared from chemical reagents of analytical grade, $NiCl_2.6H_2O$, $CoSO_4.7H_2O$, $ZnSO_4.7H_2O$ and $Cd(NO_3)_2.4H_2O$, respectively, and diluted to desired concentrations for single-metal removal experiments. Multi-metal solution was prepared from stock solution by adding equal volume of each metal solution to attain working concentration (100 mg/L) for biosorption studies as multi-metal removal system.

All the experimental transition metals present in single- and multi-metal removal system were estimated by ion chromatography (IC) method (Metrohm, Ion Chromatography, Switzerland). The Nucleosil 5SA IC cation column, the standard tartaric acid/citric acid eluent as mobile phase, PAR reagent for UV detector and the MagIC Net™ chromatography software were used that control all system components and monitor various parameters relevant to the chromatographic analysis.

2.3 *Batch Biosorption Studies*

The batch biosorption experiments were carried out both as single-metal and multi-metal removal system using dried activated tannery sludge as biosorbent. The biosorbent (0.2 g) were suspended in 25 mL of 100 mg/L initial metal solution (Ni(II), Co(II), Zn(II) and Cd(II) as single- and multi-metal) by shaking (150 rpm) at 30 °C for a period of contact time up to 24 h at the dissolution pH of each metal which varies between pH 4.9 and 5.4. The influence of environmental parameters such as effects of biosorbent dose (0.25–12 g/L), effects of initial concentrations (25–600 mg/L) and effects of contact time were determined to find out the equilibrium state of the biosorption. At the end of each experiment, solutions were separated from the biosorbent by filtration through Whatman No. 42 filter paper for analysis of each metal ion left in solution after biosorption.

The differences in metal concentration before and after biosorption were used to find out the adsorption (%) (Eq. 1).

$$\text{Biosorption}(\%) = (C_i - C_e / C_i) \times 100 \tag{1}$$

The adsorption capacity was obtained using Eq. 2.

$$q_e = \left[V(C_i - C_e)\right] / m \tag{2}$$

where q_e is the metal uptake (mg/g), C_i is the initial metal concentration (mg/L), C_e is the concentration after biosorption, m is the biosorbent amount (g) and V is the volume of metal solution (mL).

2.4 *Surface Characterization of Biosorbent*

Surface chemistry of the biosorbent before and after biosorption and presence of surface adsorbed metals were determined using FESEM-EDX and XPS analytical techniques.

2.4.1 FESEM-EDX Analysis

Surface morphology and microstructure of the biosorbent before and after biosorption both single- and multi-metal system were examined using field emission scanning electron microscopy (FESEM, Zeiss, Germany) and the elemental composition present on surface of the biosorbent were established by energy-dispersive X-ray analysis (EDX).

2.4.2 XPS Analysis

X-ray photoelectron spectroscopy technique was employed for the samples of before and after multi-metal biosorption system of dried tannery sludge using PHI 5000 XPS analyser, Versaprobe-II, USA, with an Mg-Kα source (1253.6 eV), a hemispherical analyser with 16-channel detector and PC-Access smart soft-Versaprobe 2.4.0.9 software.

2.5 *Safe Disposal of Metal-Laden Biosorbent as Inertization in Phosphate Glass*

Immobilization and safe disposal of the metal-laden hazardous biosorbent to the environment generated after biosorption process were explored for its maximum accommodation into phosphate glass matrix formulation.

The control and multi-metal-loaded biosorbent were analysed by ICP-AES for metal content determination. The biosorbents were dried at 100 °C for 6 h to remove its moisture content before incorporated into glass matrix. Glass batches were prepared by mixing defined quantities phosphate-based glass matrix with 5, 15 and 25% of control and multi-metal-laden biosorbent. Each formulation was melted in a resistive heating furnace (bottom loading-raising hearth furnace, Deltech-Model DT-31-BL-810-KC-E3504) at 1300 °C in 30 mL alumina crucible for 30 min. The glass melt was transferred into a preheated stainless steel mould and then annealed in a muffle furnace at 360 °C for 2 h followed by controlled cooling till room temperature. The amorphous phases of prepared glasses were characterized by X-ray diffraction (Philips 1710 diffractometer) patterns in the 2θ range between 10 and 90°.

The prepared glass samples were cut into blocks polished and measured its dimensions and surface area using micrometre digital vernier caliper. The glass blocks were then weighed (Mettler Toledo ±0.00001 g), and the dissolution experiments were carried out at hanging position in deionized water heated at 75 °C for 6 h per day. The dissolution rates were monitored at every seventh day interval over 1 month period of thermal cycle and were analysed for leaching of metals using ICP-AES technique. Thereafter the glass samples were removed, rinsed with acetone, dried at 100 °C in oven and weighed.

3 Results and Discussion

3.1 *Biosorbent Characterization*

The physicochemical characterizations of dried activated sludge of tannery industry, i.e. the biosorbent, were specified as surface pH of 6.55, organic matter 24.9%, bulk density 0.44 g/cm^3, porosity 83.4%, ash content 75.1%, particle size 150 μm, surface area (BET) of 23.63 m^2/g, zeta potential value −25.8 to −26.7 mV and mobility of −2.019 to −2.096 μm cm/Vs (Ramrakhiani et al. 2016b).

3.2 *Biosorption Process Optimization Parameters*

Biosorption efficiency of the prepared biosorbent for removal of the metal ions of Ni(II), Co(II), Zn(II) and Cd(II) as single- and multi-metal systems were carried out due to its complex nature, low cost and ease of accessibility (Table 1).

Influence of different experimental variables such as pH, biosorbent dosage, initial metal concentrations and contact time were studied to find out the equilibrium phase and maximum metal uptake both in the single- and multi-metal removal system.

The maximum biosorption of metal ions onto dried activated tannery sludge was found at original initial pH of metal solution both in single- and multi-metal systems. Original initial pH of metal solution was recorded for solutions of nickel chloride, 5.34; cobalt sulphate, 5.35; zinc sulphate, 4.90; and cadmium nitrate, 5.29; while in multi-metal solution, the pH was 5.21. Therefore, the further studies for biosorption were carried out without any pH adjustment. Several studies described pH value of 5.0–6.0 as the optimum pH for Cd(II), Cu(II), Pb(II), Ni(II) and Zn(II) adsorption using different biosorbents (Ramrakhiani et al. 2011). The rate of

Table 1 Biosorption efficiency of Ni(II), Co(II), Zn(II) and Cd(II), as single- and multi-metal systems using dried activated Tannery sludge

Metal biosorption system	Biosorption efficiency (%)	Metal uptake (q_e) (mg/g)
Single		
Ni	95.98	12.001
Co	88.43	11.66
Zn	99.46	12.412
Cd	99.06	12.325
Multiple		
Ni	95.62	2.988
Co	93.86	2.933
Zn	98.84	3.088
Cd	99.99	3.125

biosorption increased as solid/liquid ratio increased and 8.0 g/L of biosorbent dosage was selected as optimum solid/liquid ratio for further studies.

In the biosorption of Zn(II) and Cd(II) as single-metal system, the time required for equilibrium was 10–30 min; after that no significant changes were observed with 99.73% and 99.24% biosorption, respectively. Again in multi-metal system, the equilibrium state was achieved at 1.0 h of contact time. Thus, it can be concluded that biosorption process of Zn (II) and Cd (II) was rapid and spontaneous which may be due to binding with superficial functional groups of the biosorbent, while the kinetics studies on the biosorption of Ni(II) and Co(II) followed gradual removal. During the first phase within 30 min of contact time, it showed 49.86% and 47.07% biosorption in single-metal system and 36.86% and 61.24% biosorption in multi-metal system, respectively. The equilibrium phases were attained at 20–22 h of contact time. Thus, Ni(II) and Co(II) removal process were slow and gradual that require longer time for maximum biosorption. The lower rate of removal may be governed by ion exchange process (Areco et al. 2012; Ramrakhiani et al. 2016b).

3.3 *Characterization Using FESEM and EDX*

FESEM micrographs of control and the multi-metal-loaded biosorbents are presented in Fig. 1 a and b, respectively. A clear agglomeration was observed on the surface multi-metal-laden biosorbent when compared to that of control biosorbent. The EDX spectra of metal-laden biosorbent revealed Ni, Co, Zn and Cd peaks in the respective spectra, while no such peak could be found on the control biosorbent surface that confirms the morphological changes due to the presence of metals on the biosorbent surface. Various fungal and algal biomasses were also showed morphological changes due to the presence of adsorbed metal (Torres et al. 2009).

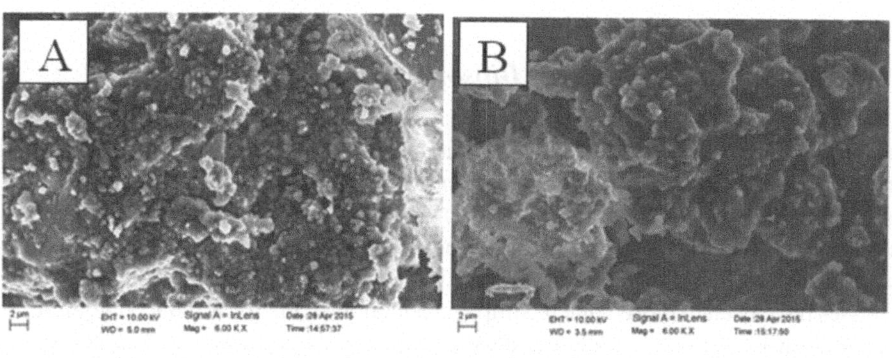

Fig. 1 FESEM micrographs. (**a**) Dried activated tannery sludge (control biosorbent). (**b**) Multi-metal (Ni(II), Cd(II), Co(II), Zn(II))-laden biosorbent

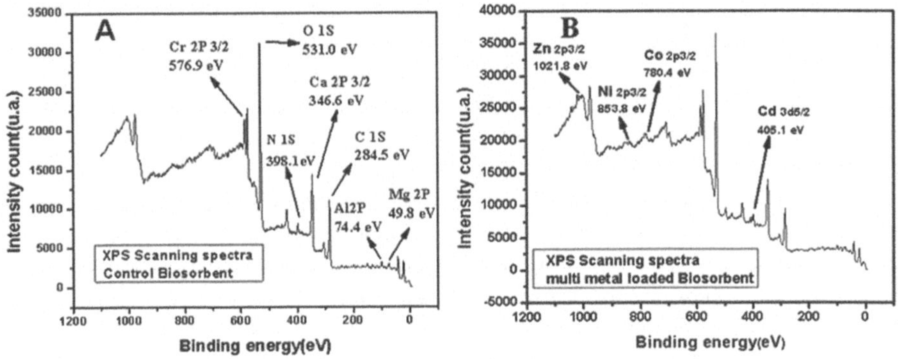

Fig. 2 Wide-scan XPS spectra: (**a**) Control biosorbent. (**b**) Multi-metal-laden biosorbent

3.4 Characterization Using XPS Spectroscopy

X-ray photoelectron spectroscopy (XPS) was used to determine the elemental constituent contained within the biosorbent for surface characterization and mechanism studies. The technique also facilitates quantitative elemental surface analysis, chemical and/or oxidation state of metal.

The surface chemistry of the biosorbent before and after multi-metal biosorption process was studied using XPS spectroscopy. The wide-scan spectra of control and multi-metal-laden biosorbent are presented in Fig. 2 a and b. The photoelectron peaks observed in control biosorbent were 49.8 eV, 74.4 eV, 284.5 eV, 346.6 eV, 398.1 eV, 530 eV and 576.9 eV which are attributed to the Mg(2p), Al(2p), C(1s), Ca(2p), N(1s), O(1s) and Cr(2p). The XPS spectra of multi-metal-laden biosorbent found peaks at 853.1 eV (Ni2p), 780.4 eV (Co2p), 1,021.8 eV (Zn2p) and 405.1 eV (Cd2p).

3.5 Management of Spent Biosorbent

Immobilization of heavy metal waste in the phosphate glass matrix in present investigation is considered as a safer option to prevent the further risk of the environmental contamination (Vijayaraghavan and Yun 2008). Moreover, glasses are the appropriate way for such inertization due to high mechanical, chemical thermal resistance and very slow dissolution rates in water. Vitrifications of nuclear waste in glasses have been explored due to simple production technology, reduced volume of resulting waste form and high tolerance to radiation damage (Weng et al. 2003).

After evaluating elemental composition of solid residues in the control and multi-metal-laden biosorbent samples by the ICP-AES analysis, both the biosorbent samples were incorporated upto 5, 15 and 25 wt.% in the phosphate glass matrix.

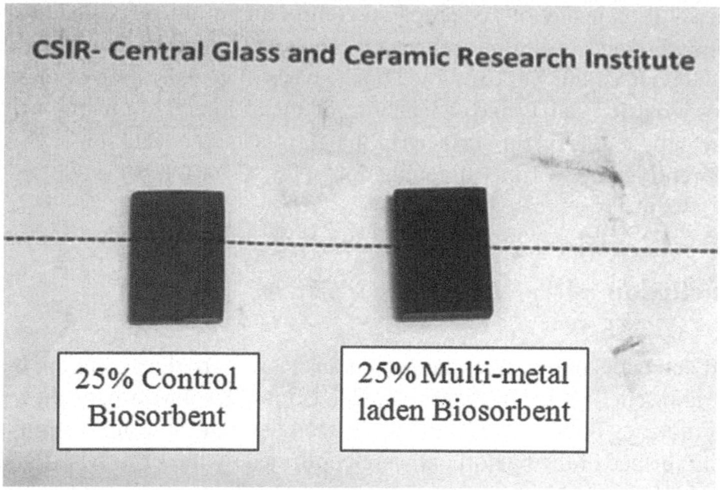

Fig. 3 Photograph of prepared phosphate glass with 25% accommodated spent biosorbent. (**a**) Control biosorbent incorporation. (**b**) Multi-metal-laden biosorbent incorporation

Fig. 4 XRD spectra of the prepared phosphate glass samples with 25 wt.% incorporation of the control and multi-metal-laden biosorbent

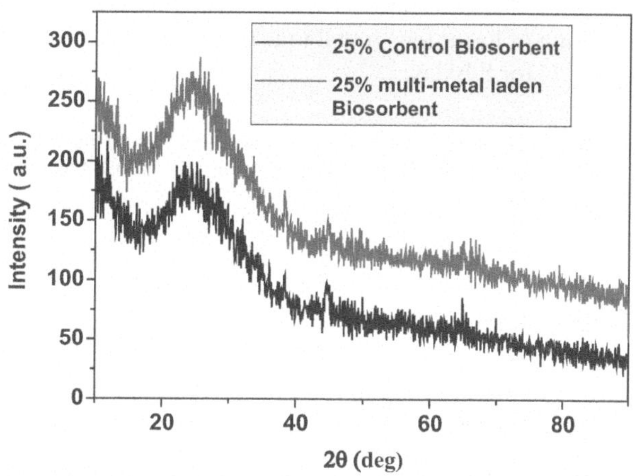

The study showed that good glasses could be prepared with up to 25% integration of the spent biosorbent in the phosphate glass matrix (Fig. 3).

All the prepared glass samples were analysed by XRD technique to detect the possible crystalline phase, if any. The XRD spectra of glass having up to 25% incorporation of the control and the multi-metal-laden spent biosorbent showed non-crystalline phase, which confirmed good glass formation (Fig. 4). Although the present study revealed up to 25% incorporation of the biosorbent, however, the possibility of incorporating higher concentration of waste into glass matrix and preparation of glass of practical interest is under progress.

The dissolution behaviour of prepared control and multi-metal-laden spent biosorbent incorporated in the phosphate glasses was studied in deionized water for 1 month. Glass leaching was examined on the basis of visualization of glass surface, weight loss and ICP analysis for leaching of metal ions. No change in the glass weight, no depositions on the surfaces of the glasses and no leaching of heavy metal ions were reported up to 1 month incubation at 75 °C for 6 h per day.

4 Conclusion

The dried activated sludge from tannery industry was used as complex biosorbent for removal of Ni(II), Co(II), Zn(II) and Cd(II) ions. The biosorbent showed >98% removal efficiency both as single and multi-metal system. The adsorptions of metal ions on the surface of the biosorbent were confirmed by the FESEM-EDX and XPS analysis.

The management of spent biosorbent as in phosphate glass formulation were successfully achieved for up to 25% accommodation of control and multi-metal-laden biosorbent. Non-crystalline phase of prepared glass was established by XRD analysis. Results of leaching test on the prepared phosphate glass showed no leaching of hazardous metal ions up to 1 month of incubation under deionized water at 75 °C for 8 h/day.

Acknowledgement The financial support to LR from the Department of Science and Technology (DST), Government of India, vide grant no. SR/WOS-A/CS-15/2014 dated 08.08.2014 is gratefully acknowledged.

References

Areco MM, Hanela S, Duran J, Afonso MS (2012) Biosorption of Cu(II), Zn(II), Cd(II) and Pb(II) by dead biomasses of green alga *Ulva lactuca* and the development of a sustainable matrix for adsorption. J Hazard Mater 213–214:123–132

Lee WE, Juoi J, Ojovan MI, Stennett MC, Hyatt NC (2006) Immobilization of radioactive waste in glasses, glass composite materials and ceramics. Adv Appl Ceram 105(1):3–12

Li H, Lin Y, Guan W, Chang J, Xu L, Guo J, Wei G (2010) Biosorption of Zn(II) by live and dead cells of *Streptomyces ciscaucasicus* strain CCNWHX 72-14. J Hazard Mater 179:151–159

Magalhães JM, Silva JE, Castro FP, Labrincha JA (2004) Effect of experimental variables on the inertization of galvanic sludges in clay-based ceramics. J Hazard Mater 106-B:139–147

Mandal AK, Biswas K, Annapurna K, Guha C, Sen R (2013) Preparation of alumino-phosphate glass by microwave radiation. Mater Res 28(14):1955–1961

Park D, Yun YS, Park JM (2010) The past, present, and future trends of biosorption. Biotechnol Bioprocess Eng 15:86–102

Paulino ALG, Cunha AJD, Alfaya RVDS, Alfaya AADS (2013) Chemically modified natural cotton fiber: a low-cost biosorbent for the removal of the Cu(II), Zn(II), Cd(II), and Pb(II) from natural water. Desal Water Treat 52:1–11

Ramrakhiani L, Majumder R, Khowala S (2011) Removal of hexavalent chromium by heat inactivated fungal biomass of *Termitomyces clypeatus*: surface characterization and mechanism of biosorption. Chem Eng J 171:1060–1068

Ramrakhiani L, Ghosh S, Majumdar S (2016a) Surface modification of naturally available biomass for enhancement of heavy metal removal efficiency, upscaling prospects, and management aspects of spent biosorbents: a review. Appl Biochem Biotechnol 180(1):41–78. doi:10.1007/s12010-016-2083-y

Ramrakhiani L, Majumdar S, Ghosh S (2016b) Heavy metal biosorption in multi component system on dried activated sludge: investigation of adsorption mechanism by surface characterization. Oral presentation in International Conference on Advances in Bioprocess Engineering and Technology (ICABET 2016), January 20–22, 2016, Sponsored by TEQIP II, Organized by Heritage Institute of Technology, Kolkata, India Abs No O-57, p 116

Torres P, Fernandes HR, Olhero S, Ferreira JMF (2009) Incorporation of wastes from granite rock cutting and polishing industries to produce roof tiles. J Eur Ceram Soc 29(1):23–30

Vijayaraghavan K, Yun YS (2008) Bacterial biosorbents and biosorption. Biotechnol Adv 26:266–291

Weng CH, Lin DF, Chiang PC (2003) Utilization of sludge as brick materials. Adv Environ Res 7:679–685

Part V
Applied Nanotechnology

Green Synthesis of Magnetic Iron Nanoparticle Using *Moringa oleifera* Lam Seeds and Its Application in Textile Effluent Treatment

K.G. Ahila, M. Vasanthy, and C. Thamaraiselvi

Abstract In the present investigation, inexpensive and easily available *Moringa oleifera* seed was utilized for the preparation of nanoparticles by coprecipitation method, followed by the treatment of textile dye effluent. The sample-treated nanoparticles showed significant reduction of color (90%), pH (7.6), COD (89%), and TDS (85%) from the effluent. Higher reduction of impurities was observed in 0.5 g of FeNP treatment. Also, pH of the treated dye effluent was neutralized. The morphological and structural features of FeNP were characterized by X-ray diffraction (XRD), Fourier transform infrared spectroscopy (FTIR), scanning electron microscopy, and energy-dispersive X-ray (EDX).

Keywords Physicochemical parameters • *Moringa oleifera* L • Magnetic Iron nanoparticle • Color • COD • TDS

1 Introduction

Textile industry contributes a major place in the Indian economy. It utilizes large quantity of water and chemical additives for carrying out various textile processes and releases highly polluted and colored wastewater into the environment with partial treatment or without treatment (Akshaya et al. 2012). Dyes are synthetic chemicals which are ionic and are aryl ring-structured aromatic organic compounds.

K.G. Ahila • C. Thamaraiselvi
Department of Biotechnology, Mother Teresa Women's University, Kodaikanal, Tamil Nadu, India

M. Vasanthy (✉)
Department of Environmental Biotechnology, Bharathidasan University, Trichy, Tamil Nadu, India
e-mail: drvasanthy02@gmail.com

S.K. Ghosh (ed.), *Utilization and Management of Bioresources*,
DOI 10.1007/978-981-10-5349-8_30

Chromophore groups in the dye provide color to the dye. Most of the synthetic dyes are nonbiodegradable, and small quantity in water is visible. Dyes used for coloring fabrics, fibers, cotton, polyester, nylon, etc. are frequently used in textile, paper, rubber, plastic, cosmetics, food, and pharmaceutical industries (Sachin et al. 2010). Dye-containing wastewater discharged into the aquatic system prevents sunlight penetration to the aquatic system and affects the photosynthetic process of aquatic flora. And the water is unfit for consumption (Mahony et al. 2002; Crini 2006). These dyes are carcinogenic, genotoxic, mutagenic, or teratogenic in nature (Nimrat et al. 2004). Several physical, chemical, and biological techniques are available for textile effluent treatment. Among these methods, treatments with nanoparticles show higher reduction of impurities from textile dye effluent. Also the synthesis of iron nanoparticle is of low cost and has higher efficiency. Iron-related materials are widely used in various applications such as biosensing, medicine, environmental cleanup, etc. Broadly used iron nanomaterials in the removal of environmental contaminants include zero-valent iron, iron oxides, iron phosphate, and iron-based bimetallic particles. Green-synthesized iron nanoparticles can be synthesized through plant extracts and microbes (Homer et al. 2016).

The chemical and physical methods used for nanoparticle preparation have some demerits which are expensive and harmful to environment and produce noxious by-products. And also, the reducing and capping agents, solvents used for the synthesis, are dangerous to the environment (Bharde et al. 2006; Nadagouda et al. 2009). Plants are used as alternative for capping and reducing agents and are eco-friendly for human and other organisms (Aromal and Philip 2012).

Moringa oleifera L. is the most important economically valued and medicinal plant which is originated from Asia and has spread to Africa. It belongs to Moringaceae family (Fuglie 2001). Its seeds are widely used as coagulant for the treatment of contaminated water. In the present study, magnetic iron nanoparticle was synthesized using *Moringa* seeds, and the synthesized particles were used for the treatment of textile dye wastewater.

2 Materials and Methods

2.1 Sample Collection

Textile dyeing effluent was collected in a pre-cleaned polyethylene container from a dyeing unit located at Karur, Tamil Nadu, India. The sample was stored at 4 °C for further studies.

2.2 Characterization of Textile Dye Effluent

The physicochemical parameters like color, odor, pH, temperature, electrical conductivity, TS, TDS, TSS, COD (Cat. No: MSW-439), DO, alkalinity, acidity, chloride, sulfate, nitrate, phosphate, and fluoride were analyzed by standard procedures (APHA 2012).

2.3 Collection of Plant Material

Moringa oleifera Lam seeds were obtained from local market, and the seeds were washed thoroughly with sterile deionized water after removing the pod and were air dried. The seed shell was removed, ground well, and sieved through ASTM (200) sieve. The extract was prepared by mixing the appropriate amount of seed powder in deionized water and was boiled for 10 min. Further, the extract was filtered and used for the study.

2.4 Green Synthesis of Magnetic Iron Nanoparticles (FeNP)

Magnetic FeNP was prepared by coprecipitation method using ferric chloride (0.37 g) and ferrous sulfate (0.954 g) with the addition of *Moringa oleifera* Lam seed as a reducing agent. The required volume of chemicals dissolved in sterile deionized water was taken in glass bottles and was mixed thoroughly at 80 °C for 10 min. Freshly prepared boiled seed extract was added to the solution in 1:1 ratio and was heated for 10 min at 60–70 °C. Followed by this, 2 ml of ammonia solution was added, and a dark orange color change into black-colored precipitate indicates the formation of magnetic iron nanoparticles (Li et al. 2011).

2.5 Purification of Magnetic FeNP

The excess ammonia from the solution was removed by washing the precipitate several times with deionized water and ethanol to neutralize the pH, and the supernatant was air-dried where liquid has evaporated. The pellet was washed with acetone and was dried at 60–70 °C (Senthilkumar and Sivakumar 2014; Sophie et al. 2008).

2.6 Characterization of Iron Nanoparticles

The synthesized magnetic FeNP was characterized by various techniques. The crystalline nature of FeNP was analyzed by X-ray diffraction (XRD, Bruker D8 Advance model). The functional groups and capping agents were characterized by FTIR spectroscopy (PerkinElmer Spectrum). The FTIR spectrum was obtained in the mid-IR region of 40–4000 cm^{-1}. SEM analysis was done using software-controlled scanned electron microscope, and the size and morphology were determined by SEM, and the elemental composition of iron nanoparticle was analyzed by energy-dispersive X-ray spectrometry (EDX). UV-visible spectrophotometer (Elico- SL 218 model) was employed to measure the color reduction.

2.7 Treatment of Dyeing Effluent

The purified magnetic nanoparticles measured in different dosages, such as 0.1 g, 0.2 g, 0.3 g, 0.4 g, 0.5 g, and 0.6 g, were added to a series of six Erlenmeyer flask containing 100 ml of dyeing effluent. The contents were mixed properly and were kept for 1 h. After treatment, the samples were centrifuged at 3000 rpm for 10 min, and the color, pH, COD, and TDS were determined.

Percentage of decolorization was calculated by the following formula (Olukanni et al. 2006).

$$\text{Decolorization}\left(\%\right) = A_0 - A_t / A_0 \times 100$$

where

A_0 = absorbance of the untreated dye effluent
A_t = absorbance of the treated dye effluent

3 Results and Discussion

3.1 Physical Characterization of Dyeing Effluent

The physical parameters such as color, odor, pH, temperature, total solids, total dissolved solids, and total suspended solids in the raw effluent was found to be dark red, unpleasant alkaline odor, 10.5, 40 °C, 6900 mg/l, 6500 mg/l, and 400 mg/l, respectively (Table 1). The color of the sample is due to the presence of higher concentration of reactive dye Red F3B. The colored wastewater influences the growth of aquatic biota and diminishes the recreation capability and inclination to chelate metal ions (Barakat et al. 2010). Total solids and dissolved solids were found to be

Table 1 Physicochemical characteristics of fabric dyeing effluent

Sl. No.	Parameters	Results (mg/l)	CPCB standard
1.	pH	10.5	5.5–9.0
2.	Temperature	40 °C	45 °C
3.	Color	Dark red	Colorless
4.	Odor	Alkaline odor	Odorless
5.	TS	6900	–
6.	TDS	6500	500
7.	TSS	400	–
8.	Acidity	283	–
9.	Alkalinity	916.6	6.0
10.	COD	2912	250
11.	DO	4.05	–
12.	Fluoride	1.84	–
13.	Phosphate	6.4	200
14.	Sulfate	1253	250
15.	Nitrate	96	–
16.	Chloride	4050	250

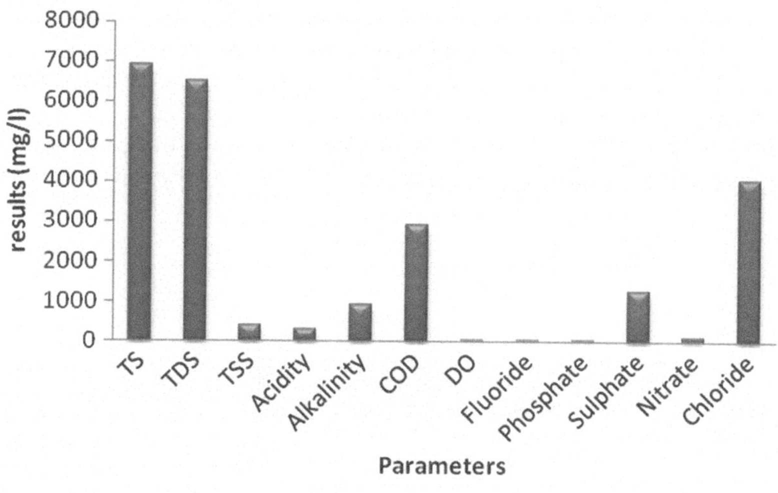

Fig. 1 Graphical representation of physicochemical characteristics of fabric dyeing effluent

higher than CPCB standard. The chemical parameters include acidity, alkalinity, COD, DO, chloride, fluoride, phosphate, sulfate, and nitrate levels in the effluent was 283 mg/l, 916.6 mg/l, 2912 mg/l, 4.05 mg/l, 4050 mg/l, 1.84 mg/l, 6.4 mg/l, 1250 mg/l, and 96 mg/l, respectively. It stressed the need of the treatment before going to discharge into natural bodies. The results are depicted in Table 1 and Fig. 1.

4 Synthesis of Magnetic Iron Nanoparticles

4.1 Characterization of Magnetic FeNP

The XRD results pattern confirmed the synthesis of iron nanoparticle, and it showed the characteristic peaks with intensity 24,223°, 35,879°, 40,936°, 49,924°, 50,103°, 63,555°, 64,611°, and 77,597° (Fig. 2). From these peaks, magnetic nanoparticle synthesis was confirmed. Hariani et al. (2013) reported that six characteristic peaks in the XRD patterns of the Fe_3O_4 at 30,2050°, 35,5150, 43,3250, 53,7110, 57,2150, and 62,9450 were corresponding to the (220), (311), (400), (422), (511), and (440) crystal planes of a pure Fe_3O_4 with a spinal structure. FTIR analysis showed the functional groups in the plant extract which are responsible for reducing and capping of iron nanoparticles (Fig. 3). The graph revealed the presence of seven peaks in ranges between 4000 and 400 cm^{-1}. The peaks in the range of 3400.89 cm^{-1} showed the presence of hydroxyl group with hydrogen bond containing OH stretch. The peaks obtained from the FTIR graph was found to be 2923.12 cm^{-1}, 1669.44 cm^{-1}, 1402.00 cm^{-1}, 1073.38 cm^{-1}, 697.16 cm^{-1}, and 604.87 cm^{-1} corresponding to alkanes with C–H stretching, C–O stretching with aldehyde groups, C–H stretching, N–H stretching, and O–H stretching and alkyne group with C–H;C–H bend and alkyl halide groups, respectively. The OH groups and N–H groups present in the seed powder might have band with the iron molecules and may be responsible for the synthesis of nanoparticles, and also it may act as reducing and capping agent. EDX characterization result was depicted in Fig. 4. The EDS spectrum showed the strong peaks of Fe and O. The composition of Fe_3O_4 synthesized through coprecipitation illustrate that the content of Fe was 71.09% and O was 28.91%.

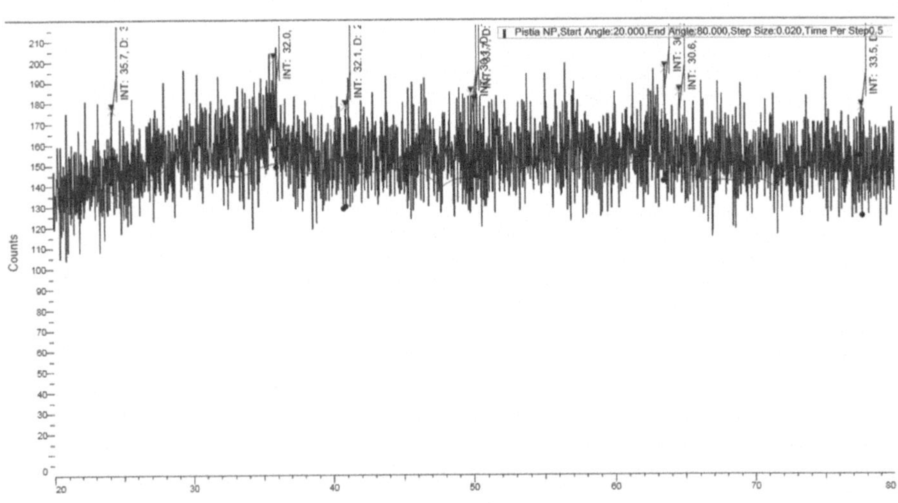

Fig. 2 XRD analysis of *Moringa* seed magnetic FeNP

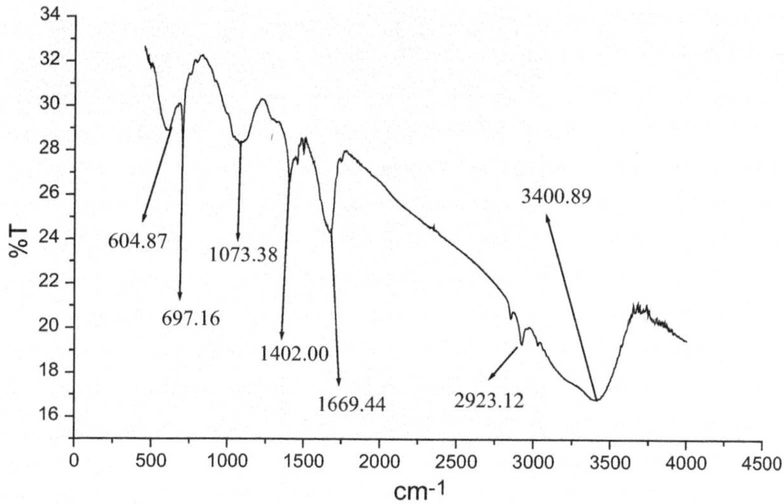

Fig. 3 FTIR analysis of *Moringa* seed magnetic FeNP

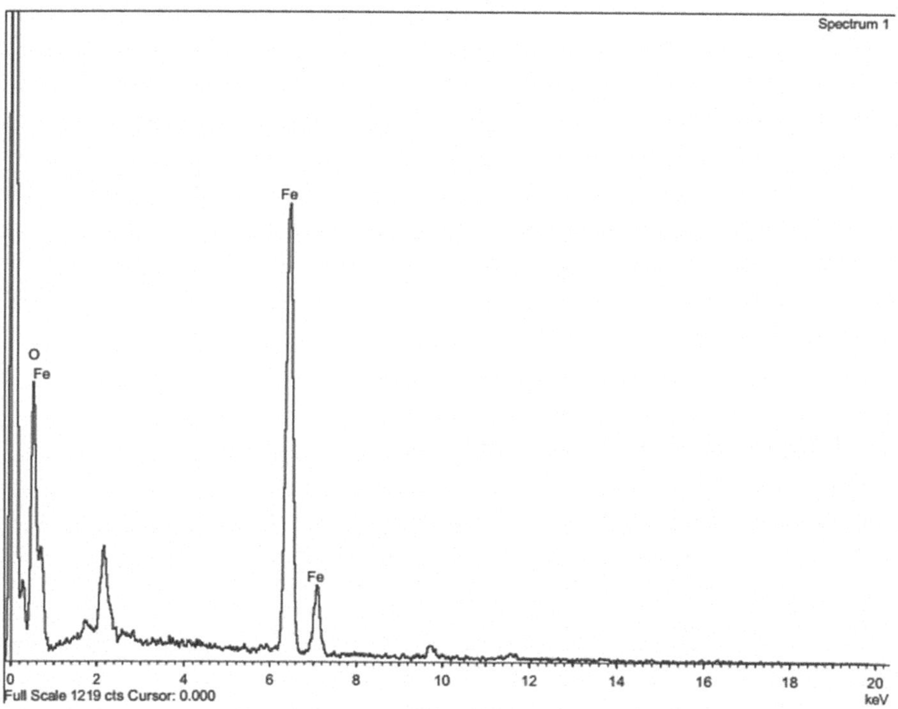

Fig. 4 EDX analysis of *Moringa* seed magnetic FeNP

Fig. 5 SEM analysis of *Moringa* seed magnetic FeNP

Table 2 Treatment of dyeing effluent using magnetic FeNP

Sl. No.	Dosage (g)	pH	Color (%)	TDS (%)	COD (%)
1.	0.1	10.2	72	68	75
2.	0.2	10	76.2	71	79
3.	0.3	9	79	75.4	81
4.	0.4	8	83.6	80.1	85
5.	0.5	7.6	90	85	89
6.	0.6	7.6	90	85	89

It indicates the purity of nanoparticles. The purity percentage (73.36% of Fe and 21.02% of O) was coincided with Hariani et al. (2013). The SEM analysis confirms the size of nanoparticle was 50 nm. The results are illustrated in the Fig. 5. All the analysis confirmed the synthesis of iron nanoparticle.

5 Treatment of Dyeing Effluent

Biosynthesized nanoparticle was applied to the dyeing effluent to reduce the TDS, COD, color, and pH. Different dosages were used and 0.5 g showed better reduction of pollutants. The treated sample showed 90% of color and 85% of TDS were reduced by the magnetic nanoparticle use uniform term it is Green nanoparticle or biosynthesized nanoparticle or magnetic nanoparticle. The pH of the treated sample got reduced to 7.2, and it signifies the pH-neutralizing capacity of FeNP. The maximum COD reduction (89%) was observed with the dosage of 0.5 g of iron nanoparticles. The results were represented in Table 2. Good floc formation was observed during the treatment of textile dye effluent with nanoparticles and may be concluded

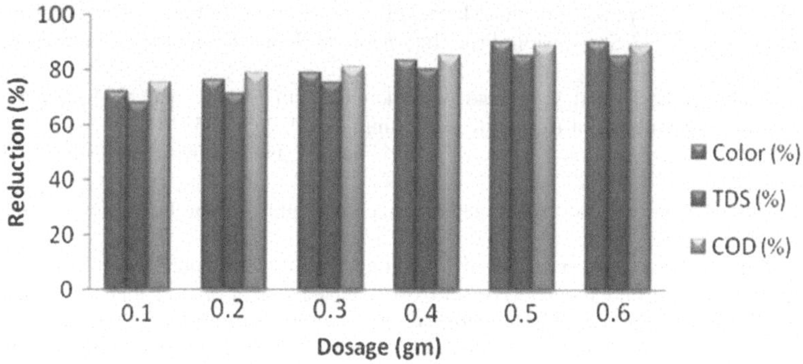

Fig. 6 Graphical representation of treatment of fabric dyeing effluent using MFeNP

that this particle has a considerable coagulating property. It was obtained by the presence of coagulating protein, and Okuda et al. (2001) reported that the IR analysis of *Moringa* seed showed the presence of amino group and hydroxide group (Fig. 6).

6 Conclusion

About 90% of color and 89% of COD were removed when the textile dye effluent was treated with 0.5 g of nanoparticles. The removal might have been achieved by the adsorption followed by flocculation process. Magnetic iron nanoparticles could be separated from the treated sample using external magnetic force and can be reused for other treatment also. Hence, it is an effective technology, and it is a low-cost method for the treatment of textile dye effluent when compared with currently used chemical coagulants.

Acknowledgment The authors acknowledge the DST-INSPIRE program and UGC authority for providing financial support for continuing the present research.

References

Akshaya KV, Bhunia P, Dash RR (2012) Decolorization and COD reduction efficiency of magnesium over iron based salt for the treatment of textile wastewater containing Dcazo and Aunthraquinone dyes. Int J Chem Mol Nucl Mater Metallugical Eng 6(6):17–24

APHA (2012) Standard methods for the examination of water and wastewater. American Public Health Association, Washington, DC

Aromal SA, Philip D (2012) Green synthesis of gold nanoparticles using Trigonellafoenum-graecum and its size-dependent catalytic activity. Spectrochim Acta – Part A: Mol Biomol Spectrosc 97:1–5

Barakat H (2010) Adsorption and photodegradation of Procion Yellow H–Exl Dyes in textile wastewater over TiO2 Suspension. Proc. International Water Technology Conference, Cairo, Egypt

Bharde A, Rautaray D, Bansal V, Ahmad A, Sarkar I, Yusuf SM, Sanyal M, Sastry M (2006) Extracellular biosynthesis of magnetite using fungi. Small 2(1):135–141

Crini G (2006) Non-conventional low cost adsorbents for dye removal. Bioresources Technol 97:1061–1085

Fuglie LJ (2001) The miracle tree: Moringaoleifera: natural nutrition for the tropics. Church World Service, Dakar, p 68

Genuino HomerC, Mazrui Nashaat, Seraji MohammadS, Luo Zhu, and Hoag George E (2016) Green synthesis of iron nanomaterials for oxidative catalysis of organic environmental pollutants, catalysis for remediation and environmental concerns. Amsterdam, Elsevier

Hariani PL, Faizal M, Ridwan M, Setiabudidaya D (2013) Synthesis and properties of Fe3O4 nanoparticles by co-precipitation method to removal Procion dye. Int J Environ Sci Dev 4(3):336–340

Li X, Xu H, Chen Z, Chen G (2011) Biosynthesis of nanoparticles by micro-organisms and their applications. J Nanomater 2011(1):1–16

Nadagouda MN, Hoag G, Collins J, Varma RS (2009) Green synthesis of Au nanostructures at room temperature using biodegradable plant surfactants, Cryst. Growth Des 9(11):4979–4983

Nimrat S, Sawangchit P, Vuthiphandchai (2004) Removal of malachite green employing physical and biological processes. Sci Asia 30:351–357

O'Mahony T, Guibal E, Tobin J, M. (2002) Reactive dye biosorption by Rhizopusarrhizus biomass. Enzyme Microb Technol 31:456–463

Okuda T, Baes AU, Nishijima W, Okada M (2001) Isolation and characterization of coagulant extracted from Moringaoleifera seed by salt solution. Water Res 35(2):405–410

Olukanni OD, Osuntoki AA, Gbenlo GO (2006) Textile effluent biodegradation potentials of textile effluent adapted and non- adapted bacteria. Afr J Biotechnol 5:1980–1984

Sachin MK, Gaikwad RW, Misal SA (2010) Low cost sugarcane bagasse ash as an adsorbent for dye removal from dye effluent. Int J Chem Eng Appl 1(4):309–318

Senthilkumar SR, Sivakumar T (2014) Green tea (*Camellia sinensis*) mediated synthesis of zinc oxide (ZnO) nanoparticles and studies on their antimicrobial activities. Int J Pharm Pharm Sci 6(6):461–465

Sophie L, Delphine F, Marc P, Alain R, Caroline R, Luce VE, Robert NM (2008) Magnetic iron oxide nanoparticles: synthesis, stabilization, vectorization, physicochemical characterization, and biological applications. Chem Rev 108:2064–2110

Index

S.K. Ghosh (ed.), *Utilization and Management of Bioresources*,
DOI 10.1007/978-981-10-5349-8

The manufacturer's authorised representative in the EU is Springer Nature Customer Service Centre GmbH, Europaplatz 3, 69115 Heidelberg, Germany. If you have any concerns regarding our products, please contact ProductSafety@springernature.com

Printed and bound by CPI Group (UK) Ltd, Croydon, CR0 4YY
25/09/2026
02222896-0010